Emerging Infectious Diseases

Emerging Infectious Diseases

Clinical Case Studies

Edited by

Önder Ergönül

Füsun Can

Lawrence Madoff

Murat Akova

AMSTERDAM • BOSTON • HEIDELBERG • LONDON
NEW YORK • OXFORD • PARIS • SAN DIEGO
SAN FRANCISCO • SINGAPORE • SYDNEY • TOKYO
Academic Press is an imprint of Elsevier

Academic Press is an imprint of Elsevier
32 Jamestown Road, London NW1 7BY, UK
225 Wyman Street, Waltham, MA 02451, USA
525 B Street, Suite 1800, San Diego, CA 92101-4495, USA

British Library Cataloguing-in-Publication Data
A catalogue record for this book is available from the British Library

Library of Congress Cataloging-in-Publication Data
A catalog record for this book is available from the Library of Congress

ISBN: 978-0-12-416975-3

For information on all Academic Press publications
visit our website at www.store.elsevier.com

Typeset by MPS Limited, Chennai, India
www.adi-mps.com

Printed and bound in United States of America

14 15 16 17 18 10 9 8 7 6 5 4 3 2 1

Dedication

We would like to dedicate our efforts to our medical students and trainees, who have inspired the authors by their interest and enthusiasm in the study of emerging infectious diseases.

Contents

10. Crimean-Congo Hemorrhagic Fever

Önder Ergönül

11. Phlebotomus Fever—Sandfly Fever

Koray Ergunay

12. Chikungunya Fever

Giovanni Rezza

15. Human Bocavirus

Tina Uršič and Miroslav Petrovec

16. Norovirus Gastroenteritis

Momoko Mawatari and Yasuyuki Kato

17. Enterohemorrhagic *Escherichia coli* (EHEC): Hemorrhagic Colitis and Hemolytic Uremic Syndrome (HUS)

Jakob P. Cramer

24. Lyme Borreliosis

Daša Stupica, Gerold Stanek and Franc Strle

25. *Plasmodium knowlesi*

Jana Preis and Larry Lutwick

26. Measles

Alpay Azap and Filiz Pehlivanoglu

List of Contributors

Jaffar A. Al-Tawfiq Saudi Aramco Medical Services Organisation, Saudi ARAMCO, Dhahran, KSA

Sevtap Arikan-Akdagli Hacettepe University Medical School, Department of Medical Microbiology, Ankara, Turkey

Tatjana Avšič Županc Institute of Microbiology and Immunology, Faculty of Medicine, University of Ljubljana, Ljubljana, Slovenia

Alpay Azap Ankara University, Infectious Diseases and Clinical Microbiology Department, Ankara, Turkey

Daniel G. Bausch Tulane Health Sciences Center, New Orleans, LA, USA

M. Bulent Ertugrul Adnan Menderes University Medical School, Department of Infectious Diseases and Clinical Microbiology, Aydin, Turkey

Remi N. Charrel Aix Marseille Université, IRD French Institute of Research for Development, EHESP French School of Public Health, EPV UMR_D 190 "Emergence des Pathologies Virales", & IHU Méditerranée Infection, APHM Public Hospitals of Marseille 13385, Marseille, France

Charles Chiu Department of Laboratory Medicine, University of California, San Francisco, CA, USA; Department of Medicine, Division of Infectious Diseases, University of California, San Francisco, CA, USA, UCSF-Abbott Viral Diagnostics and Discovery Center, San Francisco, CA, USA

Jakob P. Cramer Section Tropical Medicine and Infectious Diseases, University Medical Center Hamburg-Eppendorf and Bernhard Nocht Institute for Tropical Medicine, Bernhard-Nocht-Strasse, Hamburg, Germany

Stefano Di Bella National Institute for Infectious Diseases "Lazzaro Spallanzani," Rome, Italy

Laurent Dortet Associated National Reference Center for Antibiotic Resistance, K.-Bicêtre, France; Medical and Molecular Microbiology Unit, Department of Medicine, Faculty of Science, University of Fribourg, Switzerland

Cecilia Melina Drapeau Department of Medical Microbiology, King's College Hospital, London, UK

Hideki Ebihara Laboratory of Virology, Division of Intramural Research, National Institute of Allergy and Infectious Diseases, National Institutes of Health, Rocky Mountain Laboratories, Hamilton, MT, USA

Önder Ergönül School of Medicine, Koç University, Istanbul, Turkey

Koray Ergunay Hacettepe University Faculty of Medicine, Department of Medical Microbiology, Virology Unit, Ankara, Turkey

Joseph Fair Metabiota, Incorporated, San Francisco, CA, USA

Heinz Feldmann Laboratory of Virology, Division of Intramural Research, National Institute of Allergy and Infectious Diseases, National Institutes of Health, Rocky Mountain Laboratories, Hamilton, MT, USA

Pierre Formenty Emerging and Epidemic Zoonotic Diseases Team (CED/EZD), World Health Organization

Donald S. Grant Kenema Government Hospital, Ministry of Health and Sanitation, Kenema, Sierra Leone

Gilda Grard Unité des Maladies Virales Emergentes, Centre International de Recherches Médicales de Franceville, Franceville, Gabon

Jan Heyckendorf Medical Clinic Research Center Borstel, German Center for Infection Research (DZIF) Tuberculosis Unit, Center for Infection and Inflammation (ZIEL), University of Lübeck, Parkallee, Borstel, Germany

Dr. Gene Khai Lin Huang Infectious Diseases Registrar, Infectious Diseases Department, Austin Health, Melbourne, Australia

Prof. Dr. Paul Johnson Deputy Director, Infectious Diseases Department, Austin Health, Melbourne, Australia; Professor, University of Melbourne, Department of Microbiology and Immunology, Department of Medicine, Austin & Northern Clinical School, Melbourne, Australia; Director, World Health Organization Collaborating Centre for Mycobacterium ulcerans, Melbourne, Australia

Yasuyuki Kato Division of Preparedness and Emerging Infections, Disease Control and Prevention Center, National Center for Global Health and Medicine, Toyama, Shinjuku-ku, Tokyo, Japan

Humarr Khan Kenema Government Hospital, Ministry of Health and Sanitation, Kenema, Sierra Leone

Miša Korva Institute of Microbiology and Immunology, Faculty of Medicine, University of Ljubljana, Ljubljana, Slovenia

Christoph Lange Medical Clinic Research Center Borstel, German Center for Infection Research (DZIF) Tuberculosis Unit, Center for Infection and Inflammation (ZIEL), University of Lübeck, Parkallee, Borstel, Germany

Eric Leroy UMR MIVEGEC (IRD 224 - CNRS 5290 - UM1 - UM2) Institut de Recherche pour le Développement, Montpellier, France; Unité des Maladies Virales Emergentes, Centre International de Recherches Médicales de Franceville, Franceville, Gabon

Larry Lutwick Division of Infectious Diseases, Department of Medicine, Western Michigan University Homer Stryker MD School of Medicine, Kalamazoo, MI, USA

Julia Martensen Medical Clinic Research Center Borstel, German Center for Infection Research (DZIF) Tuberculosis Unit, Center for Infection and Inflammation (ZIEL), University of Lübeck, Parkallee, Borstel, Germany

Keita Matsuno Laboratory of Virology, Division of Intramural Research, National Institute of Allergy and Infectious Diseases, National Institutes of Health, Rocky Mountain Laboratories, Hamilton, MT, USA

Momoko Mawatari Division of Preparedness and Emerging Infections, Disease Control and Prevention Center, National Center for Global Health and Medicine, Toyama, Shinjuku-ku, Tokyo, Japan

Ziad A. Memish Global Center for Mass Gathering Medicine, Ministry of Health, Riyadh, Kingdom of Saudi Arabia; Ministry of Health, Al-Faisal University, Riyadh, KSA

Yasutaka Mizuno Department of Infection Control and Infectious Diseases, Tokyo Medical University Hospital, Tokyo, Japan

Itaru Nakamura Department of Infection Control and Infectious Diseases, Tokyo Medical University Hospital, Tokyo, Japan

Patrice Nordmann INSERM U914, South-Paris Medical School, K.-Bicêtre, France; Medical and Molecular Microbiology Unit, Department of Medicine, Faculty of Science, University of Fribourg, Switzerland, Associated National Reference Center for Antibiotic Resistance, K.-Bicêtre, France

Janusz T. Paweska Center for Emerging and Zoonotic Diseases, National Institute for Communicable Diseases of the National Health Laboratory Service and School of Pathology, Faculty of Medical Sciences, University of Witwatersrand, South Africa

Filiz Pehlivanoglu Haseki Training and Research Hospital, Infectious Diseases and Clinical Microbiology Clinic, Istanbul, Turkey

Nicola Petrosillo National Institute for Infectious Diseases "Lazzaro Spallanzani," Rome, Italy

Miroslav Petrovec Institute of Microbiology and Immunology, Faculty of Medicine, University of Ljubljana, Zaloška, Ljubljana, Slovenia

Laurent Poirel INSERM U914, South-Paris Medical School, K.-Bicêtre, France; Medical and Molecular Microbiology Unit, Department of Medicine, Faculty of Science, University of Fribourg, Switzerland, Associated National Reference Center for Antibiotic Resistance, K.-Bicêtre, France

Jana Preis Division of Infectious Diseases, New York Harbor Health Care System, Brooklyn Campus, and State University of New York, Downstate Medical School, Brooklyn, New York, NY, USA

Giovanni Rezza Department of Infectious, Parasitic and Immunomediated Diseases, Istituto Superiore di Sanità, Rome, Italy

Pierre E. Rollin Viral Special Pathogens Branch, Centers for Disease Control and Prevention, Atlanta, GA, USA

John Schieffelin Tulane Health Sciences Center, New Orleans, LA, USA

Gerold Stanek Institute for Hygiene and Applied Immunology, Medical University of Vienna, Austria

Franc Strle Department of Infectious Diseases, University Medical Centre Ljubljana, Ljubljana, Slovenia

Daša Stupica Department of Infectious Diseases, University Medical Centre Ljubljana, Ljubljana, Slovenia

Tina Uršič Institute of Microbiology and Immunology, Faculty of Medicine, University of Ljubljana, Zaloška, Ljubljana, Slovenia

Preface

The concept of "emerging infections" was introduced to the scientific literature about 20 years ago and has been reflected in daily language at an increasing rate. Though there is no standard and short definition, emerging infections could be defined as:

1. A recognized infection spreading to new areas, species or populations.
2. The discovery that a known disease is caused by an infectious agent.
3. A previously unrecognized infection appearing in areas where the habitat is changing (e.g. deforestation).
4. A new infection resulting from mutations in a known microorganism.
5. An "old" infection re-emerging because it has become resistant to treatment, as a result of a breakdown in public health initiatives or due to changes in the host population.

Emerging infections are usually a threat to public health that requires a collaborative effort to combat. This collaborative action includes basic scientists, clinicians from medical and veterinary fields, public health experts and media forces.

Clinicians recognize a new disease entity but this might only be a first step for starting an enormous scientific effort in microbiology, epidemiology and other related fields. This effort could open new windows, may result in revolutionary discoveries that could inform clinical practice. By keeping the order of this cycle, we present you emerging infections based on clinical case studies. We hope that the book will be useful to fill the gaps in this cycle of basic sciences, public health and clinical practice.

Önder Ergönül, Füsun Can, Larry Madoff and Murat Akova

Chapter 1

Severe Fever with Thrombocytopenia Syndrome Associated with a Novel Bunyavirus

Keita Matsuno, Heinz Feldmann and Hideki Ebihara
Laboratory of Virology, Division of Intramural Research, National Institute of Allergy and Infectious Diseases, National Institutes of Health, Rocky Mountain Laboratories, Hamilton, MT, USA

CASE PRESENTATION

A 40-year-old female farmer, who lived in a rural, hilly area outside Chizhou city, in Anhui Province, China, was admitted to a hospital in Shanghai (400 km/78 miles from Chizhou) in the middle of May 2012. She had no other known underlying medical conditions but a history of schistosomiasis in 2008. There was no history of previous drug or food allergies or blood transfusions.

On May 16, 2012, the patient had sudden onset of fever with a peak temperature of 40°C accompanied by muscle pain. She was examined and subsequently admitted to a local hospital, where treatment with intravenous antibiotics was started. Following no improvement, the patient was transferred on May 19 to a hospital in Chizhou, where treatment was changed to aztreonam and ribavirin. With no changes in her clinical status and fever remaining high, the patient was transferred to a hospital in Shanghai on May 21, where she presented with continuing fever, chills, body aches, and diarrhea (five to six times a day). The suspected diagnosis of SFTS was made and the patient was admitted to the intensive care unit (ICU). Her treatment was continued with cefotiam, levodropropizine, and intensive care support. The condition of the patient deteriorated and she became apathetic on May 22. Subsequent treatment included vancomycin and platelet substitution therapy, but her condition further deteriorated. The following day she presented with a stiff neck, accompanied by

Emerging Infectious Diseases. DOI: http://dx.doi.org/10.1016/B978-0-12-416975-3.00001-7

enlarged lymph nodes in the neck, axillar and mediastinum, and abnormal brain waves on electroencephalogram. On May 25, the patient was in critical condition with shortness of breath, hypotension (88/50 mmHg), oliguria, severe acidosis, an abnormal flow index, but no skin bleeding or blood stasis. Thereafter, she was treated with prednisone, platelet and plasma transfusions, and hemodialysis before she died on the same day (May 25th) with multi-organ failure including kidney and lung, disseminated intravascular coagulation, and shock.

Laboratory findings showed decreased white blood cell counts throughout the disease progression, which increased on the day of her death. The platelet count (PLT) decreased over time along with concomitant prolonged thrombin time (TT) and activated partial thromboplastin time (APTT). Multi-organ dysfunction (including liver and kidney failure) was evident by elevated blood urea nitrogen (BUN) and hepatic transaminases (alanine aminotransferase, ALT; aspartate aminotransferase, AST), lactate dehydrogenase (LDH), and creatine kinase (CK). The presence of microscopic hematuria and proteinuria was also documented. On May 25 the results of an arterial blood gas were consistent with severe metabolic acidosis and respiratory compensation (pH: 7.17; PCO_2: 14.0 mmHg; PO_2: 110 mmHg; HCO_3^-: 5.10 mmol/L; CO_2^-Ct: 5.5 mmol/L; Beecf: −23.4 mmol/L; SBC: 8.5; Beb: −21.1 mmol/L; SO_2: 97%).

SFTSV infection was confirmed by quantitative real-time reverse transcriptase polymerase chain reaction (RT-PCR) following death. The patient had no known exposure to tick bites, but contact history with birds, rodents, and other wild animals was reported. Her activities during the past 2 weeks prior to disease onset involved fieldwork collecting cotton, rice, and tea. She had no known exposure to an SFTSV case or a person with similar illness and no similar cases were found in and around her residence. No secondary SFTSV cases were found among her contacts (adapted from reference[1]).

1. WHY THIS CASE WAS SIGNIFICANTLY IMPORTANT AS AN EMERGING INFECTION

Severe fever with thrombocytopenia syndrome (SFTS) was discovered in 2009 in Central China as a newly emerged clinical syndrome with clinical and epidemiological similarity to human anaplasmosis.[2] The causative agent of SFTS was identified as a novel phlebovirus, SFTS virus (SFTSV). Currently, SFTS cases have been reported from China, Japan, and South Korea with case fatality rates ranging from 10 to 30%. A milder but similar disease has been reported from the United States caused by Heartland virus, which is a recently related discovered phlebovirus.[3]

2. WHAT IS THE CAUSATIVE AGENT?

SFTSV is tentatively classified as a novel member of the genus *Phlebovirus*, family *Bunyaviridae*. As other bunyaviruses, SFTSV is an enveloped spherical-shaped particle, about 80–100 nm in diameter, carrying a genome of three segmented negative-stranded RNA molecules. The large RNA (L) segment is 6368 nucleosides in length and encodes for the RNA-dependent RNA polymerase (RdRp), which is responsible for viral RNA transcription and replication. The medium RNA (M) segment is 3378 nucleosides in length and encodes a membrane glycoprotein precursor, which is cleaved into the two mature membrane glycoproteins (Gn and Gc). Finally, the small RNA (S) segment is 1744 nucleosides in length encoding the nucleocapsid protein (N) and a non-structural protein (NSs) in an ambisense orientation (Figure 1.1).[2] Based on phylogenetic analysis of SFTSV isolates/sequences, the Chinese isolates currently form five distinct genetic lineages.[4] The Japanese isolates mostly belong to lineage D, whereas the Korean isolate clusters belong to lineage A.

3. WHAT IS THE FREQUENCY OF THE DISEASE IN ENDEMIC REGIONS?

Since the discovery in 2009,[2] SFTS cases have been reported from China,[1,5–7] Japan,[8] and South Korea[9] (Figure 1.2). As of September 2013, China has reported more than 600 laboratory confirmed cases, of which approximately 10% were fatal.[7] Most of them have been identified in rural, hilly areas of Central and Southern China (i.e., Henan, Hubei, Anhui, Shandong, Jiangsu, Zhejiang, Liaoning, Yunnan, Guangxi, Jiangxi, and Shaanxi provinces). The oldest confirmed cases were among a family cluster reported in September 2006 from Anhui Province.[10] Incidence rates of SFTS in Hubei and Shandong Province were reported to be $0.33/10^4$ and $5/10^5$, respectively.[11] SFTS cases are largely found among farmers and people living in rural, hilly areas with a higher risk for elderly females.[12] The antibody

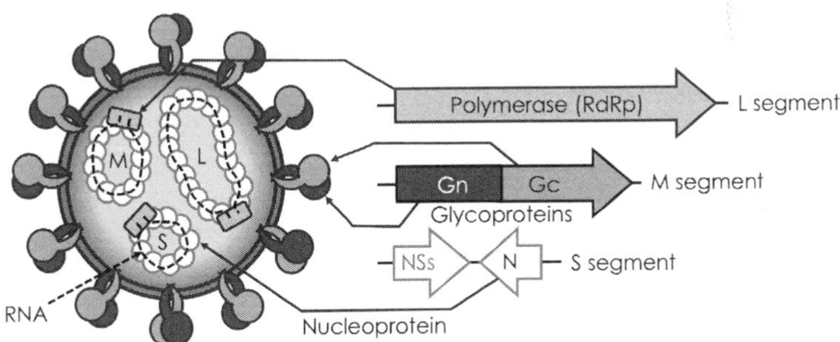

FIGURE 1.1 Schematic diagram of a SFTSV particle. SFTSV, a bunyavirus, presents as pleomorphic enveloped particles. The genome is tri-segmented, negative-sense, single-stranded RNA encoding for the nucleoprotein (N) and non-structural protein (NSs) (small (S) segment), the two glycoproteins (Gn and Gc) (medium (M) segment), and the RNA-dependent RNA polymerase (RdRp) (large (L) segment).

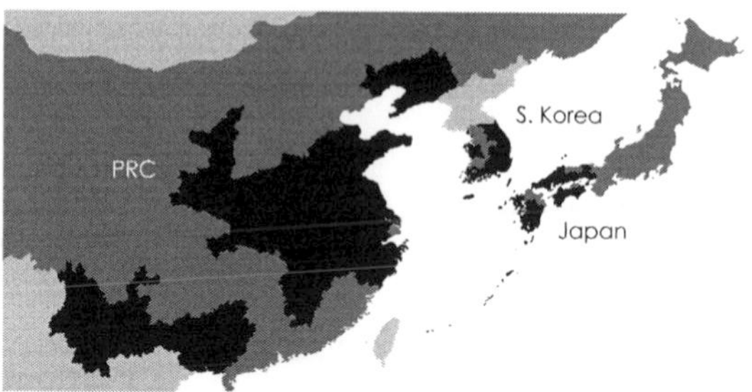

FIGURE 1.2 Endemic areas of SFTS. The map shows the endemic areas of SFTS in black. PRC = People's Republic of China; S. Korea = South Korea.

prevalence rate in healthy individuals from endemic areas ranges from 0.27 to 3.8%.[13,14] Antibody prevalence in endemic areas is high in domestic animals such as sheep, cattle, chickens, and dogs (up to 50% and higher).[15]

As of September 2013, 39 confirmed SFTSV infections and 18 deaths have been reported from Japan, mainly from the southwestern provinces of Hyogo, Shimane, Okayama, Hiroshima, Yamaguchi, Tokushima, Ehime, Saga, Nagasaki, Miyazaki, and Kagoshima. The oldest confirmed case was identified in November 2005 from Nagasaki Province.[8] Although no reports on human seroprevalence in Japan have been reported, specific antibodies against SFTSV were found in Japanese wildlife (Shika deer and wild boar) and hunting dogs.[16]

In South Korea, 15 deaths among 317 suspected SFTS cases have been reported, of which only 27 cases have been laboratory confirmed (reported by the Korean CDC). The distribution of confirmed infections is nationwide with cases identified in 10 provinces: Gangwon, Gyeongsangbuk, Daegu, Ulsan, Busan, Gyeongsangnam, Jeollanam, Jeju, Chungcheongnam, and Incheon provinces.[17]

4. HOW IS THE VIRUS TRANSMITTED?

Many (but not all) SFTS cases reported exposure to or bites of ticks during the incubation period, and case numbers in China and Japan peak during tick season.[8,14] Thus, exposure to or bite of an infected tick is thought to be the primary transmission route of SFTSV. Ticks of the species *Haemaphysalis longicornis* are widely distributed in China, Japan, and South Korea and have been found to carry SFTSV (viral RNA positive rate in China: 2.1–5.4%).[2,9,18,19] SFTSV RNA has also been detected in ticks of the species *Rhipicephalus microplus* (viral RNA positive rate in China: 0.6%),[19] *H. kitaokai*, *H. megaspinosa*, *H. flava*, and *Amblyomma testudinarium*,[16] which, therefore, could serve as additional vectors for transmission. There is no evidence for mosquitoes as potential vectors for SFTSV.[2]

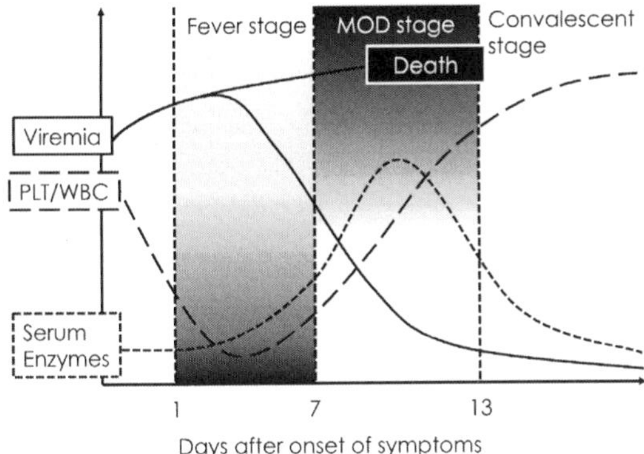

FIGURE 1.3 SFTS disease progression. The graph shows the SFTS disease progression, which can be separated into three stages: the fever stage, multi-organ dysfunction (MOD) stage, and convalescent stage.

Several small clusters of direct human-to-human transmission have been reported from China.[6,10,20,21] Each of these reported SFTS clusters began with a fatal index case but there have been no fatalities among secondary cases despite varying disease severity. Direct contact with blood of the fatal index case is the likely source and route of transmission. Higher viral loads in fatal compared to non-fatal cases have been reported,[22–24] supporting this transmission route. High prevalence of antibodies has been reported in certain domestic animal species indicating that contact with fresh meet, blood, and organs, particular during slaughter, might be another possible route of transmission.

5. WHAT ARE THE CLINICAL MANIFESTATIONS OF THE DISEASE?

The incubation period following tick exposure or unprotected contact with patients' blood or secretions/excretions is approximately 15 days on average (5–20 days).[25] SFTS disease progression can be divided into three stages (Figure 1.3): the fever stage (days 1–7 post onset of symptoms), the multi-organ dysfunction (MOD) stage (days 7–13 post onset of symptoms), and the convalescent stage (following approximately day 13 post onset of symptoms).[2,7,12,22] SFTS patients commonly present with fever, fatigue, nausea, vomiting, diarrhea, lymphadenopathy, and headache (Table 1.1). Additional symptoms such as anorexia, abdominal pain, malaise, myalgia, arthralgia, cough, and chills have been described together with hemorrhagic (e.g., conjunctival congestion, gingival bleeding, and/or melena) and central nervous system (e.g., confusion, apathy, slurred speech, dizziness, lethargy, convulsion, and/or coma) manifestations.

TABLE 1.1 Clinical Manifestations of SFTS Cases

Symptom	Number (Total Number)[a]	%
Fever	403 (409)	98.5
Anorexia	338 (389)	86.9
Fatigue	338 (409)	82.6
Body sores	289 (409)	78.3
Coarse breathing sounds	182 (350)	71.4
Chills	225 (409)	70.9
Nausea	39 (92)	70.7
Dizziness	219 (409)	68.9
Myalgia	206 (409)	57.3
Vomiting	223 (389)	55.0
Diarrhea	22 (92)	53.5
Abdominal pain/tenderness	189 (409)	52.0
Lymphadenopathy	12 (92)	50.4
Headache	129 (330)	46.2
Malaise	10 (112)	42.4
Cough	52 (409)	39.1
Sputum production	35 (151)	37.4
Muscle tremor	5 (92)	35.6
Lethargy	19 (151)	33.9
Arthralgia	202 (258)	33.6
Gingival bleeding	170 (238)	25.0
Confusion	183 (258)	23.9
Apathy	164 (238)	23.2
Melena	89 (238)	18.6
Throat congestion	80 (238)	13.0
Petechiae	5 (20)	12.7
Coma	6 (79)	12.6
Conjunctival congestion	20 (59)	8.9
Convulsion	21 (59)	8.5
Hematemesis	5 (59)	7.6
Slurred speech	11 (59)	5.4

[a] *Total number of cases were derived from four reports.*[2,5,22,26]

The fever stage is characterized by a rather non-specific prodrome including fever (temperature >38°C), headache, myalgia, arthralgia, dizziness, and malaise, which may persist for a week. During this stage patients commonly already display thrombocytopenia, leukocytopenia, and viremia (10^5–10^6 copies/mL). The MOD stage is characterized by marked thrombocytopenia, marked leukocytopenia, elevated liver enzymes, mucosal hemorrhages, hemorrhagic rash, disseminated intravascular coagulopathy, and central nervous system manifestations such as confusion. Elevated levels of ALT, AST, LDH, CK, and CK-MB can already be noticed during the late phase of the fever stage, but are manifest during the MOD stage. In fatal cases, viremia increases by up to 10^{10} copies/mL (average 10^8 copies/mL) and serum levels of AST (>400 U/L), LDH (>800 U/L), CK (>1000 U/L), and CK-MB (>50 U/L) dramatically increase indicating multi-organ failure.[22–24] The convalescent stage can last up to 30 days and more and is characterized by decreasing levels of viremia, diminishing clinical symptoms and normalization of clinical laboratory parameters.[7,23]

6. WHAT ARE THE MECHANISMS OF PATHOGENESIS?

Age and sex have been associated with disease outcome of SFTS and elderly females seem to have a higher risk for severe/fatal disease,[12] even though social and/or behavioral factors cannot be excluded as explanations for this finding.

Important factors in the pathogenesis of SFTS likely include high viral load, inflammatory responses similar to a systemic inflammatory response syndrome (SIRS), coagulation abnormalities, and multi-organ dysfunction. Uncontrolled upregulation of several cytokines (e.g., IL-1RA, IL-6, IL-10, G-CSF, IP-10, and MCP-1) appears to correlate with severity of disease suggesting that SIRS is a significant factor in SFTS pathogenesis.[23,24] Elevated serum levels of ALT, AST, LDH, CK, and CK-MB are commonly observed,[7] indicating pathological lesions and potential failure of liver and kidney. Other common clinical abnormalities are thrombocytopenia and prolonged APTT indicative of coagulation disorders including disseminated intravascular coagulopathy (DIC) resulting in hemorrhagic manifestations.[22–25,27]

As of today, experimental SFTSV infection and disease progression have been described only in rodents. Adult mice, rats, and Syrian hamsters are susceptible to SFTSV infection, but do not develop signs of severe illness.[28–30] Following intracerebral inoculation of SFTSV, newborn mice and rats develop severe disease, which appears uniformly lethal.[28,30] Adult mice (C57BL/6) are susceptible to multiple routes of infection with SFTSV, resulting in a mild, self-limited disease that recapitulates the hematological manifestations of SFTS. The infection is characterized by an acute viremia accompanied by virus replication in several organs (i.e., liver, kidney, and spleen), transient thrombocytopenia, leukocytopenia, and elevated levels of AST. Following decreased viral loads, increases in BUN and ALT are observed in association with transient liver necrosis and impaired renal function. This is accompanied by an increase in SFTSV-specific cellular and humoral immune responses. Knockout of interferon α/β receptor or inhibition of mouse adaptive immune responses results in more severe clinical signs with weight loss and case fatality rates of up to 100%[29,30] indicating that impairment of the immune system is associated with severe clinical disease as has been suggested for severe human disease.

7. HOW DO YOU DIAGNOSE THE INFECTION?

SFTS should be considered in patients with a syndrome characterized by acute fever (temperatures $\geq$38°C), thrombocytopenia (PLT <100 K/L) and leukopenia (WBC <2.5 K/L), in particular in association with a history of tick exposure/bite in an endemic area. Elevated serum levels of liver and kidney enzymes and gastrointestinal symptoms may serve as supportive diagnostic parameters. For laboratory diagnosis (virology), whole blood and serum samples should be collected.

Molecular assays targeting viral RNA are the first choice for diagnosis during the acute stage of SFTSV infections. For this, multiple platforms have been developed and evaluated including one-step real-time RT-PCR,[31–33] reverse transcription loop-mediated isothermal amplification assay (RT-LAMP),[34,35] and RT-cross priming amplification (RT-CPA).[36] In case this is the only technology applied, molecular testing might be confirmed by an assay detecting a second target or the same target on two consecutive samples. Serological assays largely based on enzyme-linked immunosorbent assay (ELISA) technology for the detection of SFTSV-specific IgM and IgG antibodies have been developed and should be considered for convalescent stage samples and as confirmatory assays for acute diagnosis.[37] Specific IgM antibodies are usually detectable by 4–5 days after infection and remain present for up to 4–5 months with a peak between 2 and 3 weeks after infection. Plaque reduction neutralization test (PRNT) utilizing convalescent stage specimens and virus isolation utilizing acute stage specimens serve as confirmatory assays, but need to be performed under high biocontainment conditions not necessarily available at diagnostic centers.

China uses the following criteria for the laboratory diagnosis of SFTS cases: (1) viral RNA positive in serum or whole blood specimens, (2) virus-specific IgM positive during acute phase disease, (3) a four-fold increase or conversion of virus-specific IgG in paired sera (acute and convalescent phase specimens), or (4) isolation of virus from patient samples.[7]

8. HOW DO YOU DIFFERENTIATE THE INFECTION FROM SIMILAR CLINICAL MANIFESTATIONS?

H. longicornis, the main potential vector for SFTSV, is widely distributed in East Asia and Australia.[38] In this area, SFTS needs to be considered for the diagnosis of an acute febrile disease. Although thrombocytopenia and lymphocytopenia are the most common findings associated with SFTSV infections,[2,5] those hematological parameters are also seen with other frequent or rare diseases. The number one differential diagnosis is human anaplasmosis, but other arboviral or zoonotic infectious diseases should be considered as well, such as Crimean-Congo hemorrhagic fever, dengue fever, hemorrhagic fever with renal syndrome, tick-borne encephalitis, ehrlichiosis, Lyme disease, Tsutsugamushi fever (Scrub typhus), rickettsiosis (*Rickettsia heilongjiangensis*), leptospirosis, Q fever, bacterial sepsis, trypanosomiasis, acute viral hepatitis, influenza, and bacterial meningitis. Therefore, rapid and sensitive laboratory diagnosis is absolutely critical (see above). For differential diagnosis, multiplex real-time RT-PCR assay would be beneficial.[39]

9. WHAT ARE THE THERAPEUTIC APPROACHES?

Currently, treatment largely consists of supportive care as there is no antiviral/therapeutic approach established. The use of the antiviral ribavirin has been reported,[1,10,21,22,27,40] but a significant therapeutic effect is not obvious. The use of Suramin, an anti-trypanosomal drug, has been suggested,[41] but human efficacy data are missing.

Supportive care includes fluid and electrolyte substitution and transfusions of fresh frozen plasma and platelets, but the effects are limited with no convincing improvement being documented. Antibiotic coverage should be initiated to prevent or combat secondary bacterial infections, especially the use of tetracycline, and new quinolones are indicated in case of tick exposure/bite to prevent possible tick-borne bacterial (co-)infections (discussed above). These antibiotics might also be beneficial due to their ability to downregulate host inflammatory responses.[42,43]

10. WHAT ARE THE PREVENTIVE AND INFECTION CONTROL MEASURES?

As for specific treatment, no vaccine is currently available for the prophylaxis of SFTSV infections. Therefore, the main measures for protection are education, prevention of tick exposure/bite, and prevention of exposure to blood and excretions/secretions of patients. SFTSV RNA has been detected in throat swabs, urine, and feces in some cases[23] indicating the potential of virus shedding, but the presence of infectious virus in those specimens has not been further evaluated. Remarkably, there is no evidence for air-borne transmission of SFTSV.

Overall, basic personal protective devices (PPE) such as gloves, mask, and eye and skin protection are recommended to prevent exposure/infection in medical centers and families of confirmed cases. The general population should use insect repellent, wear long sleeve shirts, and avoid areas of tick endemicity during the tick season.

ACKNOWLEDGMENTS

Work on severe fever with thrombocytopenia syndrome (SFTS) virus is funded by the Intramural Research Program of the National Institute for Allergy and Infectious Diseases and partially through the US–China Biomedical Collaborative Research Program.

REFERENCES

1. Pan H, Hu J, Liu S, et al. A reported death case of a novel bunyavirus in Shanghai, China. *Virol J* 2013;**10**:187.
2. Yu X-J, Liang M-F, et al. Fever with thrombocytopenia associated with a novel bunyavirus in China. *N Engl J Med* 2011;**364**:1523–32.
3. McMullan LK, Folk SM, Kelly AJ, et al. A new phlebovirus associated with severe febrile illness in Missouri. *N Engl J Med* 2012;**367**:834–41.
4. Lam TT-Y, Liu W, Bowden TA, et al. Evolutionary and molecular analysis of the emergent severe fever with thrombocytopenia syndrome virus. *Epidemics* 2013;**5**:1–10.
5. Bian-li X, Liu L, Huang X, et al. Metagenomic analysis of fever, thrombocytopenia and leukopenia syndrome (FTLS) in Henan Province, China: discovery of a new bunyavirus. *PLoS Pathog* 2011;**7**:e1002369.
6. Chang-jun B, Xii-ling G, Qi X, et al. A family cluster of infections by a newly recognized bunyavirus in eastern China, 2007: further evidence of person-to-person transmission. *Clin Infect Dis* 2011;**53**:1208–14.
7. Dexin L. A highly pathogenic new bunyavirus emerged in China. *Emerg Microbes Infect* 2013;**2**:1–4.
8. Takahashi T, Maeda K, Suzuki T, Ishido A, Shigeoka T, Tominaga T, et al. The first identification and retrospective study of severe fever with thrombocytopenia syndrome in Japan. *J Infect Dis* 2014;**209**:816–27.
9. Kim K-H, Yi J, Kim G, Choi SJ, Jun KI, Kim N-H, et al. Severe fever with thrombocytopenia syndrome, South Korea, 2012. *Emerg Infect Dis* 2013;**19**:1892–4.
10. Liu Y, Li Q, Hu W, et al. Person-to-person transmission of severe fever with thrombocytopenia syndrome virus. *Vector Borne Zoonotic Dis* 2012;**12**:156–60.
11. Zhang X, Liu Y, Zhao L, et al. An emerging hemorrhagic fever in China caused by a novel bunyavirus SFTSV. *Sci China Life Sci* 2013;**56**:697–700.
12. Xiong W-Y, Feng Z-J, Matsui T, Foxwell AR. Risk assessment of human infection with a novel bunyavirus in China. *Western Pac Surveill Response J* 2012;**3**:61–6.
13. Zhang W, Zeng X, Zhou M, et al. Seroepidemiology of severe fever with thrombocytopenia syndrome bunyavirus in Jiangsu province. *Dis Surveill* 2011;**26**:676–8.
14. Li L, Guan X-H, Xue-sen X, et al. Epidemiologic analysis on severe fever with thrombocytopenia syndrome in Hubei province, 2010. *Chin J Epidemiol* 2012;**33**:168–72.

15. Niu G, Li J, Liang M, et al. Severe Fever with Thrombocytopenia Syndrome Virus among Domesticated Animals, China. *Emerg Infect Dis* 2013;**19**:756–63.
16. Infectious Agents Surveillance Report. ＜速報＞重症熱性血小板減少症候群（SFTS）ウイルスの国内分布調査結果（第一報）. National Institute of Infectious Diseases, Japan; 2013.
17. Korea Centers for Disease Control & Prevention, South Korea. [참고자료] SFTS 바이러스 확진환자 현황('13.9.16 현재) [document on the Internet]. KCDC; 2013 Sept. 16 [cited 2013 Nov. 22]. Available from: http://www.cdc.go.kr/CDC/intro/CdcKrIntro0201.jsp?menuIds=HOME001-MNU0005-MNU0011&cid=21575.
18. Jiang X-L, Wang X, Li J-D, et al. Isolation, identification and characterization of SFTS bunyavirus from ticks collected on the surface of domestic animals. *Chin J Virol* 2012;**28**:252–7.
19. Zhang Y-Z, Zhou D-J, Qin X-C, et al. The ecology, genetic diversity, and phylogeny of Huaiyangshan virus in China. *J Virol* 2012;**86**:2864–8.
20. Chang-jun B, Qi X, Wang H. A novel bunyavirus in China. *N Engl J Med* 2011;**365**:862–3.
21. Gai Z, Liang M, Zhang Y, et al. Person-to-person transmission of severe fever with thrombocytopenia syndrome bunyavirus through blood contact. *Clin Infect Dis* 2011;**54**:249–52.
22. Gai Z-T, Zhang Y, Liang M-F, et al. Clinical progress and risk factors for death in severe fever with thrombocytopenia syndrome patients. *J Infect Dis* 2012;**206**:1095–102.
23. Zhang Y-Z, Yong-wenn H, Dai Y-A, et al. Hemorrhagic fever caused by a novel Bunyavirus in China: pathogenesis and correlates of fatal outcome. *Clin Infect Dis* 2012;**54**:527–33.
24. Sun Y, Jin C, Zhan F, et al. Host cytokine storm is associated with disease severity of severe fever with thrombocytopenia syndrome. *J Infect Dis* 2012;**206**:1085–94.
25. Deng B, Zhang S, Geng Y, et al. Cytokine and chemokine levels in patients with severe fever with thrombocytopenia syndrome virus. *PLoS One* 2012;**7**:e41365.
26. Jian-li H, Chang-jun B, Xian Q, et al. Clinical and epidemiological characteristics on 20 cases of SFTSV infection. *Chinese J Zoonosis* 2012;**28**:302–5.
27. Lu QB, Cui N, Li H, et al. Case fatality and effectiveness of ribavirin in hospitalized patients with severe fever with thrombocytopenia syndrome in China. *Clin Infect Dis* 2013;**57**:1292–9.
28. Chen X-P, Cong M-L, Li M-H, et al. Infection and pathogenesis of Huaiyangshan virus (a novel tick-borne bunyavirus) in laboratory rodents. *J Gen Virol* 2012;**93**:1288–93.
29. Jin C, Liang M, Ning J, et al. Pathogenesis of emerging severe fever with thrombocytopenia syndrome virus in C57/BL6 mouse model. *Proc Natl Acad Sci USA* 2012;**109**:10053–8.
30. Liu Y, Wu B, Paessler S, Walker DH, Tesh RB, Yu X-J. The pathogenesis of SFTSV infection in IFNα/β knockout mice: insights into the pathologic mechanisms of a new viral hemorrhagic fever. *J Virol* 2014;**88**:1781–6.
31. Sun Y, Liang M, Qu J, et al. Early diagnosis of novel SFTS bunyavirus infection by quantitative real-time RT-PCR assay. *J Clin Virol* 2012;**53**:48–53.
32. Li Z, Cui L, Zhou M, et al. Development and application of a one-step real-time RT-PCR using a minor-groove-binding probe for the detection of a novel bunyavirus in clinical specimens. *J Med Virol* 2013;**85**:370–7.
33. Yuan-yuan C, Shen-yong Z, Hong-ling W, et al. SFTSV RNA detection in sera of patients suffering from fever with thrombocytopenia syndrome. *J Shandong Univ (Health Sci)* 2012;**50**:118–21.
34. Yang G, Li B, Liu L, Huang W, Zhang W, Liu Y. Development and evaluation of a reverse transcription loop-mediated isothermal amplification assay for rapid detection of a new SFTS bunyavirus. *Arch Virol* 2012;**157**:1779–83.

35. Xu H, Zhang L, Shen G, et al. Establishment of a novel one-step reverse transcription loop-mediated isothermal amplification assay for rapid identification of RNA from the severe fever with thrombocytopenia syndrome virus. *J Virol Methods* 2013;**194**:21–5.
36. Cui L, Ge Y, Qi X, et al. Detection of severe fever with thrombocytopenia syndrome virus by reverse transcription-cross-priming amplification coupled with vertical flow visualization. *J Clin Microbiol* 2012;**50**:3881–5.
37. Jiao Y-J, Zeng X, Guo X, et al. Preparation and evaluation of recombinant severe fever with thrombocytopenia syndrome virus nucleocapsid protein for detection of total antibodies in human and animal sera by double-antigen sandwich enzyme-linked immunosorbent assay. *J Clin Microbiol* 2012;**50**:372–7.
38. You M, Xuan X, Tsuji N, et al. Identification and molecular characterization of a chitinase from the hard tick Haemaphysalis longicornis. *J Biol Chem* 2003;**278**:8556–63.
39. Li Z, Qi X, Zhou M, et al. A two-tube multiplex real-time RT-PCR assay for the detection of four hemorrhagic fever viruses: severe fever with thrombocytopenia syndrome virus, Hantaan virus, Seoul virus, and dengue virus. *Arch Virol* 2013;**158**:1857–63.
40. Chen H, Hu K, Zou J, Xiao J. A cluster of cases of human-to-human transmission caused by severe fever with thrombocytopenia syndrome bunyavirus. *Int J Infect Dis* 2013;**17**:e206–8.
41. Jiao L, Ouyang S, Liang M, et al. Structure of severe fever with thrombocytopenia syndrome virus nucleocapsid protein in complex with suramin reveals therapeutic potential. *J Virol* 2013;**87**:6829–39.
42. Salvatore CM, Techasaensiri C, Tagliabue C, et al. Tigecycline therapy significantly reduces the concentrations of inflammatory pulmonary cytokines and chemokines in a murine model of Mycoplasma pneumoniae pneumonia. *Antimicrob Agents Chemother* 2009;**53**:1546–51.
43. Kuwahara K, Kitazawa T, Kitagaki H, Tsukamoto T, Kikuchi M. Nadifloxacin, an antiacne quinolone antimicrobial, inhibits the production of proinflammatory cytokines by human peripheral blood mononuclear cells and normal human keratinocytes. *J Dermatol Sci* 2005;**38**:47–55.

Chapter 2

Bas-Congo Virus: A Novel Rhabdovirus Associated with Acute Hemorrhagic Fever

Gilda Grard[1], Joseph Fair[2], Charles Chiu[3,4,5] and Eric Leroy[1,6]

[1]Unité des Maladies Virales Emergentes, Centre International de Recherches Médicales de Franceville, Franceville, Gabon, [2]Metabiota, Incorporated, San Francisco, CA, USA, [3]Department of Medicine, Division of Infectious Diseases, University of California, San Francisco, CA, USA, [4]Department of Laboratory Medicine, University of California, San Francisco, CA, USA, [5]UCSF-Abbott Viral Diagnostics and Discovery Center, San Francisco, CA, USA, UMR MIVEGEC (IRD 224 - CNRS 5290 - UM1 - UM2), [6]Institut de Recherche pour le Développement, Montpellier, France

CLINICAL PRESENTATION

A 32-year-old male nurse from Mangala village in the Bas-Congo province of Democratic Republic of Congo (DRC) fell ill on June 13, 2009. Symptoms included epistaxis, ocular and oral hemorrhages, hematemesis, and bloody diarrhea. Two days later, the patient developed fever >39°C, anorexia, headache, fatigue, and abdominal pain. He was transferred that same day to the regional general hospital of Boma, approximately 30 km from Mangala village, where a serum sample was taken for analysis. Supportive care was then provided, including fluid resuscitation, blood transfusion, and empiric antibiotics. Initial laboratory tests were negative for malaria, tuberculosis, dengue, and bacterial sepsis. The patient recovered spontaneously from the episode of presumptive acute hemorrhagic fever.

EPIDEMIOLOGICAL CONTEXT

Within the past 3 weeks prior to becoming ill with hemorrhagic fever, the nurse had cared directly for two teenagers who had presented to the Mangala village health center with hemorrhagic symptoms. The first patient was a 15-year-old boy seen on May 25, 2009 with malaise, epistaxis, conjunctival injection, gingival bleeding, hematemesis, and diarrhea with blood.

Emerging Infectious Diseases. DOI: http://dx.doi.org/10.1016/B978-0-12-416975-3.00002-9

Although no fever was documented, he died rapidly within 2 days after onset of symptoms. The second patient was a 13-year-old girl who attended the same public school, although there was no history of any direct contact between them. She presented on June 5, 2009 with headache, fever >39°C, abdominal pain, epistaxis, conjunctival injection, mouth bleeding, hematemesis, and diarrhea with blood. Despite receiving symptomatic treatment for fever and quinine for malaria, she died within 3 days after onset of symptoms.

All three cases lived approximately 50 meters from one another in the same neighborhood in Mangala village. Household contacts for each of the three patients were closely monitored for 21 days, the duration required upon suspicion of viral hemorrhagic fever, and no other cases were observed. Notably, although DRC is an endemic country for Ebola and Marburg viral hemorrhagic fever, no previous outbreaks had ever been reported in the Bas-Congo province.

LABORATORY INVESTIGATION FOR VIRAL HEMORRHAGIC FEVER (VHF) DIAGNOSIS

The only available serum sample from the surviving 32-year-old nurse was received by the Centre International de Recherches Médicales de Franceville (CIRMF) on June 29, 2009 for viral hemorrhagic fever laboratory testing. The patient sample tested negative for Ebola virus, Marburg virus, Crimean-Congo hemorrhagic fever virus, yellow fever virus, Rift Valley fever virus, and dengue virus.

DISCOVERY OF BAS-CONGO VIRUS, A NOVEL RHABDOVIRUS, IN PATIENT SERUM

Extracted RNA from the patient serum sample was then subjected to metagenomic investigation, leading to the detection of viral genome sequences from a previously unrecognized novel rhabdovirus, subsequently named Bas-Congo virus (BASV). Likely due to the break in cold chain encountered during sample collection in DRC, the virus failed to grow on cell cultures and in suckling mice brain, but in-depth genetic analysis by next-generation sequencing and *de novo* genome assembly allowed full genomic characterization of the virus, phylogenetic confirmation of its membership in the *Rhabdoviridae* family, and development of a virus neutralization serological assay for BASV using a "pseudotyped" construct. Using the serological test, specific BASV-neutralizing antibodies were detected in serum from the surviving nurse and from an asymptomatic close contact, another nurse who took care of him upon onset of hemorrhagic symptoms in Mangala village and assisted in transporting him to the hospital in Boma. These three cases and associated research investigations were previously published in 2012.[1]

1. WHY THIS CASE WAS SIGNIFICANTLY IMPORTANT AS AN EMERGING INFECTION

Bas-Congo virus is a newly discovered rhabdovirus found in serum from a patient with acute hemorrhagic fever. Rhabdoviruses that are known to be pathogenic for humans have previously been associated with encephalitic syndromes (rabies virus and Chandipura virus)[2,3] or influenza-like syndromes (vesicular stomatitis virus).[4] This case represents the first time that a member of the family *Rhabdoviriadae* has been associated with acute hemorrhagic fever in humans.

2. WHAT IS THE CAUSATIVE AGENT?

2.1 BASV as the Etiologic Agent of an Acute Hemorrhagic Fever Syndrome

Upon discovery of a new microorganism, establishing a causal and unambiguous relationship to the disease requires several lines of evidence. This is particularly true when the suspected agent is assigned to a group that has never been associated with the type of disease under investigation. This is the case for the association of BASV with hemorrhagic fever syndrome, as was also the case with Sin Nombre hantavirus and acute pulmonary syndrome during the 1993 Four Corners outbreak in the United States.[5,6] Evidence of causality can be obtained by detecting the virus in geographically and temporally linked clusters of sick individuals with similar clinical presentations, and/or inoculation of the virus (either naturally cultured or artificially generated by reverse genetics) in a healthy animal model to induce the observed pathology.

To date, the lines of evidence supporting BASV as the etiologic agent of the hemorrhagic fever cases described here include:

1. BASV was the only credible pathogen found in the serum of the acutely ill patient
2. The viral copies per ml were 1.09×10^6 RNA copies/ml, titers similar to those observed in survivors of Ebola virus infection and correlated with disease severity
3. The close epidemiological clustering of the three patients who lived in the same neighborhood and presented with similar symptoms over a 3-week time frame
4. Clearance of viral RNA and detection of specific neutralizing antibodies in the serum of the surviving patient after convalescence.

2.2 BASV Taxonomy

The *Rhabdoviridae* family (order *Mononegavirales*) is associated with an extremely diverse host range and is currently divided into eight genera.[7,8] The genera *Cytorhabdovirus* and *Nucleorhabdovirus* are associated with plant infections, the genera *Perhabdovirus* and *Novirhabdovirus* with fish infections, and the genera *Sigmavirus* with insect infections. The genus *Lyssavirus* is associated with infections of bats and carnivores and includes the highly pathogenic rabies viruses. Finally, the genera *Ephemerovirus*, *Tibrovirus*, and *Vesiculovirus* form together, with about 120 unclassified viruses, the dimarhabdovirus supergroup (dipteran-mammal-associated rhabdovirus), the genus *Vesiculovirus* including human pathogens and viruses responsible for hemorrhagic disease in fish. Phylogenetic (Figure 2.1) and whole genome analysis of BASV identifies it is a member of the dimarhabdovirus supergroup, most closely related to the genera *Ephemerovirus* and *Tibrovirus*.[1] Viruses from these genera are usually associated with arthropod and cattle infections but have not previously been known to infect humans. Additionally, the phylogenetic branching pattern and the high amino acid divergence (≈70%) of BASV suggest it represents a distinct viral group that has yet to be taxonomically classified.

2.3 Virus Description

Rhabdoviruses are enveloped viruses with single-stranded negative-sense RNA genomes. Virions are bullet shaped, from 100 to 430 nm long and 45 to 100 nm wide.[7,9] The glycoprotein G is anchored in the lipid bilayer and is responsible for virus entry into cells (Figure 2.2A). Viral RNA is encapsulated with the nucleocapsid protein (N). This ribonucleocapsid is associated with the viral polymerase (L) through interaction with the viral phosphoprotein (P). The ribonucleocapsid is packaged with matrix protein (M), which gives the virion its characteristic form.[9] The near-complete BASV genome sequence is 11,892 nucleotides long (GenBank accession number JX297815). The genome organization of BASV displays eight open reading frames (ORFs) in the following order (Figure 2.2B): 3′- nucleoprotein (N), phosphoprotein (P), matrix protein (M), U1 protein (U1), U2 protein (U2), glycoprotein (G), U3 protein (U3), and RNA-dependent RNA polymerase (L). The N-P-M-G-L genome organization is common to rhabdoviruses while the insertion of U1, U2, and U3 genes is shared only with members of the genus *Tibrovirus*. Functions of the proteins U1 and U2 are not known, although U3 protein is hypothesized to be a candidate viroporin.[10]

3. WHAT IS THE FREQUENCY OF THE DISEASE?

To date, BASV has only been found in the Bas-Congo province of Democratic Republic of Congo, Central Africa (Figure 2.3). Infection was confirmed for one surviving patient, who presented with hemorrhagic fever symptoms, and suspected for two other closely epidemiologically linked patients, who died rapidly from fulminant hemorrhagic fever before samples could be collected. Infection by BASV was retrospectively confirmed in

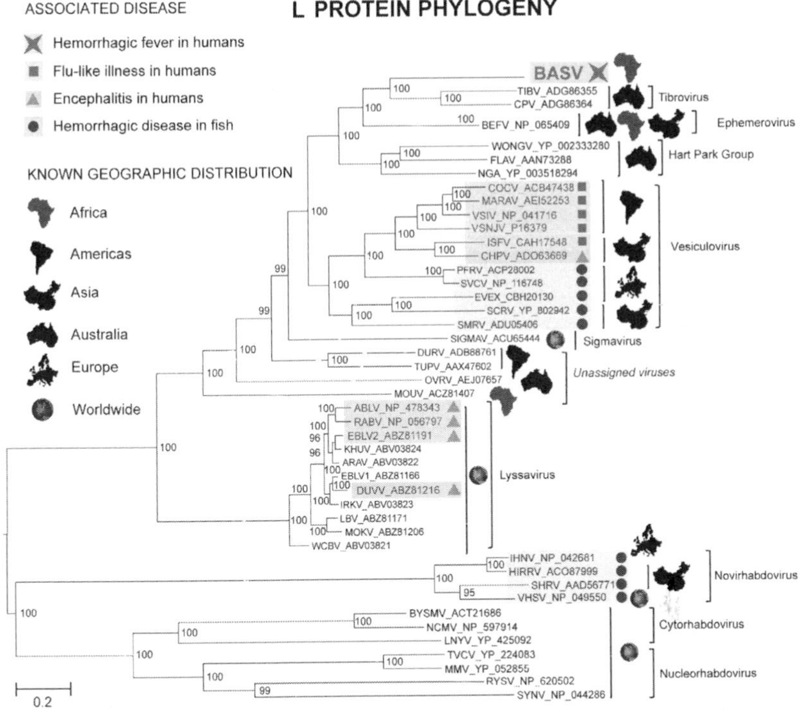

FIGURE 2.1 Phylogenetic analysis of the L proteins of rabdovirus. Bayesian tree topologies were assessed with MrBayes V.32 software (20,000 sampled trees; 5000 trees discarded as burn-in). Original figure was published by Grard et al.[1] under the creative commons license (CC-BY, available at: http://creativecommons.org/licenses/by/3.0/) and slightly modified for taxonomic updates. Virus abbreviations are listed as follows, in alphabetical order: ABLV, Australian bat lyssavirus; ARAV, Aravan virus; BEFV, bovine ephemeral fever virus; BYSMV, barley yellow striate mosaic virus (BYSMV); CHPV, Chandipura virus; CPV, Coastal Plains virus; COCV, cocal virus; DURV, Durham virus; DUVV, Duvenhage virus; EBLV1, European bat lyssavirus 1; EBLV2, European bat lyssavirus 2; EVEX, eel virus European X virus; FLAV, Flanders virus; HIRRV, Hirame rhabdovirus; IHNV, infectious hematopoietic necrosis virus; IRKV, Irkut virus; ISFV, Isfahan virus; KHUV, Khujand virus; LBV, Lagos bat virus; LNYV, lettuce necrotic yellow virus; MARAV, Maraba virus; MMV, maize mosaic virus; MOKV, Mokala virus; MOUV, Moussa virus; NCMV, northern cereal mosaic virus; NGAV, Ngaingan virus; OVRV, Oak Vale rhabdovirus; PFRV, pike fry rhabdovirus; RABV, rabies virus; RYSV, rice yellow stunt rhabdovirus; SIGMAV, sigma virus; SCRV, *Siniperca chuatsi* rhabdovirus; SHRV, snakehead virus; SMRV, *Scophthalmus maximus* rhadovirus; SVCV, spring viremia of carp virus; SYNV, Sonchus yellow net virus; TIBV, Tibrogargan virus; TUPV, Tupaia virus; TVCV, tomato vein clearing virus; VHSV, viral hemorrhagic septicemia virus; VSIV, vesicular stomatitis virus, Indiana; VSNJV, vesicular stomatitis virus, New Jersey; WCBV, West Caucasian bat virus; WONGV, Wongabel virus.

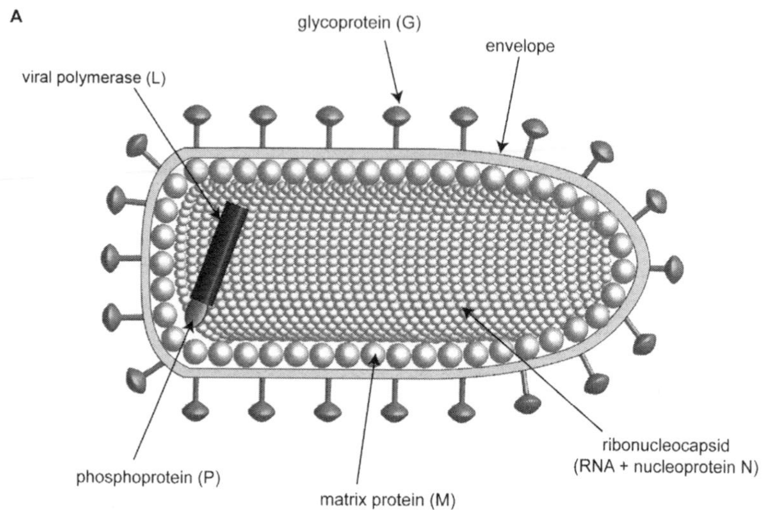

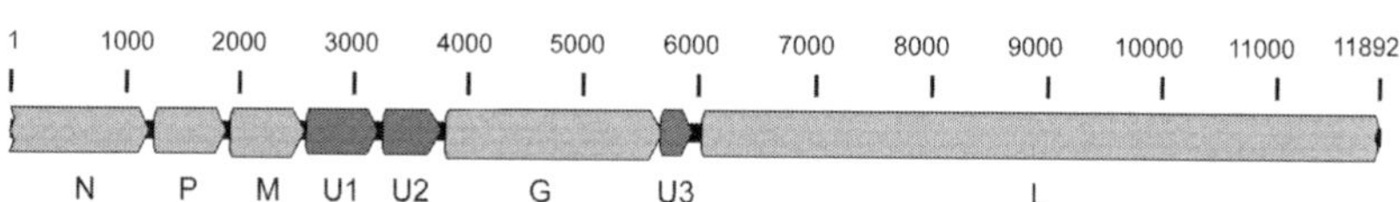

FIGURE 2.2 Panel A. Schematic representation of the rhabdovirus virion structure. Panel B. Schematic representation of the Bas-Congo virus genome organization.

another asymptomatic contact, a nurse who cared directly for the surviving patient, by seroneutralization studies. The proportion of asymptomatic vs. symptomatic infections with severe or possibly mild clinical manifestations is not known. A preliminary epidemiological study that included (1) molecular screening of 50 serum samples from patients presenting with hemorrhagic fever syndrome in DRC and (2) serological screening of 50 plasma samples from blood donors in the Kasai-Orientale province of DRC failed to detect additional cases of BASV infection. These results suggest that human infection with BASV is infrequent or that the virus has only emerged very recently. Larger epidemiological studies and surveillance of active cases of unknown hemorrhagic fever are needed to establish the frequency and impact of BASV-associated disease.

4. HOW IS THE VIRUS TRANSMITTED?

To date, the source of infection and the possible ways of transmission have not been established. Waterborne or airborne transmission appears unlikely

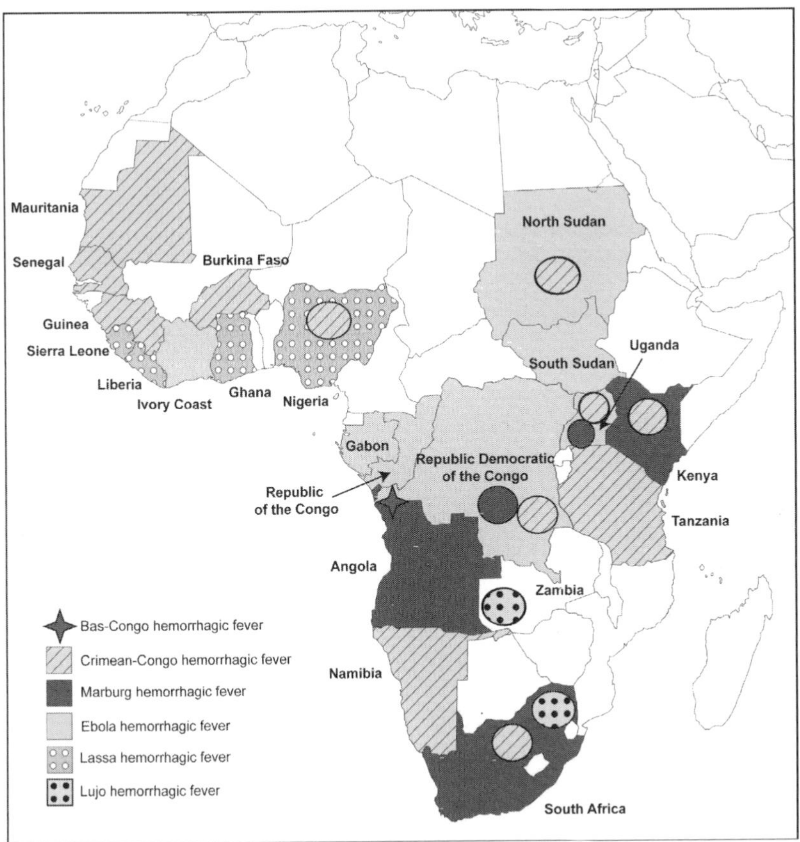

FIGURE 2.3 Map of viral hemorrhagic fever outbreaks in Africa. Yellow fever and dengue fever have a wide geographic distribution throughout sub-Saharan Africa, as well as Rift Valley fever which extends to the south of Africa and are not shown. Mangala village, located in Democratic Republic of the Congo, is represented by a red star. Original figure was published by Grard et al.[1] under the creative commons license (CC-BY, available at: http://creativecommons.org/licenses/by/3.0/) and slightly modified.

due to the small number of reported cases. The phylogenetic position of BASV in the dimarhabdovirus supergroup is consistent with a possible arthropod-borne transmission. Additionally, the epidemiological data also suggest potential human-to-human transmission. Indeed, the surviving nurse directly cared for the first two patients who presented with hemorrhagic fever symptoms, and he in turn was taken care of by the serologically confirmed asymptomatic contact. However, none of the household contacts became ill, suggesting that human-to-human transmission may be limited. This putative pattern of transmission for BASV is also seen with Crimean-Congo hemorrhagic fever virus, which is an arboviral VHF agent associated with additional cases of nosocomial human-to-human transmission.[11]

5. WHAT ARE THE CLINICAL MANIFESTATIONS?

Clinical symptoms shared by all three cases were an abrupt onset of disease, with hemorrhagic manifestations not limited to gastrointestinal sites but also affecting mucosa: epistaxis (nose bleeding), ocular hemorrhage or conjunctival injection (eye bleeding), oral hemorrhage (mouth bleeding), hemorrhagic vomiting, and hemorrhagic diarrhea. Associated unspecific symptoms included one or more of the following: fever, malaise, headache, abdominal pain, fatigue, and anorexia. Notably, fever may be concomitant with early symptoms or appear several days later, potentially explaining why one patient died without documented fever. As revealed by the retrospective serologic study, asymptomatic infection from BASV may occur. In addition, BASV-associated clinical presentations with subclinical, mild, or unrelated symptoms, such as rashes, arthralgias, hepatitis, or neurologic disorders, cannot be excluded at the present time. Many hemorrhagic fever viruses, such as yellow fever[12], are associated with a variety of associated clinical manifestations ranging from mild to severe.

6. HOW DO YOU DIAGNOSE?

6.1 Molecular Diagnosis

Diagnosis of severe viral hemorrhagic fevers classically includes detection of viral RNA or antigens in serum or blood samples taken during the acute phase of the disease. BASV was first detected by random metagenomic analysis from RNA extracted from a serum sample taken during the acute phase of the disease. Conventional reverse transcription (RT)-PCR and real-time RT-PCR systems were subsequently implemented in the lab for viral RNA detection in human sera and are currently used for diagnostic investigations of suspected VHF cases.

6.2 Viral Isolation

Virus isolation is usually the gold standard technique for diagnosis of viral infection. However, as the time of virus cultivation is several days long, this approach is not suitable for emergency detection of highly pathogenic VHF viruses, which require immediate outbreak response and management. With BASV, attempts were made to grow the virus on several cell lines with unsuccessful results: Vero (green monkey kidney), LLC-MK2 (rhesus monkey kidney), BHK (baby hamster kidney cells), CCL-106 (rabbit kidney), and C6/36 (*Aedes albopictus* mosquito). Since intracerebral inoculation of suckling mice also failed in growing the virus, we assume that the failure in virus cultivation resulted from viral inactivation due to breakdown in cold chain. A recent study of a viral pseudotype carrying the immunogenic BASV glycoprotein in a vesicular stomatitis virus (VSV, genus *Vesiculovirus*) backbone demonstrated the susceptibility of numerous cell lines including Vero and C6/36 to BASV infection *in vitro*.[13] In the absence of samples from additional cases of BASV, recovery of infectious virus may be possible by reverse genetics, as previously performed for rabies virus and VSV.[14,15]

6.3 Serologic Diagnosis

Serological testing for BASV during the acute phase of the disease is not recommended. To date, the kinetics of production of IgM and IgG antibodies relative to onset of symptoms and disease course are not known. In addition, we cannot exclude a limited or impaired humoral immune response in patients with fatal outcome as previously reported for Ebola hemorrhagic fever.[16] A seroneutralization assay for BASV using a pseudotyped viral construct is available for epidemiological serosurveys of BASV prevalence[13] but is currently not useful for diagnosis.

7. HOW DO YOU DIFFERENTIATE THIS DISEASE FROM SIMILAR ENTITIES?

To date, BASV has only been detected in Central Africa where numerous viral hemorrhagic fevers are encountered (Figure 2.3). In this context, other viral etiologies to consider are Ebola and Marburg viruses, Crimean-Congo hemorrhagic fever virus, yellow fever virus, dengue viruses, and Rift Valley fever virus. Lassa and Lujo viruses should also be considered despite currently occurring in a different geographic range. However, the early development of hemorrhagic signs (within the first 2 days of symptoms) seen in BASV infection appears unusual with respect to other VHF agents. Ebola, Marburg, Crimean-Congo, and Lassa hemorrhagic fevers typically begin with a non-specific prodrome that may be confused with the other infections listed in Table 2.1, and hemorrhagic manifestations generally appear by day 3 to 5 after symptom onset.[11,17–19]

In addition to viral hemorrhagic fevers, many other infections are associated with acute febrile disease in tropical and subtropical regions (Table 2.1). The two most frequent infections to exclude are malaria and typhoid fever.[20] Gastrointestinal infections by bacteria such as *Shigella*, *Campylobacter*, and *Salmonella* can also be accompanied by bleeding, and are part of the differential diagnosis. Other causes of acute febrile illness frequently encountered in Central Africa include typhus, plague, relapsing fever, leptospirosis, anthrax, and viral hepatitis.[19]

8. WHAT IS THE THERAPEUTIC APPROACH?

To date the therapeutic approach is limited to supportive care.

TABLE 2.1 BASV Infection Differential Diagnosis

Viral infections	
Hemorrhagic fever viruses	
Family *Filoviridae*	Zaire ebolavirus Bundibugyo ebolavirus Tai Forest ebolavirus Sudan ebolavirus Marburg marburgvirus
Family *Flaviviridae*	Dengue virus Yellow fever virus
Family *Bunyaviridae*	Crimean-Congo hemorrhagic fever virus Rift Valley fever virus
Family *Arenaviridae*	Lassa virus Lujo virus
Other viral infections Measles Hemorrhagic varicella Rubella Hemorrhagic smallpox Viral hepatitis Chikungunya fever	
Parasitic infections	
Malaria African trypanosomiasis	
Bacterial infections	
Gram-negative bacteria Gram-negative bacterial septicemia Shigellose Salmonellose Typhoid fever Septicemic plague Cholera Q fever Meningococcemia Typhus Murin typhus Leptospirose Siphilis Relapsing fever Leptospirosis	
Gram-positive bacteria Staphylococcal or streptococcal toxic shock syndrome Anthrax Leptospirosis	

9. WHAT ARE THE PREVENTIVE AND INFECTION CONTROL MEASURES?

As the source of infection and way(s) of transmission are not clearly established, instructions to prevent and control the infection cannot be precisely defined. However, a minimum of rules should already be considered.

Public health authorities must be informed whenever there is VHF suspicion and contact tracing should be engaged as soon as possible. Compliance with the standard guidelines for biosafety and hygiene (use of lab coat, gloves, sharps containers, hand and surface cleaning, etc.) during clinical examination and blood sampling is an absolute requirement to prevent a potential nosocomial transmission. An aerosol transmission seems unlikely but use of masks and face shields should be considered in light of the precautionary principle and would provide an additional protection in the event of accidental exposure to body fluids.

The scientific data do not allow a definitive and obvious risk assessment, so that prevention and control measures are a sensitive point with strong ethical concerns. Going beyond the minimal set of rules proposed here, in considering stronger isolation measures for patient management for example, must be collectively discussed with additional medical experts, including those experienced in the management of VHF outbreaks in the field, to reach a consensus on solid and balanced recommendations.

REFERENCES

1. Grard G, Fair JN, Lee D, Slikas E, Steffen I, Muyembe JJ, et al. A novel rhabdovirus associated with acute hemorrhagic fever in central Africa. *PLoS Pathog* [serial online]. 2012 [cited 2012 Sep];8(9):e1002924. Available from: <http://www.plospathogens.org>.
2. Gurav YK, Tandale BV, Jadi RS, Gunjikar RS, Tikute SS, Jamgaonkar AV, et al. Chandipura virus encephalitis outbreak among children in Nagpur division, Maharashtra, 2007. *Indian J Med Res* 2010;**132**:395–9.
3. Warrell MJ, Warrell DA. Rabies and other lyssavirus diseases. *Lancet* 2004;**363**:959–69.
4. Rodriguez LL. Emergence and re-emergence of vesicular stomatitis in the United States. *Virus Res* 2002;**85**:211–9.
5. Fredericks DN, Relman DA. Sequence-based identification of microbial pathogens: a reconsideration of Koch's postulates. *Clin Microbiol Rev* 1996;**9**:18–33.
6. Chiu C, Fair J, Leroy EM. Bas-Congo virus: another deadly virus? *Future Microbiol* 2013;**8**:139–41.
7. Kuzmin IV, Novella IS, Dietzgen RG, Padhi A, Rupprecht CE. The rhabdoviruses: biodiversity, phylogenetics, and evolution. *Infect Genet Evol* 2009;**9**:541–53.
8. King AMQ, Adams MJ, Carstens EB, Lefkowitz EJ. *Virus taxonomy: classification and nomenclature of viruses: ninth report of the International Committee on Taxonomy of Viruses*. San Diego: Elsevier Academic Press; 2012.
9. Lyles DS, Rupprecht CE. Rhabdoviridae. In: Knipe DM, Howley PM, editors. *Fields virology*. 5th ed. Philadelphia: Lippincott Williams & Wilkins; 2007. p. 1363–408.

10. Gubala A, Davis S, Weir R, Melville L, Cowled C, Boyle D. Tibrogargan and Coastal Plains rhabdoviruses: genomic characterization, evolution of novel genes and seroprevalence in Australian livestock. *J Gen Virol* 2011;**92**:2160–70.
11. Ergonul O. Crimean-Congo haemorrhagic fever. *Lancet Infect Dis* 2006;**6**:203–14.
12. Monath TP, Barrett AD. Pathogenesis and pathophysiology of yellow fever. *Adv Virus Res* 2003;**60**:343–95.
13. Steffen I, Liss NM, Schneider BS, Fair JN, Chiu CY, Simmons G. Characterization of the Bas-Congo virus glycoprotein and its function in pseudotyped viruses. *J Virol* 2013;**87**:9558–68.
14. Schnell MJ, Mebatsion T, Conzelmann KK. Infectious rabies viruses from cloned cDNA. *EMBO J* 1994;**13**:4195–203.
15. Stanifer ML, Cureton DK, Whelan SP. A recombinant vesicular stomatitis virus bearing a lethal mutation in the glycoprotein gene uncovers a second site suppressor that restores fusion. *J Virol* 2011;**85**:8105–15.
16. Baize S, Leroy EM, Georges-Courbot MC, Capron M, Lansoud-Soukate J, Debre P, et al. Defective humoral responses and extensive intravascular apoptosis are associated with fatal outcome in Ebola virus-infected patients. *Nat Med* 1999;**5**:423–6.
17. Yun NE, Walker DH. Pathogenesis of lassa fever. *Viruses* 2012;**4**:2031–48.
18. Bwaka MA, Bonnet MJ, Calain P, Colebunders R, De Roo A, Guimard Y, et al. Ebola hemorrhagic fever in Kikwit, Democratic Republic of the Congo: clinical observations in 103 patients. *J Infect Dis* 1999;**179**(Suppl. 1):S1–7.
19. Sanchez A, Geisbert TW, Feldmann H. Filoviridae: Marburg and Ebola viruses. In: Knipe DM, Howley PM, editors. *Fields virology*. 5th ed. Philadelphia: Lippincott Williams & Wilkins; 2007. p. 1409–48.
20. Feldmann H, Geisbert TW. Ebola haemorrhagic fever. *Lancet* 2011;**377**:849–62.

Chapter 3

Hantavirus Infections

Tatjana Avšič Županc and Miša Korva
Institute of Microbiology and Immunology, Faculty of Medicine, University of Ljubljana, Ljubljana, Slovenia

CASE PRESENTATION

A 35-year-old male, professional driver was admitted to the General Hospital of Novo Mesto in Slovenia with a 3-day history of weakness, inappetence, nausea, vomiting, diarrhea, and oliguria with brownish urine. On admission, the patient complained of severe abdominal pain and blurred vision. On physical examination, the patient was generally affected, dehydrated, and febrile (39°C). The abdomen was diffusely sensitive to palpitation, but was not rigid. The patient's arterial blood pressure was 100/70 mmHg, his pulse was 92 beats per minute, and his central venous pressure was 1 cm of H_2O. Laboratory evaluation done at admission showed a hematocrit level of 0.52 l, WBC count of 14.4×10^9/L, and platelet count of 96×10^9/L. The patient's urine output was 20 ml/h, serum urea nitrogen (15 mmol/L) and creatinine (185 μmol/L) were elevated, and coagulation factors were diminished. Chest radiography showed bilateral effusion with discrete infiltrate of the right lower lobe. Based on anamnestic data (patient was from a highly endemic region), hantavirus infection was considered in etiology. Serum drawn at hospital admission revealed IgG antibody titers of 1:128 for Dobrava and 1:16 for Puumala viruses when tested by indirect immunofluorescent antibody (IFA) assay. The IgM titers by antibody-capture enzyme-linked immunosorbent assay (ELISA) were 1:12,800 for Dobrava and <1:100 for Puumala viruses. Over the next 30 h, the patient's general condition worsened with continued weakness and strong abdominal pains; he developed complete anuria (serum urea nitrogen 19.4 mmol/L, creatinine 398 μmol/L), atrial fibrillation, systemic arterial hypotension, pulmonary arterial hypertension, and hemorrhagic manifestations: hematuria, hematemesis, melena, and hemorrhagic pleural fluid. Laboratory tests showed leukocytosis (18.8×10^9/L), worsening thrombocytopenia (platelet count 14×10^9/L), and disseminated intravascular coagulation. In spite of intensive supportive therapy (dopamine, dobutamine,

Emerging Infectious Diseases. DOI: http://dx.doi.org/10.1016/B978-0-12-416975-3.00003-0

medigoksin, furosemide, and transfusions of erythrocytes, platelets, and fresh frozen plasma) his general condition deteriorated abruptly with signs of massive hemorrhages. Finally, hemodynamic instability developed, leading to circulatory collapse. Resuscitation was unsuccessful and the patient died on the third day of hospitalization (on day 6 of illness).

On gross examination of autopsy, hemorrhages in many organs were seen, especially in the gastrointestinal tract, the cardiac right atrium, the renal medulla, the adrenal glands, and the pituitary gland. The both side hemorrhagic hydrothorax (700 + 500 ml) was found, as well as hemorrhagic ascites (300 ml) in abdominal cavity. Microscopic examination showed focal hemorrhages in various organs, especially in the subepicardium and subendocardium of the right atrium, the anterior lobe of the pituitary, the submucosa of the gastrointestinal tract, and the cortex of both adrenal glands. In the kidneys, acute necrotizing hemorrhagic tubulo-interstitial nephritis was seen with severe congestion and extravasation of red cells, most intensive in the medulla.

1. WHY THIS CASE WAS SIGNIFICANTLY IMPORTANT AS AN EMERGING INFECTION

Hantaviruses cause hemorrhagic fever with renal syndrome (HFRS) in Asia and Europe and hantavirus cardio pulmonary syndrome (HCPS) in the Americas, with mortality rates from 12% (HFRS) to 50% (HCPS).[1]

2. WHAT IS THE CAUSATIVE AGENT?

Hantaviruses (genus *Hantavirus*, family *Bunyaviridae*) are enveloped RNA viruses, spherical in shape with a diameter of 80 to 120 nm (Figure 3.1).

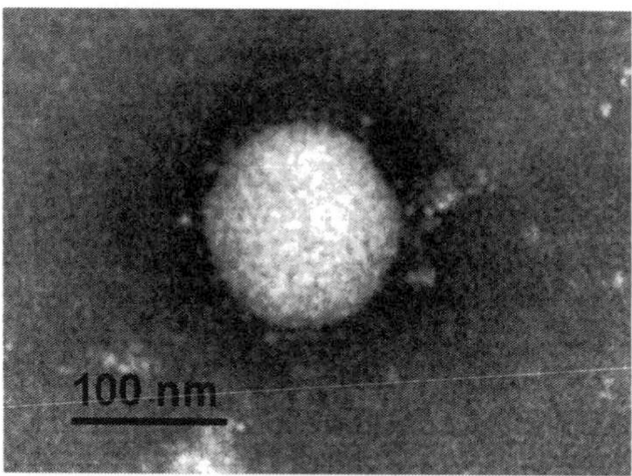

FIGURE 3.1 Electron microscopy image of Dobrava virus in a direct preparation from cell culture, negatively stained with 2% phosphotungistic acid (PTA; pH 4.5) and examined with a transmission electron microscope (JEM 1200 EX II; Jeol, Japan) at a magnification of ×120,000.

They are negative-sense, single-stranded RNA viruses with a three segmented genome. Each segment contains an open reading frame (ORF) flanked by non-coding regions at the 3′ and 5′ ends. Terminal parts of the non-coding region form a panhandle structure, the function of which has a crucial role in viral replication and transcription.[2] The three segments, small (S), medium (M), and large (L), encode the nucleoprotein (N), envelope glycoproteins (Gn and Gc), and viral RNA-depended RNA polymerase (RdRp), respectively.[3] Above that, in some hantaviruses there is an additional non-structural protein (Ns) coded in the S segment, which is associated with IFN antagonism.[4]

Hantaviruses infect many species of rodents, shrews, moles, and bats, but for now only rodent-borne hantaviruses are pathogenic for humans. Currently, 24 hantavirus species are recognized and they are divided into four distinct phylogroups: HFRS is caused by Old World hantaviruses, carried by rodents of subfamilies *Murinae* (Old World mice and rats) and *Arvicolinae* (voles and lemmings); HCPS is caused by New World hantaviruses carried by rodents of subfamilies *Sigmodontinae* and *Neotominae* (New World mice and rats); and the most ancestral clade is represented by hantaviruses found in insectivores and bats (order *Soricomorpha* and *Chiroptera*).[3,5]

3. WHAT IS THE FREQUENCY OF THE DISEASE?

Since hantaviruses are harbored by different small mammals, it is not surprising that the time and space distribution of hantavirus infections in man mirrors the distribution of the reservoir (Figure 3.2). The majority of HFRS cases, approximately 100,000 annually, are reported from Asia, where China is the most endemic country with 70–90% of all cases. A severe form of the disease is caused by hantaan virus (HTNV), the prototype strain isolated from the striped field mouse in 1976,[6] with mortality rate up to 15%, while a moderate disease course is caused by Seoul virus (SEOV), carried by rats, with mortality rate up to 2%.[1,7,8] In Europe, there are roughly 9000 HFRS

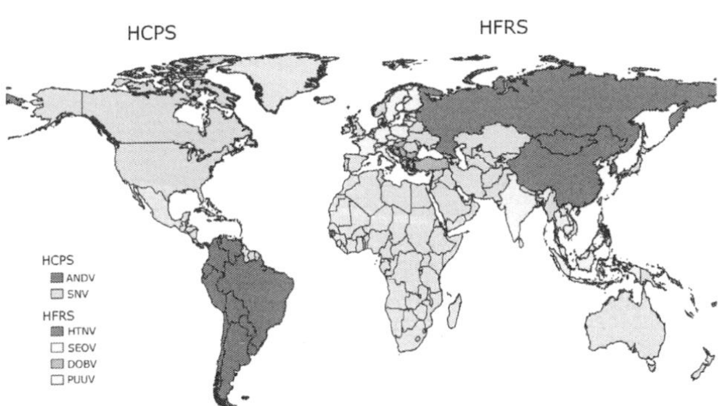

FIGURE 3.2 Distribution of pathogenic hantaviruses in the world.

cases reported yearly.[9,10] A severe HFRS, with mortality rate up to 12%, is caused by Dobrava virus (DOBV), harbored by the yellow-necked mouse, predominantly in south Europe (Balkan region).[11] The majority of cases occur in north and central Europe, where Puumala virus (PUUV), carried by the bank vole, and Dobrava–Kurkino virus, harbored by striped field mouse, are causing only a milder disease with mortality less than 0.5%.[3,12] Even though in North and South America, where the reported number of HCPS (about 200 cases) is considerably smaller, the average case fatality rate is about 40%. The predominant cause of HCPS in the United States is Sin Nombre virus (SNV), carried by deer mouse, while in South America the most important causative agent is Andes virus (ANDV), harbored by the long-tailed rice rat.[1,7] In addition to reported HFRS and HCPS cases around the world, there is probably a great number of asymptomatic, subclinical, and non-specific mild infections that remain unrecognized.[10] In line with this, in Africa only serological evidence of hantavirus infection was reported from several countries.[12]

4. HOW IS THE VIRUS TRANSMITTED?

In nature rodents act as reservoir host for pathogenic hantaviruses, with chronic and almost asymptomatic infection. The viruses are excreted in urine, feces, and saliva of infected reservoirs and it can stay infective in the environment for more than 10 days at room temperature, and even more if the temperatures are lower.[3,7,13] Most human infections occur when contaminated aerosolized rodent excreta are inhaled (Figure 3.3). Undoubtedly, the aerosol route of infection is the most common; however, infection after a rodent bite[14] and person-to-person transmission of ANDV in Argentina has been reported.[15] The risk factor of contracting hantavirus from rodents is related to closeness of contact, like farming, camping, forestry work, and cleaning barns and sheds.[1,10]

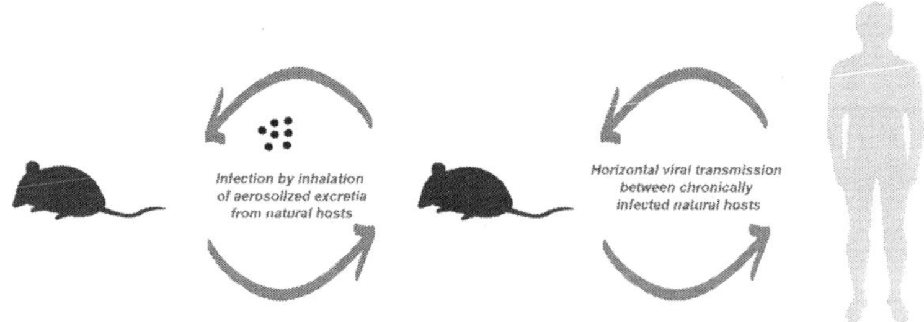

FIGURE 3.3 Hantavirus transmission route.

5. WHICH FACTORS ARE INVOLVED IN DISEASE PATHOGENESIS? WHAT ARE THE PATHOGENIC MECHANISMS?

Hantavirus disease (HFRS and HCPS) is a systemic infectious disease targeting different organs and organ systems. Infections mainly occur in pulmonary or renal microvascular endothelial cells and macrophages, albeit the viral antigen is present also in many different organs.[16,17] The central mechanism in pathogenesis is increased capillary permeability and acute thrombocytopenia leading to pleural edema and intracranial hemorrhage.[3,17,18] Although, hantaviruses non-lytically infect endothelial cells, they cause dramatic changes in barrier function of infected endothelial cells; thus, the immunopathology, caused by activation of innate and adaptive immune responses, has been suggested. Above that, a genetic predisposition toward severe forms of hantavirus disease was shown to be related to HLA-type and tumor necrosis factor-α (TNF-α) polymorphism.[3] After infection, immature dendritic cells probably play a pivot role in hantavirus dissemination through the body, since they express β_3 integrin receptors for viral entry and can also serve as vehicles for the transport of the virions through the lymphatic vessels.[18–20] Recognition of viral proteins by cell receptors induces activation of signaling cascades that lead to vigorous production of $CD8^+$, $CD4^+$ T cells, and antibody production. Hantavirus-specific antibodies persist for life, although they are not able to completely control virus replication.[3,21] Monocytes and macrophages appear to play roles both in systemic immunity to hantavirus infections and in pathogenesis through the release of cytokines and chemokines, which can act as a double-edged sword. Elevated levels of TNF-α, interleukin (IL)-6, IL-2, IL-1, and IL-10 have been linked to fever and septic shock in acute phase of the disease.[3,22] The complement system, which promotes clearance of immune complexes and opsonization of microorganisms, is also involved in hantavirus immunopathology, contributing to development of vascular leakage.[3,23]

6. WHAT ARE THE CLINICAL MANIFESTATIONS?

The incubation period for hantavirus infection ranges from 2 to 4 weeks. The infection can result in two clinical syndromes: HFRS and HCPS caused by Old World hantaviruses (HTNV, DOBV, PUUV, and SEOV) or New World hantaviruses (SINV and ADNV), respectively. The spectrum of illnesses varies from subclinical, mild, and severe courses to fatal outcome for which virus- and patient-specific determinants are responsible. The initial symptoms of all hantavirus infections are very similar, including a sudden onset of high fever, chills, malaise, myalgia, headache, and other flu-like symptoms. A typical course of HFRS can be divided into five phases: febrile, hypotensive, oliguric, polyuric, and convalescent (Figure 3.4). Backache, abdominal pains, nausea, vomiting, somnolence, and visual disturbances

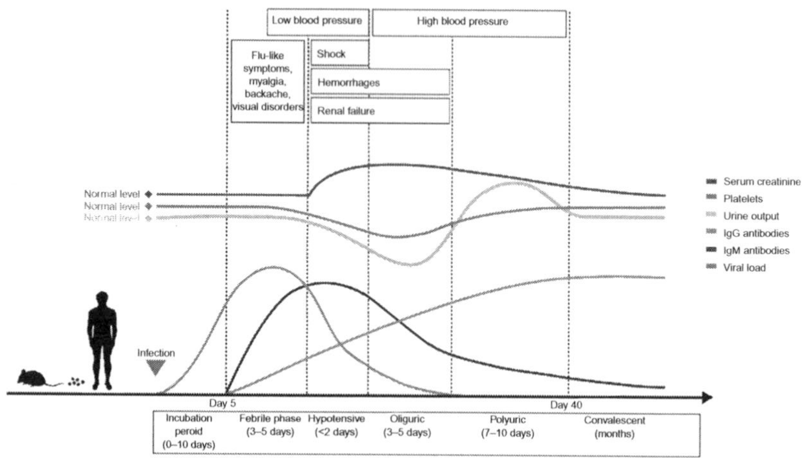

FIGURE 3.4 The clinical and laboratory course of hantavirus disease (HFRS).

(blurred vision) are frequently reported during the febrile phase. In the hypotensive stage, conjunctival hemorrhages, petechiae, and hypotension, which may progress to fulminant irreversible shock, are common (see case presentation). Thrombocytopenia, leukocytosis, and severe hemorrhagic manifestations (petechiae, ecchymoses, hematemesis, epistaxis, hematuria, melena, and fatal intracranial hemorrhages) are characteristic for this phase. In the oliguric phase, blood pressure normalizes, while oliguria or even anuria, proteinuria, azotemia, and elevated levels of serum creatinine and urea are associated with acute renal failure (Table 3.1). In the polyuric phase, renal function improves and urinary output increases. Convalescence is prolonged and full recovery is usually reached; complications are rare but can include chronic renal failure and hypertension.[24–26]

The initial HCPS symptoms resemble those of HFRS, and are generally presented through three phases: prodromal, cardiopulmonary, and convalescent. The illness progresses rapidly to severe respiratory distress followed by pulmonary edema and shock. Renal and hemorrhagic manifestations are rare in patients with HCPS (Table 3.1). Although the clinical course of HCPS varies from mild hypoxemia, cardiogenic shock due to respiratory failure is responsible for the often fatal outcome of HCPS.[27]

Although renal disease is usually assigned to HFRS and lung disease to HCPS, the increased medical knowledge about the clinical courses of HFRS and HCPS has resulted in the conclusion that both syndromes partly overlap. Namely, the number of reported HFRS cases with lung involvement and HCPS cases with renal and/or hemorrhagic involvement is continuously growing.[26] However, vascular alterations and leakage, which are responsible for organ failure, are characteristic for both clinical entities.

TABLE 3.1 Signs, Symptoms, and Laboratory Characteristics for Patients with Hantavirus Infections (HFRS and HCPS)

Finding[a]	HFRS	HCPS
Fever	++++	++++
Headache	++++	++
Chills	+++	+++
Myalgia	+++	+++
Abdominal pain	+++	+
Backache	++++	++
Nausea	+++	+
Vomiting, diarrhea	++	++
Cough	++	++++
Blurred vision	+++	–
Petechiae	+++	–
Bleedings	++	+
Hemorrhages	++	+
Hypotension	+++	++
Tachycardia	+	++++
Dyspnea	+/–	++++
Shock	++	++++
Oliguria	+++	–
Polyuria	++++	–
Leukocytosis	++++	++++
Thrombocytopenia	++++	++++
Proteinuria	++++	–
Hematuria	+++	+
S-creatinine	+++	+
Mortality	0–15%	50%

[a]*Based on reported findings.*[1,25]

7. HOW DO YOU DIAGNOSE?

The diagnosis of HFRS and HCPS is based on clinical and epidemiological data and laboratory tests. The symptoms that should alert the physician of a possible hantavirus infection are high fever, headache, abdominal and back pains, blurred vision, and pathological laboratory findings with leukocytosis, thrombocytopenia, increased serum creatinine, proteinuria and hematuria (Table 3.2). However, it is impossible to diagnose hantavirus infections solely on clinical grounds, as the early signs of the disease are non-specific.[10,24]

Laboratory diagnosis of acute hantavirus infections is based on serology as virtually all patients have IgM and usually also IgG antibodies present in serum at the onset of symptoms. The most commonly used serological tests are indirect IgM and IgG ELISAs as well as IgM capture ELISAs, which have higher specificity than indirect ELISAs. IFAs are also regularly used for diagnostics but have lower specificity.[9,28] In addition, rapid 5-min user-friendly immunochromatographic IgM-antibody tests have been developed and are available commercially.[29,30]

The hantavirus infection can also be confirmed by detection of hantavirus genome in blood or serum samples by RT-PCR. Both traditional and quantitative RT-PCRs are used to detect viremia.[31] Although the presence of viremia varies, viral RNA can usually be detected up until 14 days after onset of symptoms. It has also been suggested that higher viremia is found in more severe hantavirus infections (DOBV, SNV, ANDV), compared to milder infections, caused by PUUV.[21,32]

TABLE 3.2 Patient Findings Leading to Increased Suspicion of Hantavirus Infection

HFRS	HCPS
Acute renal failure	Acute respiratory distress
Sudden onset of fever with thrombocytopenia	Rapid onset of thrombocytopenia
Blurred vision (myopic shift)	Caught
Subconjunctival hemorrhage or presence of petechial rash	Hypocapnea
Proteinuria	Hemoconcentration
Prominent gastrointestinal symptoms	Prominent gastrointestinal symptoms
History of rodent–host exposure risk	History of rodent–host exposure risk

8. HOW DO YOU DIFFERENTIATE THIS DISEASE FROM SIMILAR ENTITIES?

Due to such diversity of clinical signs and symptoms, there is a long list of differential diagnoses: acute renal failure of another etiology, acute febrile urinary tract infection, tubulointerstitial nephritis of another etiology, acute and chronic glomerulonephritis, acute abdomen, including appendicitis, hemolytic uremic syndrome (HUS), thrombocytopenic thrombotic purpura (TTP), acute respiratory infections, atypical pneumonias, ARDS, sepsis, and cardiogenic shock. Other hemorrhagic fevers have to be considered according to the origin of the patient and the mode and risk of potential exposure (e.g., Crimean-Congo hemorrhagic fever in southern Europe, the Middle East, and Asia).[25,33] However, leptospirosis, rickettsiosis and Q fever should be included in the differential diagnosis of HFRS, particularly in south-east Europe, as these diseases are endemic in the area and are presented with similar epidemiological characteristics and almost the same symptoms and clinical laboratory findings at the onset of disease.[34]

9. WHAT IS THE THERAPEUTIC APPROACH?

At present, there is no specific treatment available for hantavirus diseases. Early treatment of HFRS patients with ribavarin can reduce the severity of symptoms. However, ribavirin did not provide an apparent clinical benefit in the treatment of HCPS patients.[1] Supportive therapy mainly comprised cardiovascular, respiratory, and renal function support, with fluid and electrolyte homeostasis being important components of care and early management of patients, which is crucial for the survival of severe cases. HFRS patients with severe renal insufficiency often need dialysis treatment, while severe HCPS patients are treated with supplemental oxygen, mechanical ventilation, and the appropriate use of pressures.[1]

10. WHAT ARE THE PREVENTIVE AND INFECTION CONTROL MEASURES?

Human behavior is a key factor in hantavirus infections. Risk factors for human infections include professions related to forestry, farming, military activities, or outdoor activities such as camping or using summer houses and hiking.[35] Further risk factors include rodents and their excreta, and cleaning and working in barns and woodsheds. Thus, prevention of hantavirus infections is based on personal preventive measures to avoid virus-contaminated dust, on proper cleaning and disinfecting of areas containing rodent droppings, and on continuous rodent control. In addition, minimizing food available to rodents around residential areas is known to effectively reduce the rodent population.[36] While there is currently no hantavirus vaccine licensed in Europe or the Americas, inactivated virus vaccines have been used in China[8] and Korea.[37] Research into a molecular vaccine is in development with candidate vaccines under phase I testing, but there are technical, regulatory, and economic problems that need to be solved before a vaccine becomes available. A comprehensive two-component vaccine is required, particularly for the highly endemic regions where different hantaviruses are causing HFRS.[38]

REFERENCES

1. Jonsson CB, Figueiredo LT, Vapalahti O. A global perspective on hantavirus ecology, epidemiology, and disease. *Clin Microbiol Rev* 2010;**23**:412–41.
2. Hussein IT, Haseeb A, Haque A, Mir MA. Recent advances in hantavirus molecular biology and disease. *Adv Appl Microbiol* 2011;**74**:35–75.
3. Vaheri A, Strandin T, Hepojoki J, et al. Uncovering the mysteries of hantavirus infections. *Nat Rev Microbiol* 2013;**11**:539–50.
4. Jaaskelainen KM, Kaukinen P, Minskaya ES, et al. Tula and Puumala hantavirus NSs ORFs are functional and the products inhibit activation of the interferon-beta promoter. *J Med Virol* 2007;**79**:1527–36.
5. Guo WP, Lin XD, Wang W, et al. Phylogeny and origins of hantaviruses harbored by bats, insectivores, and rodents. *PLoS Pathog* 2013;**9**:e1003159.
6. Lee HW, Lee PW, Johnson KM. Isolation of the etiologic agent of Korean hemorrhagic fever. *J Infect Dis* 1978;**137**:298–308.
7. Bi Z, Formenty PB, Roth CE. Hantavirus infection: a review and global update. *J Infect Dev Ctries* 2008;**2**:3–23.
8. Zhang YZ, Zou Y, Fu ZF, Plyusnin A. Hantavirus infections in humans and animals, China. *Emerg Infect Dis* 2010;**16**:1195–203.
9. Vapalahti O, Mustonen J, Lundkvist A, Henttonen H, Plyusnin A, Vaheri A. Hantavirus infections in Europe. *Lancet Infect Dis* 2003;**3**:653–61.
10. Heyman P, Vaheri A, Lundkvist A, Avsic-Zupanc T. Hantavirus infections in Europe: from virus carriers to a major public-health problem. *Expert Rev Anti Infect Ther* 2009;**7**:205–17.
11. Avsic-Zupanc T, Petrovec M, Furlan P, Kaps R, Elgh F, Lundkvist A. Hemorrhagic fever with renal syndrome in the Dolenjska region of Slovenia—a 10-year survey. *Clin Infect Dis* 1999;**28**:860–5.
12. Klempa B. Hantaviruses and climate change. *Clin Microbiol Infect* 2009;**15**:518–23.
13. Kallio ER, Klingstrom J, Gustafsson E, et al. Prolonged survival of Puumala hantavirus outside the host: evidence for indirect transmission via the environment. *J Gen Virol* 2006;**87**:2127–34.
14. Douron E, Moriniere B, Matheron S, et al. HFRS after a wild rodent bite in the Haute-Savoie—and risk of exposure to Hantaan-like virus in a Paris laboratory. *Lancet* 1984;**1**:676–7.
15. Enria D, Padula P, Segura EL, et al. Hantavirus pulmonary syndrome in Argentina. Possibility of person to person transmission. *Medicina (B Aires)* 1996;**56**:709–11.
16. Hughes JM, Peters CJ, Cohen ML, Mahy BW. Hantavirus pulmonary syndrome: an emerging infectious disease. *Science* 1993;**262**:850–1.

17. Vapalahti O, Lundkvist A, Vaheri A. Human immune response, host genetics, and severity of disease. *Curr Top Microbiol Immunol* 2001;**256**:153–69.
18. Gavrilovskaya I, Gorbunova E, Matthys V, Dalrymple N, Mackow E. The role of the endothelium in HPS pathogenesis and potential therapeutic approaches. *Adv Virol* 2012;**2012**:467059.
19. Peebles Jr RS, Graham BS. Viruses, dendritic cells and the lung. *Respir Res* 2001;**2**:245–9.
20. Schonrich G, Rang A, Lutteke N, Raftery MJ, Charbonnel N, Ulrich RG. Hantavirus-induced immunity in rodent reservoirs and humans. *Immunol Rev* 2008;**225**:163–89.
21. Korva M, Saksida A, Kejzar N, Schmaljohn C, Avsic-Zupanc T. Viral load and immune response dynamics in patients with haemorrhagic fever with renal syndrome. *Clin Microbiol Infect* 2013;**19**:E358–66.
22. Sadeghi M, Eckerle I, Daniel V, Burkhardt U, Opelz G, Schnitzler P. Cytokine expression during early and late phase of acute Puumala hantavirus infection. *BMC Immunol* 2011;**12**:65.
23. Paakkala A, Mustonen J, Viander M, Huhtala H, Pasternack A. Complement activation in nephropathia epidemica caused by Puumala hantavirus. *Clin Nephrol* 2000;**53**:424–31.
24. Linderholm M, Elgh F. Clinical characteristics of hantavirus infections on the Eurasian continent. *Curr Top Microbiol Immunol* 2001;**256**:135–51.
25. Sargianou M, Watson DC, Chra P, et al. Hantavirus infections for the clinician: from case presentation to diagnosis and treatment. *Crit Rev Microbiol* 2012;**38**:317–29.
26. Krautkramer E, Zeier M, Plyusnin A. Hantavirus infection: an emerging infectious disease causing acute renal failure. *Kidney Int* 2013;**83**:23–7.
27. Enria DA, Briggiler AM, Pini N, Levis S. Clinical manifestations of New World hantaviruses. *Curr Top Microbiol Immunol* 2001;**256**:117–34.
28. Elgh F, Lundkvist A, Alexeyev OA, et al. Serological diagnosis of hantavirus infections by an enzyme-linked immunosorbent assay based on detection of immunoglobulin G and M responses to recombinant nucleocapsid proteins of five viral serotypes. *J Clin Microbiol* 1997;**35**:1122–30.
29. Hjelle B, Jenison S, Torrez-Martinez N, et al. Rapid and specific detection of Sin Nombre virus antibodies in patients with hantavirus pulmonary syndrome by a strip immunoblot assay suitable for field diagnosis. *J Clin Microbiol* 1997;**35**:600–8.
30. Hujakka H, Koistinen V, Kuronen I, et al. Diagnostic rapid tests for acute hantavirus infections: specific tests for Hantaan, Dobrava and Puumala viruses versus a hantavirus combination test. *J Virol Methods* 2003;**108**:117–22.
31. Aitichou M, Saleh SS, McElroy AK, Schmaljohn C, Ibrahim MS. Identification of Dobrava, Hantaan, Seoul, and Puumala viruses by one-step real-time RT-PCR. *J Virol Methods* 2005;**124**:21–6.
32. Xiao R, Yang S, Koster F, Ye C, Stidley C, Hjelle B. Sin Nombre viral RNA load in patients with hantavirus cardiopulmonary syndrome. *J Infect Dis* 2006;**194**:1403–9.
33. Christova I, Younan R, Taseva E, et al. Hemorrhagic fever with renal syndrome and Crimean-Congo hemorrhagic fever as causes of acute undifferentiated febrile illness in Bulgaria. *Vector Borne Zoonotic Dis* 2013;**13**:188–92.
34. Markotic A, Kuzman I, Babic K, et al. Double trouble: hemorrhagic fever with renal syndrome and leptospirosis. *Scand J Infect Dis* 2002;**34**:221–4.
35. Watson D.C., Sargianou M., Papa A., Chra P., Starakis I., Panos G. Epidemiology of Hantavirus infections in humans: a comprehensive, global overview. *Crit Rev Microbiol* 2014;**40**(3):261–72.

36. Kraigher A., Frelih T., Korva M., Avsic T. Increased number of cases of haemorrhagic fever with renal syndrome in Slovenia, January to April 2012. Euro Surveill 2012;17.
37. Cho HW, Howard CR. Antibody responses in humans to an inactivated hantavirus vaccine (Hantavax). *Vaccine* 1999;**17**:2569–75.
38. Schmaljohn CS. Vaccines for hantaviruses: progress and issues. *Expert Rev Vaccines* 2012;**11**:511–3.

Chapter 4

Lassa Fever

Donald S. Grant[1], Humarr Khan[1], John Schieffelin[2] and Daniel G. Bausch[2]

[1]*Kenema Government Hospital, Ministry of Health and Sanitation, Kenema, Sierra Leone,*
[2]*Tulane Health Sciences Center, New Orleans, LA, USA*

CASE PRESENTATION

A 20-year-old male farmer from eastern Sierra Leone (Konabu Village, Nongowa Chiefdom) presented to the outpatient department of Kenema Government Hospital, Kenema, Sierra Leone, with a 10-day history of fever, headache, sore throat, retrosternal pain, diarrhea, vomiting, abdominal pain, and dry cough unresponsive to self-medication with various antibiotics and antipyretics. Malaria, infectious gastroenteritis, and/or pneumonia were suspected and the patient was treated with artemisinin and metronidazole. Nevertheless, his symptoms worsened, with severe prostration and hematemesis noted the following day. Lassa fever (LF), which is hyper-endemic in the area, was suspected and the patient was transferred to the hospital's Lassa Ward, an isolation unit specifically dedicated to the care of patients with this disease.

Vital signs on admission to the Lassa Ward were T = 37.8°C (axillary), P = 86 bpm; BP = 110/70 mmHg, and RR = 32 breaths/min. The patient was in obvious respiratory distress, with nasal flaring and basal crepitations, and transmitted sounds on auscultation. Abdominal examination revealed marked epigastric tenderness. Conjunctival injection, blood clots at the angles of the mouth, and oozing of blood around a left forearm IV site were noted (Figures 4.1 and 4.2). Laboratory results showed Hb = 13 g/dl, PCV = 39%, ESR = 48 mm/hr, WBC = 4800/μl (60% polymorphonucleocytes, 26% lymphocytes, 4% monocytes), and trace protein on urinalysis. A thick smear for malaria was negative. Subsequent serological testing by ELISA for Lassa virus (LASV)-specific antigen was positive, confirming a diagnosis of LF. Anti-LASV IgM and IgG antibodies were negative. Intravenous dextran solution, ribavirin, and ceftriaxone (for possible secondary bacterial infection) were begun, along with oral paracetamol and antacids, and oxygen by nasal cannula.

Emerging Infectious Diseases. DOI: http://dx.doi.org/10.1016/B978-0-12-416975-3.00004-2

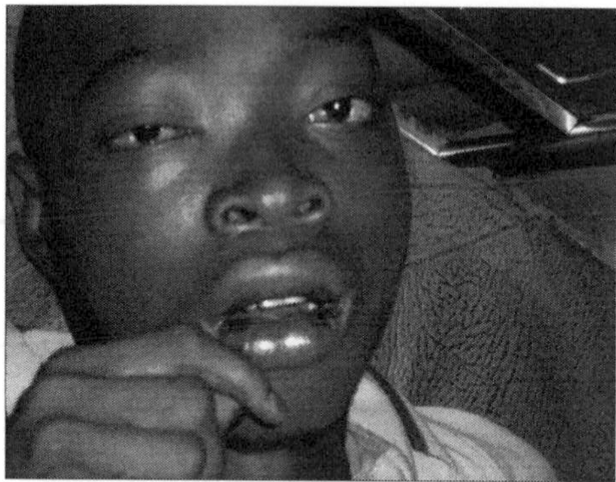

FIGURE 4.1 Patient with Lassa fever manifesting conjunctival injection and blood clots at the angles of the mouth. *Photo by Donald S. Grant.*

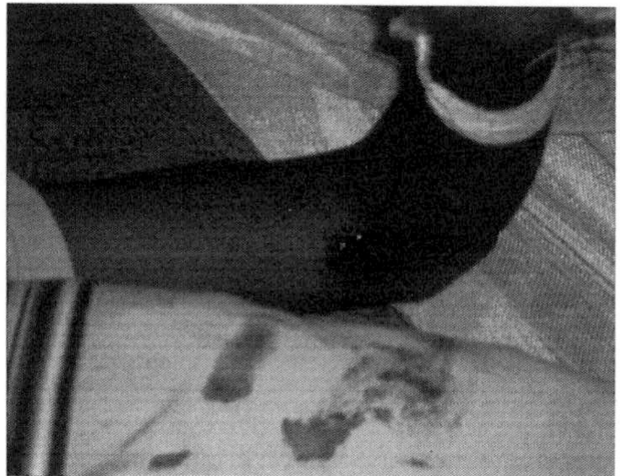

FIGURE 4.2 Patient with Lassa fever manifesting oozing of blood around the IV (recently removed) site on the left forearm. *Photo by Donald S. Grant.*

After 3 days of treatment the patient was communicative, although still with complaints of headache and weakness and unable to walk without aid. Hematemesis and bleeding from the nasal mucosal and around IV sites continued. Vital signs were T = 36.5°C (axillary), P = 90 bpm; BP = 100/70 mmHg, and RR = 25 breaths/min. Repeat ELISA testing for LF parameters was unchanged. The patient's condition stabilized over the next few days, with cessation of hematemesis but appearance of melanotic stools. He completed 10 days of IV ribavirin, at which point he was completely alert and oriented and able to walk unaided. Vital signs returned to normal. However, pallor was noted and a recheck of his Hb revealed a level of 7.6 g/dl. Ferrous fumarate and folic acid were begun. On repeat ELISA testing, LASV antigen and IgG antibody were negative but IgM antibody was positive. All symptoms subsided by day 17 of hospitalization and the patient was discharged to home. Final ELISA testing prior to discharge showed a negative LASV antigen but positive IgM and IgG antibodies. A follow-up outreach visit to the patient's home revealed signs of heavy rodent infestation. In addition, the patient admitted to hunting rodents as a source of protein.

1. WHY LASSA FEVER IS IMPORTANT AS AN EMERGING INFECTION

Lassa virus is one of more than 25 causative viruses of viral hemorrhagic fever (Table 4.1), an acute systemic illness classically involving fever, a constellation of initially non-specific signs and symptoms, and a propensity for bleeding and shock.[1] Unlike many viral hemorrhagic fevers, LF is not a rare disease that emerges only in outbreak form. After first discovery in Nigeria in 1969, it was subsequently noted in many countries across West Africa.[2–4] Yearly infections may number in tens or even hundreds of thousands, with thousands of deaths. In addition, use of LASV and other hemorrhagic fever viruses as bioweapons is a concern because of their high pathogenicity, risk of secondary spread, and tendency to cause public panic and social disruption.[5] Lassa virus is classified as biosafety level four or "high containment."

2. WHAT IS THE CAUSATIVE AGENT?

Lassa virus is a member of the *Arenaviridae* family, which includes numerous viruses known to cause viral hemorrhagic fever in Africa and South America (Table 4.1).[6] Arenaviruses have an ≈ 11 kb bisegmented genome comprised of single-stranded RNA with ambisense coding.[6] The naked RNA is not infectious. There is considerable sequence heterogeneity of LASVs across West Africa, with four recognized lineages—three in Nigeria and one in the area comprising Sierra Leone, Liberia, Guinea, and Ivory Coast (Figure 4.3).[7] There is also considerable genetic heterogeneity within lineages, especially in Nigeria.

3. WHAT IS THE FREQUENCY OF THE DISEASE?

Lassa fever is endemic exclusively in West Africa (Figure 4.4). Estimating the true incidence and mortality of LF is extremely difficult due to the non-specific clinical presentation; logistical impediments presented by civil unrest;[8] unstable governments with underdeveloped surveillance systems;[9] extensive human migration and perturbation of the physical landscape; and lack of reagents and laboratories for laboratory confirmation in West Africa.[10] Nevertheless, yearly infections are thought to number in the tens or even hundreds of thousands, with thousands of deaths. The highest incidence

TABLE 4.1 Principal Viruses Causing Hemorrhagic Fever

Virus	Disease	Principal Reservoir/Vector	Geographic Distribution of Disease	Annual Cases	Disease-to-Infection Ratio	Human-to-Human Transmissibility	Case Fatality
Filoviridae							
Ebola[a]	Ebola HF	Fruit bat ("Egyptian fruit bat" or *Rousettus aegyptiacus*, perhaps others)	Sub-Saharan Africa	–[b]	1:1	High	25–85% depending upon species[b]
Marburg	Marburg HF	Fruit bat ("Egyptian fruit bat" or *Rousettus aegyptiacus*, perhaps others)	Sub-Saharan Africa	–[b]	1:1	High	25–85%[c]
Arenaviridae[d]							
Old World							
Lassa	Lassa fever	Rodent ("natal mastomys" or "multimammate rat" or *Mastomys natalensis*)	West Africa	30,000–50,000	1:5–10	Moderate	25%
Lujo[e]	Lujo HF	Unknown. Presumed rodent	Zambia	Unknown	Unknown	Moderate to high	80%
New World							
Junín	Argentine HF	Rodent ("corn mouse" or *Calomys musculinus*)	Argentine pampas	<50	1:1.5	Low	15–30%
Machupo	Bolivian HF	Rodent ("large vesper mouse" or *Calomys callosus*)	Beni department, Bolivia	<50	1:1.5	Low	15–30%

Guanarito	Venezuelan HF	Rodent ("cane mouse" or *Zygodontomys brevicauda*)	Portuguesa state, Venezuela	<50	1:1.5	Low	30–40%
Sabiá[f]	Brazilian HF	Unknown. Presumed rodent	Rural area near Sao Paulo, Brazil?	–[f]	1:1.5	Low?	33%
Chapare[g]	Chapare HF	Unknown. Presumed rodent	Cochabamba, Bolivia	Unknown	Unknown	Unknown	Unknown
Bunyaviridae							
Old World hantaviruses							
Hantaan, Seoul, Puumala, Dobrava-Belgrade, others	HF with renal syndrome	Rodent (Hantaan: "striped field mouse" or *Apodemus agrarius*; Seoul: "Brown rat" or *Rattus norvegicus*; Puumala: "bank vole" or *Clethrionomys glareolus*; Dobrava-Belgrade: "yellow-necked field mouse" or *Apodemus flavicollis*)	Hantaan: northeast Asia; Seoul: urban areas worldwide; Puumala and Dobrava-Belgrade: Europe	50,000–150,000	Hantaan: 1:1.5, Others: 1:20	None	<1–50%, depending on specific virus
New World hantaviruses							
Sin Nombre, Andes, Laguna Negra, others	Hantavirus pulmonary syndrome	Rodents. Sin Nombre: "North American deer mouse" or *Peromyscus maniculatus*; Andes: "long-tailed colilargo" or *Oligoryzomys longicaudatus*; Laguna Negra: "little laucha" or "small vesper mouse" or *Calomys laucha*	Americas	50,000–150,000	Sin Nombre: 1:1, Others up to 1:20	None, except for Andes virus	<1–50%, depending on specific virus

(*Continued*)

TABLE 4.1 (Continued)

Virus	Disease	Principal Reservoir/Vector	Geographic Distribution of Disease	Annual Cases	Disease-to-Infection Ratio	Human-to-Human Transmissibility	Case Fatality
Rift Valley fever	Rift Valley fever	Domestic livestock/mosquitoes (*Aedes* and others)	Sub-Saharan Africa, Madagascar, Saudi Arabia, Yemen[g]	100–100,000[b,h]	1:100	None	Up to 50% in persons manifesting severe forms
Crimean-Congo HF	Crimean-Congo HF	Wild and domestic vertebrates/tick (primarily *Hyalomma* species)	Africa, Balkans, southern Russia, Middle East, India, Pakistan, Afghanistan, western China	≈500	1:1–2	High	15–30%
Flaviviridae							
Yellow fever	Yellow fever	Monkey/mosquito (*Aedes aegypti*, other *Aedes* and *Haemagogus* spp.)	Sub-Saharan Africa, South America up to Panama	5000–200,000[i]	1:2–20	No	20–50%
Dengue	Dengue HF	Human/mosquito (*Aedes aegypti* and *albopictus*)	Tropics and subtropics worldwide	Dengue HF: 100,000–200,000[h]	1:10–100 depending on age, previous infection, genetic background, and infecting serotype	None	Untreated: 10–15% Treated: <1%

Kyasanur Forest disease	Kyasanur Forest disease	Vertebrate (rodents, bats, birds, monkeys, others)/tick (*Haemophysalis* species and others)	Karnataka State, India; Yunnan Province, China; Saudi Arabia[j]	≈ 500	Unknown	Not reported, but laboratory infections have occurred	3–5%
Omsk HF	Omsk HF	Rodent/ticks (primarily *Dermacentor* and *Ixodes* species)	Western Siberia	100–200	Unknown	Not reported	1–3%

[a] *Six species or subtypes of Ebola virus with varying associated case-fatality ratios are recognized: Ebola Zaire—85%, Ebola Sudan—55%, Ebola Bundibugyo—40%, Ebola Cote d'Ivoire—0 (only one recognized case, who survived), Ebola Reston—0 (not pathogenic to humans), Lloviu—no human infections recognized. All are endemic to sub-Saharan Africa, with the exceptions of Ebola Reston virus, which is found in the Philippines, and Lloviu virus, which was detected in bats in Spain.*

[b] *Although some endemic transmission of the filoviruses (Ebola > Marburg) and Rift Valley fever virus occurs, these viruses have most often been associated with outbreaks. Filovirus outbreaks are typically less than 100 cases and have never been greater than 500.*

[c] *The case fatality ratio was 22% in the first recognized outbreak of Marburg HF in Germany and Yugoslavia in 1967 but has been consistently over 80% in outbreaks in Central Africa where the virus is endemic. Possible reasons for this discrepancy include differences in quality of care, strain pathogenicity, route and dose of infection, underlying prevalence of immunodeficiency and co-morbid illnesses, and genetic susceptibility.*

[d] *In addition to the arenaviruses listed in the table, Flexal and Tacaribe viruses have caused human disease as a result of laboratory accidents. Another arenavirus, Whitewater Arroyo, has been noted in sick persons in California but its role as a pathogen has not been clearly established.*

[e] *Discovered in 2008 in an outbreak of five cases (four of them fatal) in South Africa. The index case came to South Africa from Zambia.*

[f] *Discovered in 1990. Only three cases (one fatal) have been noted, two of them from laboratory accidents.*

[g] *Discovered in 2003 from a small outbreak in Cochabamba, Bolivia. Blood was obtained from one fatal case and Chapare virus isolated but few other details from the outbreak have been reported.*

[h] *Although Rift Valley fever virus can be found throughout sub-Saharan Africa, large outbreaks usually occur in East Africa.*

[i] *Based on estimates from the World Health Organization. Significant underreporting occurs. Incidence may fluctuate widely depending on epidemic activity.*

[j] *Numerous variants of Kyasanur Forest disease virus have been identified, including Nanjianyin virus in Yunnan Province, China, and Alkhumra virus (also spelled "Alkhurma" in some publications) in Saudi Arabia.*

Abbreviation: HF, hemorrhagic fever.

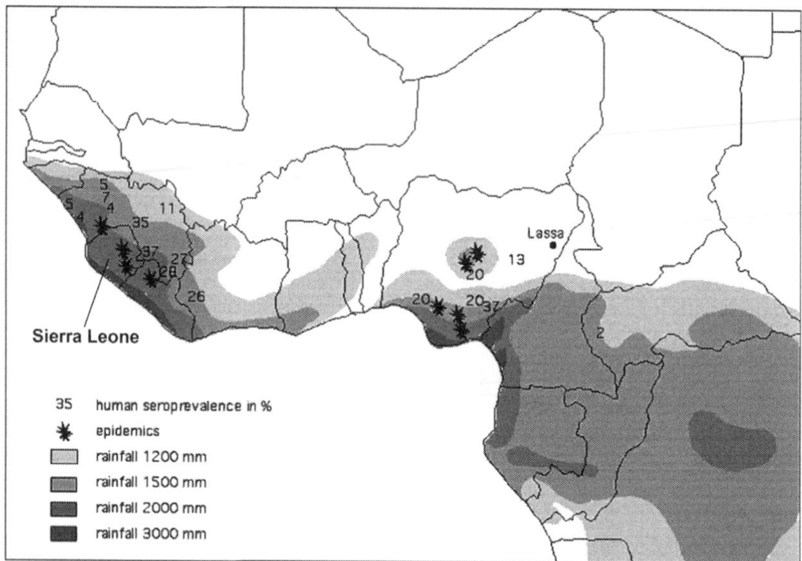

FIGURE 4.3 West and Central Africa mean annual rainfall (1951–1989), Lassa fever nosocomial outbreaks (stars) and human seroprevalence (numbers in %). Varying degrees of capacity for surveillance and diagnosis and existence of Lassa fever-focused research programs may significantly shape the observed distribution of cases. The town of Lassa in northeast Nigeria, after which the disease was named, is shown. *From Ref. 50.*

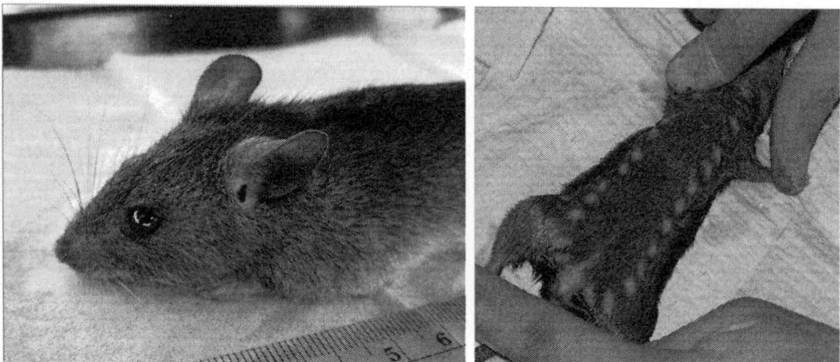

FIGURE 4.4 *Mastomys natalensis*, the reservoir for Lassa virus. The animal is commonly known as the "natal mastomys" or "multimammate rat," the latter name due to the female's multiple and prominent mammary glands, as seen in the right-hand panel. *From Ref. 19.*

appears to be in eastern Sierra Leone, northern Liberia, southeastern Guinea, and central and southern Nigeria (Figure 4.4).[11–14] However, the risk of exposure to LASV varies significantly in a given country and often among regions or even villages within endemic areas. The incidence of LF is consistently highest during the dry season (Figure 4.4), which may relate to greater stability of LASV in lower humidity,[15] seasonal fluctuations of numbers and prevalence of LASV infection in rodents,[16] and/or societal and behavioral factors.[14]

4. HOW IS THE VIRUS TRANSMITTED?

Lassa virus is maintained via chronic asymptomatic infection of the rodent *Mastomys natalensis* (common name "natal mastomys," also called the "multimammate rat") (Figure 4.5).[17] Despite the occurrence of *Mastomys* species throughout sub-Saharan Africa, LASV has not been found in rodents outside of West Africa, perhaps due to historical bottlenecks in dispersal of the virus, reservoir, or both. *M. natalensis* is almost always found in close association with humans in rural villages and surrounding cultivated fields and, less commonly, in grasslands and at the forest edge.[16] Poor quality housing, which may reflect ease of rodent access to the home, has been shown to be a risk factor for LF.[18,19] Foreign military personnel, peacekeepers, and aid workers in rural settings are occasionally infected, sometimes importing LASV back to their countries of origin.[20]

Transmission of LASV to humans is believed to occur via exposure to rodent excreta, either from direct inoculation to the mucous membranes or from inhalation of aerosols produced when rodents urinate.[11] The relative frequency of these modes of transmission is unknown. Lassa virus may be contracted when rodents are trapped and prepared for consumption, a common practice in some parts of West Africa.[21] Since LASV is easily inactivated by heating, eating cooked rodent meat should pose no danger.

Human-to-human transmission of LASV occurs through direct contact with infected blood or bodily fluids, presumably from oral or mucous membrane exposure in the context of providing care to a sick family member (community) or patient (nosocomial transmission). However, secondary attack rates are generally low as long as strict barrier nursing practices are observed (see below). Large outbreaks are almost always fueled by nosocomial transmission, usually in resource-poor regions where barrier nursing practices may not be maintained.[22] Extensive field experience has not suggested aerosol transmission between humans in natural settings,[23] although the artificial production of infectious aerosols has been shown, with obvious implications for the potential use of LASV as a bioweapon.[5] Despite delayed clearance of LASV for weeks to months from some immunologically protected sites, such as the kidney, gonads, and central nervous system, secondary transmission during convalescence has not been noted, with the exception of rare reports of sexual transmission occurring months after recovery from acute disease.[24–26] Viremia and infectivity of persons with LF generally parallels the clinical state, with highest infectivity late in the course

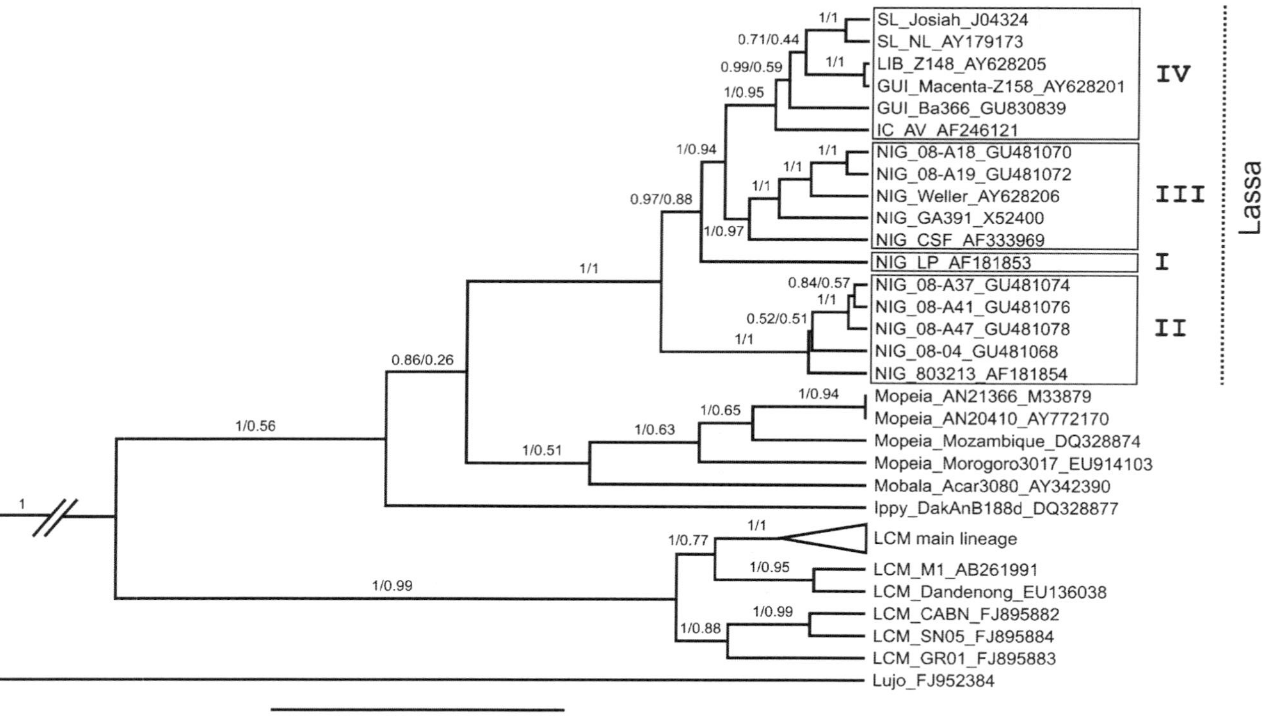

FIGURE 4.5 Phylogenetic analysis of Lassa virus and other arenaviruses of the Old World Arenavirus Complex. The phylogram is based on complete nucleotide sequences of the GP gene (1473 nucleotides). Similar results (not shown) were found with complete nucleotide sequencing of the Lassa virus NP and L genes. The origin of Lassa virus strains is indicated by the following prefixes: SL, Sierra Leone; LIB, Liberia; GUI, Guinea; IC, Ivory Coast; NIG, Nigeria. LCM = lymphocytic choriomeningitis virus. Recently discovered nonpathogenic arenaviruses are not shown. *Adapted from Ref. 51.*

of severe disease, especially when bleeding. Thus, the risk of transmission during the incubation period or from asymptomatic persons is negligible.

5. WHICH FACTORS ARE INVOLVED IN DISEASE PATHOGENESIS? WHAT ARE THE PATHOGENIC MECHANISMS?

Microvascular instability and impaired hemostasis are the hallmarks of viral hemorrhagic fever, with considerable overlap with the pathogenesis of septic shock. The pathogenesis of LF is thought to relate primarily to disruption of cellular function, as opposed to extensive cell death; patients often die without significant bleeding and histopathological lesions are usually not severe enough to account for death.[27] Severe LF appears to result from an insufficient immune response,[28–30] with higher levels of viremia and lower antibody titers in fatal cases relative to survivors.[24,26]

After inoculation, LASV first replicates in dendritic cells and other local tissues, with subsequent migration to regional lymph nodes and dissemination via the lymph and blood monocytes to a broad range of tissues and organs.[31] The liver is consistently the most affected organ. Tissue damage may ultimately be mediated through direct necrosis or indirectly through apoptosis. Impaired hemostasis may result from endothelial cell, platelet, and/or coagulation factor dysfunction and hemorrhage may occur even when platelet counts are not drastically low.[32] Disseminated intravascular coagulopathy does not appear to be part of the pathogenesis of LF, although this finding bears confirmation.[33]

Viremia usually peaks at days 4–9 of illness and clears within 2–3 weeks in survivors. In survivors, IgM antibodies begin to appear after about a week and progressively increase as virus clears, lasting at least some months.[24] IgG antibody begins to appear 2–3 weeks after onset and lasts for years.[3]

A long-standing mystery of LF is the apparent extreme range of clinical severity. Although this finding bears confirmation, mild or asymptomatic LASV infection appears to be frequent and may relate to heterogeneity in strain virulence, route and dose of inoculation, human genetic predisposition, underlying co-infections and/or premorbid conditions, or misclassification of reinfection as new infection due to waning of antibody.[24]

6. WHAT ARE THE CLINICAL MANIFESTATIONS?

Lassa fever is seen in both genders and all age groups. Most patients present with non-specific signs and symptoms difficult to distinguish from a host of other febrile illnesses common in the tropics. After an incubation period of about 1 week (range 3–21 days), illness typically begins with the gradual onset of fever and constitutional symptoms, including general malaise, anorexia, headache, chest or retrosternal pain, sore throat, myalgia, arthralgia, lumbosacral pain, and dizziness.[14,34] The pharynx may be erythemic or even exudative, a finding which has at times led to misdiagnosis of streptococcal pharyngitis. Gastrointestinal signs and symptoms occur early in the course of disease and may include nausea, vomiting, epigastric and abdominal pain and tenderness, and diarrhea. A morbilliform, maculopapular, or petechial skin rash almost always occurs in fair-skinned persons but, for unclear reasons, rarely in blacks. Conjunctival injection or hemorrhage is frequent but is not accompanied by itching, discharge, or rhinitis. A dry cough, sometimes accompanied by a few scattered rales on auscultation, may be noted, but prominent pulmonary symptoms are uncommon early in the course of the disease. Jaundice is not typical and should suggest another diagnosis.

In severe cases, patients progress to vascular instability, which may be manifested by subconjunctival hemorrhage, facial flushing, edema, bleeding, hypotension, shock, and proteinuria. Swelling in the face and neck and bleeding are particularly specific signs but are not very sensitive—seen in fewer than 20% of cases. Clinically discernible hemorrhage is seen in fewer than 20% of cases and never in the first few days of illness. Hematemesis, melena, hematochezia, metrorrhagia, petechiae, epistaxis, and bleeding from the gums and venupuncture sites may develop, but hemoptysis and hematuria are infrequent. Neurological complications, including disorientation, tremor, ataxia, seizures, and coma, may be seen, particularly in the late stages, and usually portend a fatal outcome.[35] Patients may be normothermic or even hypothermic in these late stages, so the absence of fever should not preclude consideration of LF.

Pregnant women with LF often present with spontaneous abortion and vaginal bleeding, with LASV found at high concentrations in placenta and fetal tissues.[27] Anasarca has been described in a single report of children with LF (termed the "swollen baby syndrome") but may have been related to aggressive rehydration.[36] Typical clinical laboratory findings are presented in Table 4.2. Radiographic and electrocardiographic findings are generally non-specific and correlate with the physical examination.[37,38]

Death in fatal cases usually occurs within 2–3 weeks after onset from shock and multi-organ system failure. The case-fatality rate in hospitalized cases is usually around 20%.[11] Indicators of a poor prognosis include shock, bleeding, neurological manifestations, viremia $>10^8$ $TCID_{50}$/ml (or cDNA copy number or ELISA antigen as surrogates), AST >150 IU/L, and pregnancy, especially during the third trimester when maternal and fetal mortality approach 100%.[34] Convalescence may be prolonged, with persistent myalgia, arthralgia, anorexia, weight loss, alopecia up to a year after infection, and sometimes with lasting psychological effects. Sensorineural deafness is reported in up to 25% of cases and is the only recognized permanent sequela.

TABLE 4.2 Indicated Clinical Laboratory Tests and Characteristic Findings in Patients with Lassa Fever

Test	Characteristic Findings and Comments
Leukocyte count	Early: moderate leukopenia; later: leukocytosis with left shift; granulocytosis more suggestive of bacterial infection
Hemoglobin and hematocrit	Hemoconcentration if advanced stage
Platelet count	Mild-to-moderate thrombocytopenia
Electrolytes	Sodium, potassium, and acid–base perturbations, depending upon fluid balance and stage of disease
BUN/creatinine	Renal failure may occur late in disease
Serum chemistries (AST, ALT, amylase, gamma-glutamyl transferase, alkaline phosphatase, creatinine kinase, lactate dehydrogenase, lactate acid)	Usually increased, especially in severe disease; AST > ALT; a lactate level greater than 4 mmol/L (36 mg/dL) may indicate persistent hypoperfusion and sepsis
Sedimentation rate	Normal or increased
Blood gas	Metabolic acidosis may be indicative of shock and hypoperfusion
Coagulation studies (PT, PTT, fibrinogen, fibrin split products, platelets, D-dimer)	Usually normal (except platelets, see above). DIC not common in Lassa fever
Urinalysis	Proteinuria common; hematuria may be occasionally noted; sediment may show hyaline-granular casts, and round cells with cytoplasmic inclusions
Blood culture	Useful early to exclude Lassa fever and later to evaluate for secondary bacterial infection; blood should be drawn before antibiotic therapy is instituted
Stool culture	Useful to exclude Lassa fever (in favor of hemorrhagic bacillary dysentery)
Thick and thin blood smears	May aid in the diagnosis of blood parasites (malaria and trypanosomes) and bacterial sepsis (meningococcus, capnocytophaga, and anthrax); all negative in Lassa fever unless co-infection
Rapid test, PCR, or other assay for malaria	Negative in Lassa fever unless co-infection with malaria
Febrile agglutinins or other assay for *Salmonella typhi*	Negative in Lassa fever unless co-infection with *S. typhi*

Abbreviations: ALT, Alanine aminotransferase; AST, aspartate aminotransferase; DIC, disseminated intravascular coagulation.

7. HOW DO YOU MAKE THE DIAGNOSIS?

The non-specific presentation of LF makes it extremely difficult to diagnose clinically, especially early in the course of disease before hemorrhage and other more severe manifestations develop. Therefore, prompt laboratory testing is imperative. Unfortunately, since there are presently no commercially available assays, LF diagnostics are available only in a few specialized laboratories,[10] although various recombinant-protein and virus-like particle-based assays are being developed.[39,40] The ELISA for LASV-specific antigen and IgM antibody[24] and RT-PCR[41] are the usual mainstays of LF diagnosis, with sensitivities and specificities over 90%.[42] Postmortem diagnosis may be made by immunohistochemical staining of formalin-fixed tissue.[43]

8. HOW DO YOU DIFFERENTIATE THIS DISEASE FROM SIMILAR ENTITIES?

The differential diagnosis of LF includes a broad array of febrile illnesses common in West Africa (Table 4.3). A diagnosis of LF should be considered in patients with a clinically compatible syndrome who, within 3 weeks prior to disease onset, (1) lived in or traveled to West Africa, especially in areas where LF is known to be endemic (Figure 4.4), (2) had potential direct contact with blood or bodily fluids of a person with LF during their acute illness (this group most often is comprised of healthcare workers), (3) worked in a laboratory or animal facility where LASV is handled, or (4) had sex with someone recovering from LF in the last 3 months. Recognized direct contact with rodents in West Africa should heighten suspicion but is rare even among confirmed cases. Acts of bioterrorism must be considered if LF is strongly suspected in a patient without any of the aforementioned risk factors, especially if clusters of cases are seen. It should be noted that even persons who meet the above criteria most commonly have a disease other than LF, so alternative diagnoses should always be aggressively sought, especially malaria and typhoid fever.

Consultation with infectious disease specialists with experience diagnosing and treating patients with viral hemorrhagic fever should be sought when the diagnosis is suspected. The inclusion of viral hemorrhagic fever in the differential diagnosis has the potential to induce considerable anxiety in patients, hospital staff, and the general community. Knowledge that imported LF is rare and that routinely practiced barrier nursing is protective in the vast majority of cases should offer reassurance. All confirmed cases of LF should be reported immediately to government health authorities.

TABLE 4.3 Differential Diagnosis of Lassa Fever

Disease	Distinguishing Characteristics and Comments
Parasites	
Malaria	Classically shows paroxysms of fever and chills; hemorrhagic manifestations less common; malaria smears or rapid test usually positive; co-infection (or baseline asymptomatic parasitemia) common; responds to anti-malarials
Amebiasis	Hemorrhagic manifestations other than bloody diarrhea generally not seen; amebic trophozoites identified in the stool; responds to anti-parasitics
Giardiasis	Positive stool antigen test and/or identification of trophozoites or cysts in stool; responds to anti-parasitics
African trypanosomiasis (acute stages)	Especially the east African form. Examination of peripheral blood smear/buffy coat may show trypanosomes
Bacteria (including *Spirochetes, Rickettsia, Ehrlichia,* and *Coxiella*)	
Typhoid fever	Hemorrhagic manifestations other than bloody diarrhea generally not seen; responds to antibiotics
Bacillary dysentery (including shigellosis, campylobacteriosis, salmonellosis, and enterohemorrhagic *Escherichia coli* and others)	Hemorrhagic manifestations other than bloody diarrhea generally not seen; respond to antibiotics
Capnocytophaga canimorsus	Associated with dog and cat bites, typically in persons with underlying immunodeficiency, notably asplenic patients; responds to antibiotics
Meningococcemia	Bacterial-induced DIC may mimic the bleeding diathesis of Lassa fever; bleeding within the first 24–48 hours after onset of illness and rapidly progressive illness typical; large ecchymoses typical of meningococcemia are unusual in Lassa fever; rapid serum latex agglutination tests can be used to detect bacterial antigen in meningococcal septicemia; may respond to antibiotics (critical to administer early)

(*Continued*)

TABLE 4.3 (Continued)

Disease	Distinguishing Characteristics and Comments
Staphylococcemia	Bacterial-induced DIC may mimic the bleeding diathesis of Lassa fever; may respond to antibiotics
Septic abortion	History of pregnancy and positive pregnancy test
Septicemic plague	Bacterial-induced DIC may mimic the bleeding diathesis of Lassa fever; large ecchymoses typical of plague are unusual in Lassa fever; may respond to antibiotics
Streptococcal pharyngitis	May mimic the exudative pharyngitis sometimes seen in Lassa fever
Tuberculosis	Hemoptysis of advanced pulmonary tuberculosis may suggest Lassa fever, but tuberculosis generally has a much slower disease evolution
Tularemia	Ulceroglandular and pneumonic forms more common; responds to antibiotics
Acute abdominal emergencies	Appendicitis, peritonitis, and bleeding upper gastrointestinal ulcer
Anthrax (inhalation or gastrointestinal)	Prominent pulmonary manifestations and widened mediastinum on chest X-ray in inhalation form; responds to antibiotics
Atypical bacterial pneumonia (*Legionella, Mycoplasma, Chlamydophila pneumoniae* and *psittaci*, others)	May mimic hantavirus pulmonary syndrome; exposure to birds and symptoms often not present until late in the illness in psittacosis; respond to antibiotics
Relapsing fever	Recurrent fevers and flu-like symptoms, with direct neurologic involvement and splenomegaly; spirochetes visible in blood while febrile; responds to antibiotics
Leptospirosis	Jaundice, renal failure, and myocarditis in severe cases; responds to antibiotics
Spotted fever group rickettsia (including African tick bite fever, Boutonneuse fever, Rocky Mountain spotted fever)	Incubation period of 7–10 days after tick bite, compared with 1–3 days in Crimean-Congo HF; necrotic lesions (eschar) typically seen at site of tick bite in some rickettsial diseases while there may only

(*Continued*)

TABLE 4.3 (Continued)

Disease	Distinguishing Characteristics and Comments
	be slight bruising at the bite site in Crimean-Congo HF; rash (if present) of rickettsial infection classically involves palms and soles
Q fever (*Coxiella burnetii*)	Broad spectrum of illness, including hepatitis, pneumonitis, encephalitis, and multisystem disease with bleeding; responds to antibiotics
Ehrlichiosis	Responds to antibiotics
Viruses	
Influenza	Prominent respiratory component to clinical presentation; no hemorrhagic manifestations; influenza rapid test may be positive; may respond to anti-influenza drugs
Arbovirus infection (including dengue and West Nile fever)	Encephalitis unusual, but when present may mimic the Lassa fever with significant neurologic involvement; usually less severe than Lassa fever; hemorrhage not reported
Viral hepatitis (including hepatitis A, B, and E, Epstein–Barr, and cytomegalovirus)	Jaundice atypical in HF except yellow fever; tests for hepatitis antigens positive; fulminant infection resembling Lassa fever may be seen in persons with underlying immune deficiencies
Herpes simplex or varicella-zoster	Fulminant infection with hepatitis (with/without vesicular rash); elevated transaminases and leucopoenia typical; disseminated disease may be noted in otherwise healthy persons; poor response to acyclovir drugs unless recognized early
HIV/AIDS	Seroconversion syndrome or HIV/AIDS with secondary infections, especially septicemia
Measles	Rash may mimic that seen in early stages of Lassa fever and may sometimes be hemorrhagic; prominence of coryza and upper respiratory symptoms in measles should help differentiate; vaccine preventable

(*Continued*)

TABLE 4.3 (Continued)

Disease	Distinguishing Characteristics and Comments
Rubella	Rash may mimic that seen in early stages of some Lassa fever; usually a mild disease; vaccine preventable
Hemorrhagic or flat smallpox	Diffuse hemorrhagic or macular lesions; in contrast to Lassa fever, the rash may involve the oral mucosa, palms, and soles; smallpox in the wild has been eradicated
Alphavirus infection (including chikungunya and o'nyong-nyong)	Joint pain typically a predominant feature
Non-infectious etiologies	
Heat stroke	History for extreme heat exposure; absence of sweating; bleeding not typical but DIC may occur
Idiopathic and thrombotic thrombocytopenic purpura (ITP/TTP)	Presentation usually less acute than Lassa fever; may have prominent neurologic symptoms in TTP; often respond to corticosteroids (ITP) or plasma exchange (TTP)
Drug sensitivity or overdose	Stevens–Johnson's syndrome and anticoagulant (warfarin) overdose
Industrial and agricultural chemical poisoning	Especially anticoagulants, although other symptoms of Lassa fever absent
Hematoxic snake bite envenomation	History of snake bite

Abbreviations and symbols: DIC, disseminated intravascular coagulopathy; HF, hemorrhagic fever.

9. WHAT IS THE THERAPEUTIC APPROACH?

Patients with LF should be treated in an intensive care unit following fluid and blood pressure management guidelines for septic shock.[1] Intravenous administration of the antiviral drug ribavirin should be started immediately, completing a 10-day course (Table 4.4).[44] The primary adverse effect is a dose-dependent, mild-to-moderate hemolytic anemia that infrequently necessitates transfusion and disappears with cessation of treatment.[19,44] Despite possible teratogenicity, ribavirin should still be considered in pregnant women given the extremely high maternal and fetal mortality associated with LF in pregnancy. Uterine evacuation also appears to lower maternal mortality in pregnant patients, although extreme caution is merited given the

TABLE 4.4 Ribavirin Therapy for Lassa Fever

Indication	Route	Dose	Interval
Treatment	IV	30 mg/kg (maximum 2 g)[a]	Loading dose, followed by:
	IV	15 mg/kg (maximum 1 g)[a]	Every 6 hr for 4 days, followed by:
	IV	7.5 mg/kg (maximum 500 mg)[b]	Every 8 hr for 6 days
Prophylaxis	PO	35 mg/kg (maximum 2.5 g)[a]	Loading dose, followed by:
	PO	15 mg/kg (maximum 1 g)[a]	Every 8 hr for 10 days

[a] *Reduce the dose in persons known to have significant renal insufficiency (creatinine clearance of less than 50 ml/min).*
[b] *The drug should be diluted in 150 ml of 0.9% saline and infused slowly.*
Abbreviations: IV, intravenous; PO, oral administration.

risks of nosocomial transmission.[45] Broad spectrum antibacterial and/or antiparasitic coverage should be continued until the diagnosis of LF can be confirmed. Impaired gas exchange is not typically a prominent feature of LF and intubation and mechanical ventilation should be avoided because of the risk of barotrauma and pleural-pulmonary hemorrhage. Various immune modulators (anti-TNF-α, nitric oxide inhibitors, statins, interleukins) and coagulation modifiers (recombinant activated protein C, rNAPc2) may have theoretical benefit in septic shock and viral hemorrhagic fever but have not been specifically tested in LF and should still be considered experimental.

Patients who have recovered from their acute illness can safely be sent home. However, because of the potential for delayed virus clearance from the urine and semen, abstinence or condom use is recommended for 3 months thereafter. Transmission through use of toilet facilities has not been noted, but simple precautions to avoid contact with potentially infected excretions in this setting are prudent, including separate toilet facilities and regular hand washing. Breastfeeding should be avoided during convalescence. Clinical management during convalescence includes the use of warm packs, acetaminophen, non-steroidal anti-inflammatory drugs, cosmetics, hair-growth stimulants, anxiolytics, antidepressants, nutritional supplements, and nutritional and psychological counseling as indicated.

10. WHAT ARE THE PREVENTIVE AND INFECTION CONTROL MEASURES?

Although contact precautions are generally protective, the possibly grave consequences of infection merit management of patients under specialized

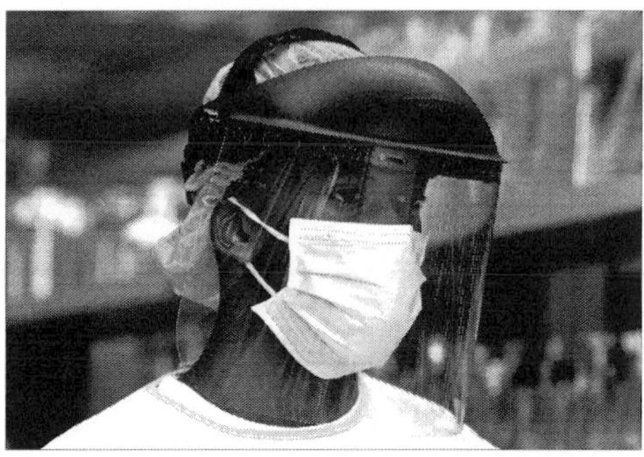

FIGURE 4.6 Face shield and surgical mask recommended for healthcare workers as part of viral hemorrhagic fever precautions. Also included in the recommended personal protective equipment are double gloves, gowns, protective aprons, and shoe covers. *From Ref. 52.*

viral hemorrhagic fever precautions (Figure 4.6).[46,47] Contacts should be monitored daily for 3 weeks. Post-exposure prophylaxis with oral ribavirin should be considered for persons with direct unprotected contact with blood or bodily fluids from a person with confirmed LF.[19] A number of experimental vaccine platforms are being explored, including some that have shown efficacy as post-exposure prophylaxis.[48] There is little long-term infection risk through shedding of LASV in the environment, where evidence indicates it is viable only for hours or days.[49] When recent contamination may have occurred, decontamination can be performed by heating and a variety of disinfectants.[1] Prevention in the community is oriented toward limiting contact with rodents by improving "village hygiene," such as eliminating unprotected storage of garbage, foodstuffs, and water, and plugging holes that allow rodents entry into homes. Rodent trapping or poisoning is generally not thought to be an effective long-term control strategy because animals from surrounding fields will likely soon recolonize the area.

REFERENCES

1. Bausch DG, Moses LM, Goba A, Grant D, Khan H. Lassa fever. In: Singh SK, Ruzek D, editors. *Viral hemorrhagic fevers*. Boca Raton, FL: Taylor and Francis Group/CRC Press; 2013.
2. Frame JD, Baldwin Jr JM, Gocke DJ, Troup JM. Lassa fever, a new virus disease of man from West Africa. I. Clinical description and pathological findings. *Am J Trop Med Hygiene* 1970;**19**:670–6.
3. Bond N, Schieffelin JS, Moses LM, Bennett AJ, Bausch DG. A historical look at the first reported cases of Lassa Fever: IgG antibodies 40 years after acute infection. *Am J Trop Med Hygiene* 2013;**88**:241–4.
4. Bausch DG, White H. ASTMH remembers Penny Pinneo, a pioneer in combating Lassa fever (1917–2012). In: ASTMH Blog: *American Society of Tropical Medicine and Hygiene*; 2012:Blog of the American Society of Tropical Medicine and Hygiene.

5. Bausch DG, Peters CJ. The viral hemorrhagic fevers. In: Lutwick LI, Lutwick SM, editors. *Beyond anthrax: the weaponization of infectious diseases*. New York: Humana Press; 2009. p. 107–44.
6. Enria DA, Mills JN, Shieh W, Bausch D, Peters CJ. Arenavirus infections. In: Guerrant RL, Walker DH, Weller PF, editors. *Tropical infectious diseases: principles, pathogens, and practice*. 3rd ed. Philadelphia: Elsevier Inc; 2011. p. 449–61.
7. Bowen MD, Rollin PE, Ksiazek TG, et al. Genetic diversity among Lassa virus strains. *J Virol* 2000;**74**:6992–7004.
8. Fair J, Jentes E, Inapogui A, et al. Lassa virus-infected rodents in refugee camps in Guinea: a looming threat to public health in a politically unstable region. *Vector Borne Zoonotic Dis* 2007;**7**:167–71.
9. Allan R. The progression from endemic to epidemic Lassa fever in war-torn West Africa. *Emerg Dis* 1998;2.
10. Khan S, Goba A, Chu M, et al. New opportunities for field research on the pathogenesis and treatment of Lassa fever. *Antiviral Res* 2008;**78**:103–15.
11. McCormick JB, Webb PA, Krebs JW, Johnson KM, Smith ES. A prospective study of the epidemiology and ecology of Lassa fever. *J Infect Dis* 1987;**155**:437–44.
12. Bloch A. A serological survey of Lassa fever in Liberia. *Bull World Health Organ* 1978;**56**:811–3.
13. Ehichioya DU, Hass M, Olschlager S, et al. Lassa fever, Nigeria, 2005–2008. *Emerg Infect Dis* 2010;**16**:1040–1.
14. Bausch DG, Demby AH, Coulibaly M, et al. Lassa fever in Guinea: I. Epidemiology of human disease and clinical observations. *Vector Borne Zoonotic Dis* 2001;**1**: 269–81.
15. Stephenson EH, Larson EW, Dominik JW. Effect of environmental factors on aerosol-induced Lassa virus infection. *J Med Virol* 1984;**14**:295–303.
16. Fichet-Calvet E, Lecompte E, Koivogui L, et al. Fluctuation of abundance and Lassa virus prevalence in Mastomys natalensis in Guinea, West Africa. *Vector Borne Zoonotic Dis* 2007;**7**:119–28.
17. Bowen MD, Peters CJ, Nichol ST. Phylogenetic analysis of the Arenaviridae: patterns of virus evolution and evidence for cospeciation between arenaviruses and their rodent hosts. *Mol Phylogenet Evol* 1997;**8**:301–16.
18. Bonner PC, Schmidt W-P, Belmain SR, Oshin B, Baglole D, Borchert M. Poor housing quality increases risk of rodent infestation and Lassa fever in refugee camps of Sierra Leone. *Am J Trop Med Hyg* 2007;**77**:169–75.
19. Kelly DJ, Barrie MB, Ross RA, Temple BA, Moses LM, Bausch DG. Housing equity for health equity: a rights-based approach to the control of Lassa fever in post-war Sierra Leone. *BMC Int Health Hum Rights* 2013;**13**:2.
20. ter Meulen J, Lenz O, Koivogui L, et al. Short communication: Lassa fever in Sierra Leone: UN peacekeepers are at risk. *Trop Med Int Health: TMIH* 2001;**6**:83–4.
21. ter Meulen J, Lukashevich I, Sidibe K, et al. Hunting of peridomestic rodents and consumption of their meat as possible risk factors for rodent-to-human transmission of Lassa virus in the Republic of Guinea. *Am J Trop Med Hygiene* 1996;**55**:661–6.
22. Fisher-Hoch SP, Tomori O, Nasidi A, et al. Review of cases of nosocomial Lassa fever in Nigeria: the high price of poor medical practice. *BMJ* 1995;**311**:857–9.
23. Carey DE, Kemp GE, White HA, et al. Lassa fever. Epidemiological aspects of the 1970 epidemic, Jos, Nigeria. *Trans R Soc Trop Med Hygiene* 1972;**66**:402–8.

24. Bausch DG, Rollin PE, Demby AH, et al. Diagnosis and clinical virology of Lassa fever as evaluated by enzyme-linked immunosorbent assay, indirect fluorescent-antibody test, and virus isolation. *J Clin Microbiol* 2000;**38**:2670–7.
25. Lunkenheimer K, Hufert FT, Schmitz H. Detection of Lassa virus RNA in specimens from patients with Lassa fever by using the polymerase chain reaction. *J Clin Microbiol* 1990;**28**:2689–92.
26. Johnson KM, McCormick JB, Webb PA, Smith ES, Elliott LH, King IJ. Clinical virology of Lassa fever in hospitalized patients. *J Infect Dis* 1987;**155**:456–64.
27. Walker DH, McCormick JB, Johnson KM, et al. Pathologic and virologic study of fatal Lassa fever in man. *Am J Pathol* 1982;**107**:349–56.
28. Baize S, Kaplon J, Faure C, Pannetier D, Georges-Courbot MC, Deubel V. Lassa virus infection of human dendritic cells and macrophages is productive but fails to activate cells. *J Immunol* 2004;**172**:2861–9.
29. Mahanty S, Hutchinson K, Agarwal S, McRae M, Rollin PE, Pulendran B. Cutting edge: impairment of dendritic cells and adaptive immunity by Ebola and Lassa viruses. *J Immunol* 2003;**170**:2797–801.
30. Mahanty S, Bausch DG, Thomas RL, et al. Low levels of interleukin-8 and interferon-inducible protein-10 in serum are associated with fatal infections in acute Lassa fever. *J Infect Dis* 2001;**183**:1713–21.
31. Hensley LE, Smith MA, Geisbert JB, et al. Pathogenesis of Lassa fever in cynomolgus macaques. *Virol J* 2011;**8**:205.
32. Roberts PJ, Cummins D, Bainton AL, et al. Plasma from patients with severe Lassa fever profoundly modulates f-met-leu-phe induced superoxide generation in neutrophils. *Br J Haematol* 1989;**73**:152–7.
33. Fisher-Hoch S. Pathophysiology of shock and haemorrhage in viral haemorrhagic fevers. *Southeast Asian J Trop Med Public Health* 1987;**18**:390–1.
34. McCormick JB, King IJ, Webb PA, et al. A case-control study of the clinical diagnosis and course of Lassa fever. *J Infect Dis* 1987;**155**:445–55.
35. Solbrig MV, McCormick JB. Lassa fever: central nervous system manifestations. *J Trop Geograph Neurol* 1991;**1**:23–30.
36. Monson MH, Cole AK, Frame JD, Serwint JR, Alexander S, Jahrling PB. Pediatric Lassa fever: a review of 33 Liberian cases. *Am J Trop Med Hygiene* 1987;**36**:408–15.
37. Ketai L, Alrahji AA, Hart B, Enria D, Mettler Jr. F. Radiologic manifestations of potential bioterrorist agents of infection. *AJR: Am J Roentgenol* 2003;**180**:565–75.
38. Cummins D, Bennett D, Fisher-Hoch SP, Farrar B, McCormick JB. Electrocardiographic abnormalities in patients with Lassa fever. *J Trop Med Hyg* 1989;**92**:350–5.
39. Branco LM, Matschiner A, Fair JN, et al. Bacterial-based systems for expression and purification of recombinant Lassa virus proteins of immunological relevance. *Virol J* 2008;**5**:74.
40. Saijo M, Georges-Courbot MC, Marianneau P, et al. Development of recombinant nucleoprotein-based diagnostic systems for Lassa fever. *Clin Vaccine Immunol* 2007;**14**:1182–9.
41. Olschlager S, Lelke M, Emmerich P, et al. Improved detection of Lassa virus by reverse transcription-PCR targeting the 5′ region of S RNA. *J Clin Microbiol* 2010;**48**:2009–13.
42. Niedrig M, Schmitz H, Becker S, et al. First international quality assurance study on the rapid detection of viral agents of bioterrorism. *J Clin Microbiol* 2004;**42**:1753–5.
43. Zaki SR, Shieh WJ, Greer PW, et al. A novel immunohistochemical assay for the detection of Ebola virus in skin: implications for diagnosis, spread, and surveillance of Ebola

hemorrhagic fever. Commission de Lutte contre les Epidemies a Kikwit. *J Infect Dis* 1999;**179**(Suppl. 1):S36–47.

44. McCormick JB, King IJ, Webb PA, et al. Lassa fever. Effective therapy with ribavirin. *New Engl J Med* 1986;**314**:20–6.
45. Price ME, Fisher-Hoch SP, Craven RB, McCormick JB. A prospective study of maternal and fetal outcome in acute Lassa fever infection during pregnancy. *BMJ* 1988;**297**:584–7.
46. CDC and WHO. *Infection control for viral haemorrhagic fevers in the African health care setting*. Atlanta: Centers for Disease Control and Prevention; 1998.
47. Interim Infection control guidelines for care of patients with suspected or confirmed filovirus (Ebola, Marburg) haemorrhagic fever. WHO, 2008. (Accessed October 13, 2013, at <http://www.who.int/csr/bioriskreduction/filovirus_infection_control/en/>.)
48. Geisbert TW, Jones S, Fritz EA, et al. Development of a new vaccine for the prevention of Lassa fever. *PLoS Med* 2005;**2**:e183.
49. Sagripanti JL, Lytle CD. Sensitivity to ultraviolet radiation of Lassa, vaccinia, and Ebola viruses dried on surfaces. *Arch Virol* 2011;**156**:489–94.
50. Fichet-Calvet E, Rogers DJ. Risk maps of Lassa fever in West Africa. *PLoS Negl Trop Dis* 2009;**3**:e388.
51. Ehichioya DU, Hass M, Becker-Ziaja B, et al. Current molecular epidemiology of Lassa virus in Nigeria. *J Clin Microbiol* 2011;**49**:1157–61.
52. Bausch DG, Feldmann H, Geisbert TW, et al. Outbreaks of filovirus hemorrhagic fever: time to refocus on the patient. *J Infect Dis* 2007;**196**(Suppl. 2):S136–41.

Chapter 5

Alkhurma Hemorrhagic Fever

Pierre E. Rollin[1] and Ziad A. Memish[2]
[1]*Viral Special Pathogens Branch, Centers for Disease Control and Prevention, Atlanta, GA, USA,* [2]*Global Center for Mass Gathering Medicine, Ministry of Health, Riyadh, Kingdom of Saudi Arabia*

CASE PRESENTATION

A 32-year-old male Egyptian butcher working in Saudi Arabia was presented to hospital with fever, generalized body aches, and vomiting of 5 day duration. He then developed irritability, convulsions with hematemesis, and melena. On examination the patient was unconscious and in bad general condition. His neck was lax, with no sign of meningeal irritation. His temperature was 38.5°C, blood pressure 90–160 mmHg, and pulse rate 112/min and regular. Abdominal examination showed no abnormality. Skin examination showed ecchymotic patches at injection sites. Laboratory examination revealed the following: leucopenia (leucocytes 2.9 × 109/L), thrombocytopenia (platelets 38 × 109/L), and elevated levels of liver enzymes (serum glutamic oxalacetic transaminase (AST) 680 units/L, serum glutamic pyruvic transaminase (ALT) 1950 units/L), creatinine phosphokinase (18,200 units/L), and blood urea (186 mmol/L). The patient received intensive supportive treatment; however, he became deeply comatose and died of irreversible shock with renal and respiratory failure 2 days after admission. The diagnosis was confirmed by virus isolation from a blood sample taken 6 days after the onset of symptoms.

This is a published case report.[1]

1. WHY THIS CASE WAS SIGNIFICANTLY IMPORTANT AS AN EMERGING INFECTION

Alkhurma hemorrhagic fever (AHF) is a viral infection recently described in Saudi Arabia, associated in severe forms with severe hemorrhagic and neurologic manifestations. Mortality in hospitalized patients varied between 1 and 20%. The recent laboratory confirmation of the disease in tourists visiting

Emerging Infectious Diseases. DOI: http://dx.doi.org/10.1016/B978-0-12-416975-3.00005-4

Egypt extends the geographical range of the virus and suggests that infections due to AHF virus, a tick-borne flavivirus, are underreported. The persistence of the virus within the tick population (vertical and horizontal transmission such as co-feeding) and the role of livestock infection in the transmission of the virus to humans are unknown. Their understanding will have great implications on public health, prophylaxis, and health education measures.

2. WHAT IS THE CAUSATIVE AGENT?

Alkhurma hemorrhagic fever virus (AHFV) was first isolated from specimens collected from a patient in Mecca, Saudi Arabia, in 1995.[1] The patient (a butcher) presented a rapid, fatal hemorrhagic fever and Crimean-Congo hemorrhagic fever was clinically suspected. AHFV is a variant of Kyasanur Forest disease virus (KFDV), which is endemic in the Karnataka State in India and a member of the tick-borne encephalitis group (family *Flaviviridae*, genus *Flavivirus*, enveloped, segmented, negative-strand RNA virus).[2,3] Phylogenetic analysis of full-genome sequences of AHFV and KFDV revealed a low diversity and a slow rate of molecular evolution, similar to other tick-borne flaviviruses.[3,4] The suggested divergence between AHFV and KFDV, around 770 years ago, argues against a recent introduction of KFDV in Saudi Arabia. The virus is classified as a BSL-3 or BSL-4 agent and its manipulation required high-containment laboratories and trained personnel.

3. WHAT IS THE FREQUENCY OF THE DISEASE?

Since the first description in the Mecca region, several hundred human cases have been reported in other western Governorates of Saudi Arabia: Jeddah, Jizan, and Najran (Figure 5.1). Antibodies have been found in people living in other Governorates of Saudi Arabia, and in United Arab Emirates.[5, unpublished data] Three human cases have also been reported in tourists visiting a camel market in Al-Shalateen, along the Red Sea in southern Egypt.[6,7] Because of the large geographic range of the tick vector, it is expected that the distribution of the virus is more extensive, with probable unsuspected human cases in Yemen and Sudan. The monthly distribution of human cases, with a peak in spring and summer, is certainly related to the ecology of the reservoirs. In the early description of the disease, the case fatality was reaching 25%. Further studies in Najran added subclinical infection to the spectrum of the disease, lowering the case fatality rate to 1–2%.[1,8,9]

4. HOW IS THE VIRUS TRANSMITTED?

Epidemiological, veterinary and entomological aspects, and the cycle of transmission of AHFV are still poorly understood. AHFV is a zoonotic

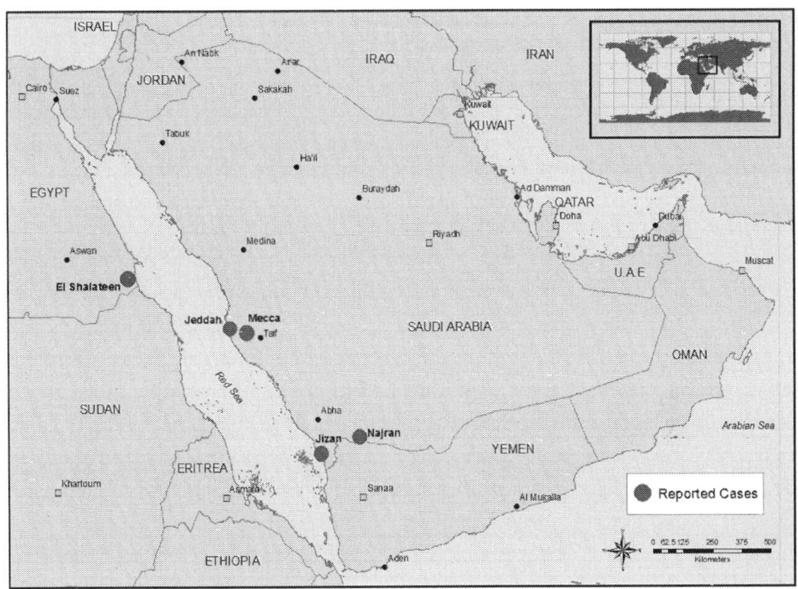

FIGURE 5.1 Geographic range of Alkhurma virus (the large black dots correspond to the region on known endemicity of the virus).

virus, and has been detected and isolated from adult soft ticks (*Ornithodoros savignyi*) and hard ticks (*Hyalomma dromedari*) collected in western Saudi Arabia.[10,11] Both ticks have a very wide distribution. People can be infected through tick bite or when crushing infected ticks. Soft ticks feed and stay attached to the host only for a short period of time and this could be a reason for them being unnoticed by patients. All the epidemiological studies identified contacts with domestic animals or livestock as risk factors for human infections. There is no known disease in animals after infection by AHFV, but no experimental infections have been done, and as with Crimean-Congo hemorrhagic fever virus, another tick-borne virus, the animal could develop a viremia without obvious clinical symptoms. If this is the case, slaughtering animals during this phase of infection could certainly be the source of human infection. Mosquito transmission has been hypothesized but no detection or isolation in wild-caught mosquitoes has been reported and the taxonomic classification of the virus makes it unlikely. No transmission through non-pasteurized milk consumption has been described, although other viruses from the tick-borne encephalitis group have been transmitted to humans through this route. No nosocomial transmissions are reported in healthcare workers but standard precautions are recommended when handling AHFV patients or their biological specimens.

5. WHICH FACTORS ARE INVOLVED IN DISEASE PATHOGENESIS? WHAT ARE THE PATHOGENIC MECHANISMS?

Very little is known about the pathogenesis of AHFV in humans and no animal models of the human disease are available. The virus is present in the blood during the clinical phase of the disease, then disappears and is replaced by IgM and IgG antibodies. AHFV certainly affect the liver, as reflected by the elevation of the liver enzymes, and this tropism may be partly associated with hemostasis abnormalities and bleeding tendencies in patients. Although the central nervous system is affected in a certain number of patients, no cerebrospinal fluid (CSF) examination or imaging evaluation of the brain has been reported. The neurotropism during human infection by members of the tick-borne flavivirus group is well known: KFDV can be isolated from the CSF[12] and intrathecal production of antibodies is also described in tick-borne encephalitis.[13]

6. WHAT ARE THE CLINICAL MANIFESTATIONS?

It is difficult to have a definite duration for the incubation of AHF, because of the permanent exposure to the potential risk factors, but in the Italy-imported patients, the incubation after apparent tick bite was as short as 2 to 4 days, comparable with the reported 2–7 days for KFD. The disease appears to be diphasic at least in some patients, and starts as a non-specific flu-like syndrome with fever, anorexia, malaise, diarrhea, vomiting, followed by either neurological or hemorrhagic manifestations in the severe forms (Table 5.1).[1,6–9,14–16] Laboratory abnormalities such as thrombocytopenia, leukopenia, elevated creatine phosphokinase, LDH, and liver enzymes are nearly always present in hospitalized patients (Table 5.2).[1,6–8,14–16] No data are available on CSF even in neurological forms. Multi-organ failures precede the fatal outcome. No sequelae have been reported after clinical recovery. Rhabdomyolysis has been reported in one patient.[7] As reported by Alzahrani and colleagues,[9] milder clinical forms that do not require hospitalization or even doctor consultation exist.

7. HOW DO YOU DIAGNOSE?

In Saudi Arabia, a clinical, laboratory, and epidemiologic case definition has been put in place (Table 5.3). ALK virus infection can be confirmed at the acute phase of the disease by direct detection of virus in blood, plasma, or serum either by virus isolation or by detection of the viral RNA. The same techniques can be used to detect ALK virus from viremic animals or infected ticks. At later stages of human disease, in serological surveys, or in animal samples, IgM and IgG antibodies may be detectable in the blood, plasma,

TABLE 5.1 Alkhurma Hemorrhagic Fever: Clinical Signs and Symptoms

Variable	Zaki[1] n (%)	Madani[14] n (%)	Memish[15] n (%)	Charrel[16] n (%)	Madani[8] n (%)	Alzahrani[9] hospitalized n (%)	Alzahrani[9] non-hospitalized n (%)	Carletti[6] n (%)	Madani[8] n (%)	Ravanini[7] n (%)
Fever	10 (100)	20 (100)	1 (100)	16 (100)	78 (100)	11 (100)	4 (24)	1 (100)	78 (100)	1 (100)
Headache	10 (100)	15 (75)		16 (100)	67 (85.9)				67 (85.9)	1 (100)
Malaise		15 (75)		16 (100)	67 (85.9)			1 (100)	67 (85.9)	
Myalgia		15 (75)		16 (100)	64 (82.1)				64 (82.1)	
Retro-orbital pain	10 (100)	1 (5)		16 (100)	43 (55.1)				43 (55.1)	
Arthralgia		9 (45)			65 (83.3)				65 (83.3)	
Backache, body ache	10 (100)	5 (25)		16 (100)	56 (71.8)				56 (71.8)	
Nausea, vomiting	10 (100)	10 (50)		10 (62.5)	56 (71.8)			1 (100)	56 (71.8)	
Diarrhea		5 (20)		4 (25)	40 (51.3)				40 (51.3)	1 (100)
Rash	2 (20)	3 (15)		2 (12.5)	0	5 (46)	2 (12)		0	

(Continued)

TABLE 5.1 (Continued)

Variable	Zaki[1] *n* (%)	Madani[14] *n* (%)	Memish[15] *n* (%)	Charrel[16] *n* (%)	Madani[8] *n* (%)	Alzahrani[9] hospitalized *n* (%)	Alzahrani[9] non-hospitalized *n* (%)	Carletti[6] *n* (%)	Madani[8] *n* (%)	Ravanini[7] *n* (%)
Bleeding (*)	2 (20)	11 (55)	1 (100)	4 (25)	20 (25.6)	5 (46)	0		20 (25.6)	
Confusion, disorientation		1 (5)	1 (100)		8 (10.7)				8 (10.7)	
Neck stiffness					7 (9.3)	3 (27)	0		7 (9.3)	
Seizures					4 (5.1)	2 (18)	0		4 (5.1)	
Encephalitis	1 (10)	4 (20)			10 (12.8)				10 (12.8)	
Mortality	2 (20)	5 (25)	0	5 (31)	1 (1.3)				1 (1.3)	

May include epistaxis, gum bleeding, melena, hematemesis, and genital bleeding.

TABLE 5.2 Alkhurma Hemorrhagic Fever: Laboratory Data

Variable	Zaki[1] *n* (%)	Madani[14] *n* (%)	Memish[15] *n* (%)	Charrel[16] *n* (%)	Carletti[6] *n* (%)	Madani[8] *n* (%)	Ravanini[7] *n* (%)
Leukopenia	10 (100)	13 (65)	1 (100)	16 (100)	1 (100)	57/65 (87.7)	1 (100)
Thrombocytopenia	10 (100)	15 (75)	1 (100)	16 (100)	1 (100)	30/65 (46.2)	1 (100)
High bilirubin		6 (30)				5/39 (12.8)	
High hemoglobin		1 (5)				5/62 (8.1)	
High INR		9 (45)				5/21 (23.8)	1 (100)
High PTT		15 (75)				10/19 (52.6)	1 (100)
High CPK	10 (100)	19 (5)	1 (100)	16 (100)		21/46 (41.7)	1 (100)
High LDH		17 (85)	1 (100)			9/36 (25)	1 (100)
High AST	10 (100)	20 (100)	1 (100)	16 (100)	1 (100)	54/63 (85.7)	1 (100)
High ALT	10 (100)	16 (80)	1 (100)	16 (100)	1 (100)	43/64 (67.2)	1 (100)

TABLE 5.3 Alkhurma Hemorrhagic Fever: Saudi Arabia Case Definition[17]

Clinical case definition (human cases):
Suspected: Case meets the clinical *and* exposure criteria
Probable: suspected case with clinical laboratory data (e.g., thrombocytopenia, leucopenia, elevation of liver enzymes, elevated CPK or LDH) and IgM detected by capture ELISA
Confirmed: Probable case *and* laboratory criteria listed below

Clinical criteria:
Unexplained acute febrile illness (fever. 38°C) with one of the three following features:
- Hemorrhagic manifestations not related to injury (bleeding under the skin, in internal organs or from body orifices; and positive tourniquet test)
- Liver involvement (jaundice, hepatomegaly)
- Neurological involvement (severe headache, altered mental status, and/or seizures)

Laboratory criteria (one or more of the following laboratory findings):
- AHFV RNA detected by real-time or conventional RT-PCR
- Virus isolation/identification using cell culture or suckling mice
- Four-fold antibody (IgM and or IgG) rise in paired serum samples using ELISA or IFA
- Neutralization test—preferably plaque reduction for paired sera

Exposure criteria (one or more of the following exposures before onset of symptoms):
- Recent contact with animal, blood, or other animal products
- Recent exposure to or bite by tick
- Contact with blood or body fluid from a confirmed human case
- Work in a laboratory that handles AHFV specimens/isolates

and serum by ELISA.[18] Viral antigen can certainly be detected by immunohistochemistry in formalin-fixed tissues and paraffin-embedded blocks. As usual, when an infectious disease etiology is suspected, a combination of several techniques should be used for laboratory confirmation.

7.1 Viral Isolation

All isolation attempts require high-containment laboratory and trained personnel. Virus can be isolated from acute patient samples using several cell types, but most often Vero or BHK21 cells; a cytopathic effect is easily seen after 9–10 days. If newborn suckling mice are available, the animals, after intracerebral inoculation, show paralysis and die after 5–7 days.[1,4]

7.2 Molecular Detection: RT-PCR

Conventional and real-time reverse transcriptase polymerase chain reactions (RT-PCR) are available for detecting ALK RNA from acute human samples (first week post-onset), or from infected ticks. A real-time kit is commercially available (TIB MolBiol LightMix® Kit Alkhurma Virus, Cat. No. 40-0581-16). The virus being present in circulating macrophages, using blood or buffy-coat will increase the sensitivity of the molecular detection.[19]

7.3 Serological Diagnosis

The interpretation of the serological assays needs caution when several flaviviruses are circulating in the same geographical zone. Several IgM and IgG ELISAs have been developed and are used in Saudi Arabian laboratories for identification of acute human cases or surveys.[5,9]

8. HOW DO YOU DIFFERENTIATE THIS DISEASE FROM SIMILAR ENTITIES?

Clinical diagnosis is difficult or impossible without etiologic laboratory confirmation. Crimean-Congo hemorrhagic fever and Rift Valley fever should be suspected in patients fulfilling the clinical criteria of the proposed case definition, coming from the AHFV endemic area, and exposed to identical risk factors (exposition to livestock, working in slaughterhouses). The initial laboratory data (white blood and platelets count, liver function tests) may also be identical. Molecular assays are very specific and sensitive. But the well-known antibody cross-reactivity within flaviviruses (i.e., dengue and maybe Kadam viruses) circulating in the same geographical range make any diagnosis based on a single serological assay suspicious.[20,21]

9. WHAT IS THE THERAPEUTIC APPROACH?

There is no standard specific treatment for ALK disease. Patients receive supportive therapy. This consists of balancing the patient's fluids and electrolytes, maintaining their oxygen status and blood pressure, and treating them for any complicating infections.

10. WHAT ARE THE PREVENTIVE AND INFECTION CONTROL MEASURES?

No treatment or specific prophylaxis being available, prevention and awareness are the only recommended measures. Controlling vectors is difficult and interrupting the cycle of the virus within the reservoirs is impossible. In the endemic region, it is recommended, as for other tick-borne pathogens, if possible to avoid tick-infected areas and limit contact with tick-infested livestock or domestic animals. At the individual level, people should use tick repellents on skin and clothes, check skin for attached ticks and remove them as soon as possible, and prevent tick infestation on domestic animals (tick collars) and livestock (dipping). Applications of pesticides and insecticides around the habitations and animal living areas could have some effect in reducing the tick populations. Although difficult in practice, people working with animals or animal products on farms or in slaughterhouses should avoid unprotected contact with blood of potentially viremic animals. No human-to-human or nosocomial infections have been described, but standard infection control practices are recommended for healthcare workers.

REFERENCES

1. Zaki AM. Isolation of a flavivirus related to the tick-borne encephalitis complex from human cases in Saudi Arabia. *Trans. Roy Soc Trop Med Hyg* 1997;**91**:179–81.
2. Charrel RN, Zaki AM, Attoui H, et al. Complete coding sequence of the Alkhurma virus, a tick-borne flavivirus causing severe hemorrhagic fever in humans in Saudi Arabia. *Biochem Biophys Res Comm* 2001;**287**:455–61.
3. Charrel RN, Zaki AM, Fakeeh M, et al. Low diversity of Alkhurma hemorrhagic fever virus, Saudi Arabia, 1994–1999. *Emerg Infect Dis* 2005;**11**:683–8.
4. Dodd KA, Bird BH, Khristova ML, et al. Ancient Ancestry of KFDV and AHFV revealed by complete genome analyses of viruses isolated from ticks and mammalian hosts. *PLoS Negl Trop Dis* 2011;**5**:e1352.
5. Memish ZA, Albarrak A, Almazroa MA, et al. Seroprevalence of Alkhurma and other hemorrhagic fever viruses, Saudi Arabia. *Emerg Infect Dis* 2011;**17**:2316–8.
6. Carletti F, Castilletti C, Di Caro A, et al. Alkhurma hemorrhagic fever in travelers returning from Egypt, 2010. *Emerg Infect Dis* 2010;**16**:1979–82.
7. Ravanini P, Hasu E, Huhtamo E, et al. Rhabdomyolysis and severe muscular weakness in a traveler diagnosed with Alkhurma hemorrhagic fever virus infection. *J Clin Virol* 2011;**52**:254–6.
8. Madani TA, Azhar EI, Abuelzein EME, et al. Alkhumra (Alkhurma) virus outbreak in najran, Saudi Arabia: epidemiological, clinical, and laboratory characteristics. *J Infect* 2011;**62**:67–76.
9. Alzahrani AG, Al Shaiban HM, Al Mazroa MA, et al. Alkhurma hemorrhagic fever in humans, Najran, Saudi Arabia. *Emerg Infect Dis* 2010;**16**:1882–8.
10. Charrel RN, Fagbo S, Moureau G, Alqahtani MH, Temmam S, de Lamballerie X. Alkhurma hemorrhagic fever virus in *Ornithodoros savignyi* ticks. *Emerg Infect Dis* 2007;**13**:153–5.
11. Mahdi M, Erickson BR, Comer JA, et al. Kyasanur Forest disease virus Alkhurma subtype in ticks, Najran Province, Saudi Arabia. *Emerg Infect Dis* 2011;**17**:945–7.
12. Bhatt PN, Work TH, Varma VGR, Trapido H, Narasimha Murthy DP, Rodrigues FM. Isolation of Kyasanur forest disease virus from infected humans and monkeys of Shimoga district, Mysore State. *Indian J Med Sci* 1966;**20**:316–20.
13. Gunther G, Haglund M, Lindquist L, Skoldenberg B, Forsgren M. Intrathecal IgM, IgA and IgG antibody response in tick-borne encephalitis. Long-term follow-up related to clinical course and outcome. *Clin Diag Virol* 1997;**8**:17–29.
14. Madani TA. Alkhumra virus infection, a new viral hemorrhagic fever in Saudi Arabia. *J Infect* 2005;**51**:91–7.

15. Memish ZA, Balkhy HH, Francis C, Cunningham G, Hajeer AH, Almuneef MA. Alkhurma haemorrhagic fever: case report and infection control details. *Br J Biomed Sci* 2005;**62**:37–9.
16. Charrel RN, de Lamballerie X, Zaki AM. Human cases of hemorrhagic fever in Saudi Arabia due to a newly discovered flavivirus, Alkhurma hemorrhagic fever virus. In: Lu Y, Essex M, Roberts B, editors. *Emerging infections in Asia*. New York: Springer; 2008. p. 179–92.
17. Memish ZA, Fagbo SF, Assiri AM, et al. Alkhurma viral hemorrhagic fever virus: proposed guidelines for detection, prevention, and control in Saudi Arabia. *PLoS Negl Trop Dis* 2012;**6**:e1604.
18. Memish ZA, Charrel RN, Zaki AM, Fagbo SF. Alkhurma haemorrhagic fever—a viral haemorrhagic disease unique to the Arabian Peninsula. *Int J Antimicrob Agents* 2010;**36** (Suppl. 1):S53–7.
19. Madani TA, Abuelzein EE, Azhar EI, et al. Superiority of the buffy coat over serum or plasma for the detection of Alkhumra virus RNA using real time RT-PCR. *Arch Virol* 2012;**157**:819–23.
20. Allwinn R, Doerr HW, Emmerich P, Schmitz H, Preiser W. Cross-reactivity in flavivirus serology: new implications of an old finding? *Med Microbiol Immunol* 2002;**190**:199–202.
21. Kuno G. Serodiagnosis of flaviviral infections and vaccinations in humans. *Adv Virus Res* 2003;**61**:3–65.

Chapter 6

Rift Valley Fever

Janusz T. Paweska

Center for Emerging and Zoonotic Diseases, National Institute for Communicable Diseases of the National Health Laboratory Service and School of Pathology, Faculty of Medical Sciences, University of Witwatersrand, South Africa

CASE PRESENTATION

A 68-year-old male involved in farming sheep and cattle in the North-West Province of South Africa, without recent travel history, was admitted to Kimberly Hospital in North Cape Province. He became ill 4 days after slaughtering a warthog, which was apparently killed on his farm by a dog during late summer of 2010 when many cases of Rift Valley fever (RVF) were reported throughout the country in livestock, wild animals, and humans. His early complaints included headache, malaise, fever, and myalgia followed by vomiting and bleeding from the gums. He was admitted to the hospital 5 days after exposure to a potentially sick animal, and despite supportive treatment he died the following day after admission. His major complications included hepatitis, renal failure, thrombocytopenia, and hemorrhage. On the day of admission he had raised levels of bilirubin, alanine transaminase, aspartate transaminase, creatine kinase, alkaline phosphatase, γ-glutamyl transferase, and decreased platelets. The patient had no detectable IgG and IgM antibodies against Crimean-Congo hemorrhagic fever virus (CCHFV), and also tested negative by CCHFV RT-PCR, but RVF virus (RVFV) nucleic acid was detected in his serum. RVF IgG and IgM serology was negative. The RVFV RT-PCR result was subsequently confirmed by isolation of the virus from his blood in suckling mice.

Recent re-emergence of RVF in South Africa, 34 years after the major outbreak of 1974–1976, was first associated with scattered outbreaks of the disease in adjacent parts of north-eastern parts of the country in 2008 and limited outbreaks in KwaZulu Natal and North Cape Provinces in 2009. RVFV had progressively reinfiltrated much of interior plateau of South Africa in 2010. The last human cases were reported in May of 2011. Of the total of 2014 suspected cases, 302 were laboratory confirmed. The majority

Emerging Infectious Diseases. DOI: http://dx.doi.org/10.1016/B978-0-12-416975-3.00006-6

of cases were reported in March and April each summer season, which coincides with increased mosquito activity in South Africa. Most cases were diagnosed in 2010 with a case fatality rate of 8%. No person-to-person transmission or nosocomial outbreaks were reported.

1. WHY RVF OUTBREAK WAS IMPORTANT AS AN EMERGING INFECTION

The etiologic agents of viral hemorrhagic fevers (VHFs) belong to taxonomically diverse RNA viruses of *Arenaviridae*, *Bunyaviridae*, *Flaviviridae*, and *Filoviridae* families. Timely recognition and management of VHF cases constitutes a challenging task for diagnostic laboratory and public health systems. The capacity of Rift Valley fever virus (RVFV) to cause large and severe outbreaks in animal and human populations and to cross significant natural geographic barriers, as exemplified by the virus spread over the Indian Ocean, Sahara desert, and the Red Sea in the past three decades, is of great concern for veterinary and public health authorities worldwide. RVFV is one of the most important emerging zoonotic threats, particularly to vulnerable African communities with low resilience to economic and environmental challenges.[1] In recent years, significant progress has been made on various aspects of the disease and its etiological agent; however, unpredictability of virus emergence, gaps in understanding its ecology, and particularly the enigma surrounding the mechanisms involved in inter-epizootic transmission of the virus remain a challenge for science.[2] Outbreaks of RVF prompt implementation of strict and long-term international restrictions of livestock exports with devastating economic effects on countries for which animal trade constitutes the main source of national revenue. A high death rate among pregnant and new-born ruminants affects the survival of pastoral nomads and local herders who are economically and physically dependent on milk and meat. Large outbreaks in livestock are associated with high numbers of human infections, which clinical management might pose a significant challenge in recourse-limited healthcare settings.[3] Moreover, there are no licensed vaccines or chemotherapeutics available for RVF prevention and treatment in humans. Competent mosquitoes are present in many RVF-free countries. Therefore, there is an increasing international concern that, should the virus be introduced to naïve populations of humans, domestic and wild ruminants in non-endemic regions, it would potentially cause dramatic health and socioeconomic consequences. RVFV is also regarded as a potential bioweapon, and biosecurity veterinary and health authorities are concerned about the possibility of deliberate release of this zoonotic agent.[4–6] Within the framework of the International Health Regulations,[7] members of the World Health Organization (WHO) and members of the World Organization for Animal Health (OIE) have responsibility for early detection and reporting of RVF outbreaks, together with assessment of the risk of spread to new areas.

2. WHAT IS THE CAUSATIVE AGENT?

RVFV is a member of the genus *Phlebovirus* in the family *Bunyaviridae* characterized by a tri-segmented single-stranded RNA genome of negative or ambisense polarity.[8] RVFV particles are 90–110 nm in diameter and consist of an envelope and a ribonucleocapsid (RNP). The envelope is composed of a lipid bilayer containing heterodimers of Gn and Gc glycoproteins, which are the building blocks of 122 capsomers (110 hexamers and 12 pentamers) arranged in T = 12 lattice. The viral ribonucleoproteins corresponding to each of the three genomic segments associated with numerous copies of the nucleoprotein N and the RNA-dependent RNA polymerase L are packaged into the virion.[9]

The RVFV genome of ≈12 kb consists of three single-stranded RNA segments designated large (L), medium (M), and small (S), which are used as templates to generate complementary RNA (cRNA) and messenger RNA (mRNA). The L and M segments are of negative polarity. The L segment encodes for the viral RNA-dependent RNA polymerase (L protein). The M segment encodes four proteins in a single open reading frame (ORF): the precursor to the glycoproteins Gn and Gc, and two non-structural proteins, designated NSm1 and NSm2. The S segment utilizes an ambisense strategy to code for two proteins, the nucleoprotein N and a non-structural protein NSs.[10–12] The RVFV virion composition and transcription strategy is illustrated in Figure 6.1A and B.

The glycoproteins are involved in the penetration of the virus and virions escape from infected cells. They induce the production of neutralizing antibodies, which play an important role in protection.[13] Both Gn and Gc contribute to the virion assembly process and interact with the N protein.[14] The N protein is essential for virion capsid formation, transcription, and replication.[15] The N protein is strongly immunogenic but does not elicit neutralizing antibodies. However, immunization with recombinant N protein induces a partial immune protection in animals.[16,17] The role of the M-segment encoded NSm1 and NSm2 in triggering apoptosis through the caspase 3, 8, and 9 pathways remains unclear.[18,19] The S-segment encoded NSs protein is a major factor of virulence inhibiting host innate viral defenses by different mechanisms at the transcriptional and translational levels.[14,20–23] Removal of the NSs gene results in attenuation of RVFV.[24] The NSs and NSm were also shown to play an important role in RVFV replication in mosquito vectors. Virus lacking NSs and NSm failed to infect *Aedes aegypti*, and in *Culex. quinquefasciatus* infection rates were lower than for wild-type virus.[25] Despite extensive geographic dispersion, and a wide range of susceptible arthropod vectors and vertebrate hosts, RVFV displays low genetic diversity, irrespective of the genome segments analyzed. The low genetic diversity of RVFV, 4% and 1% at the nucleotide and protein coding levels, respectively, likely reflects the evolutionary constraint imposed on arboviruses by their altering replication in mammalian

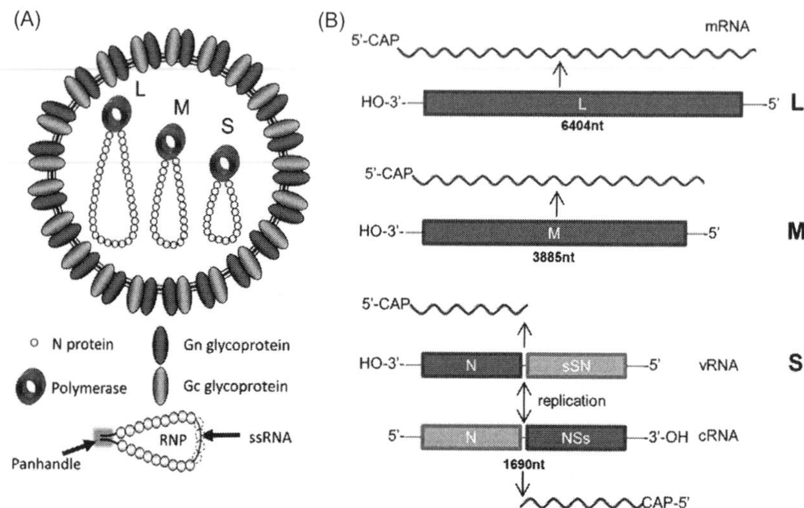

FIGURE 6.1 RVFV virion structure and transcription strategy. (A) RVFV particle with the Gn and Gc glycoproteins anchored in the viral envelope. RVFV genome consists of three single-stranded RNA segments S, M, and L which are ensapsidated by the nucleoprotein (N) into ribonucleoproteins (RNP) and associate with the viral polymerase. The terminal ends of the genomic RNA segments are complementary, permitting the ends of the RNAs to base-pair with a panhandle structure. (B) L and M segments are of negative and the S segment of ambisense polarity. Blue = genomic sense, orange = antigenomic-sense. Courtesy of Steffen Mattjijn de Boer, Central Veterinary Institute of Wageningen University and Research Center, Lelystad, The Netherlands.

and arthropod hosts.[16] Recent findings suggest that host alternation is important to maintain stability of the NSs gene, thereby promoting RVFV capacity in evasion of the innate immune response.[26]

3. WHAT IS THE BURDEN AND GEOGRAPHIC DISTRIBUTION OF RVF?

RVFV was first identified in 1930, during an outbreak associated with high rates of abortions among pregnant ewes and acute deaths of new-born lambs on a sheep farm in the Rift Valley of Kenya.[27] Subsequently, periodic epizootics occured in Kenya until the first major outbreak was recognized in South Africa in 1950–1951, which caused an estimated 500,000 abortions and 100,000 deaths of sheep. Further epizootics were subsequently reported from Namibia, Mozambique, Zimbabwe, Zambia, the Sudan, and other East African countries. In 1974–1976, South Africa experienced a second major and even more widespread outbreak of RVF during which the potential lethality of the virus for man associated with encephalitis and/or hemorrhagic fever was first recognized. By the late 1980s the geographic distribution of

RVFV had further extended, with outbreaks of the disease or evidence of virus activity reported in many countries of sub-Saharan and Central Africa.[2,28] The virus was isolated for the first time outside of continental Africa in 1979 across the Indian Ocean in Madagascar.[29] The first emergence of RVF in Egypt in 1977–1978 resulted in an estimated 200,000 human infections, and some 18,000 cases of illness of which at least 598 were fatal due to encephalitis and/or hemorrhagic fever.[30] Outbreaks of the disease in West Africa were first recorded in Mauritania and Senegal in 1987–1988.[31,32] In 2000–2001, RVFV spread across the Red Sea into the Arabian Peninsula[33] and resulted in an estimated 20,000 human infections and 886 reported cases of illness, of which 123 were fatal. Large outbreaks of RVF in East Africa in 1997–1998 resulted in an estimated 89,000 cases, of which 478 were fatal.[1] Between 2006 and 2011, resurgence of severe outbreaks of RVF was reported from East Africa,[34–36] Madagascar,[37,38] South Africa,[39–41] and for the first time from the Archipelago of Comoros, on the French Island of Mayotte.[42] Of the total of 2842 RVF cases reported in East Africa, West Africa, Madagascar, and South Africa during the period of 2006 to 2010, 613 (21.57%) were fatal.[1]

4. HOW IS RVFV MAINTAINED AND TRANSMITTED?

4.1 Vectors

Biological vectors of RVFV are members of the subgenera *Neomelaniconion* of *Aedes* and *Culex* of *Culex*.[18] There are significant differences in the ecology and transmission patterns of RVFV in endemic regions. In eastern and southern Africa, large outbreaks of RVF occur at irregular intervals of up to 15 years, after heavy rainfall and floods.[43–46] The fate of the virus during the inter-epizootic period (IEP) is not well understood, but cryptic maintenance and transmission cycles have been postulated.[2] Isolation of RVFV from *Ae. mcintoshi* mosquitoes collected during IEP in Kenya[47,48] led to the generally accepted hypothesis that the virus is maintained in nature by transovarial transmission in aedines. However, demonstration that the larvae of *Cx. pipiens*, *Ae. mcintoshi*, and *Ae. circumluteolus* become infected after feeding on liver homogenates from an experimentally inoculated hamster might be also of epidemiological importance.[49] After flooding of aedine mosquito breeding habitats (dambos), they are succeeded by *Culex* spp., which if infected upon feeding on viremic vertebrate hosts further disperse the virus. While the floodwater *Aedes* spp. tend to remain in the immediate vicinity of dambos and only feed at dusk and dawn, the more nocturnal *Culex* spp. disperse more widely to find vertebrate hosts for blood feeding.[18] Flooding contributes to a concentration of animals and humans on areas of dry land, thus further increasing the potential for virus transmission. The known RVF outbreaks in North and West Africa differ from the epidemiology of disease in sub-Saharan Africa, and occur independently of rainfall and most likely in association with vectors that breed in rivers and dams. The construction of large dam systems in North and West Africa increased not only the area of breeding sites for competent mosquito species, but also led to the concentration of people and livestock in the proximity of the dams during severe drought conditions. The main vector in the Egyptian epizootic of 1977–1978, *Cx. pipiens*, is known to be peridomestic and anthropophilic, implying that large numbers of humans can be infected by mosquito bites and subsequently serve as amplifying hosts for the infection of mosquitoes. In contrast to the Egyptian epizootic, the principal mosquito vectors of RVFV in sub-Saharan Africa tend to be zoophilic and sylvatic and therefore in this part of Africa humans become infected mostly from contact with infected animal blood and tissues.[46] In West Africa, the *Aedes* (*Aedimorphus*) spp. typically breed in the small, temporary ground pools that occur after localized rains. This region had not experienced heavy rainfall or floods and the typical vectors, *Ae. mcintoshi*, *Ae. circumluteolus*, and *Cx.* spp., were not present in high numbers during the 1987 outbreak.[50] In this region, breeding conditions are more suited for flood-breeding aedine mosquitoes of the subgenera *Aedimorphus,* including *Ae. vexans*, *Ae. ochraceus*, *Ae. dalzieli*, and *Ae. cumminsii.*[18,51] During the second large outbreak of RVF in this region in 1998–1999, there was apparently a shift in the dominant species towards *Mansonia uniformis* and *Cx. poicilipes*.[52] Irrigation for agriculture in the Tihama regions of Yemen and Saudi Arabia and the proximity of the Jizan Dam provide suitable breeding grounds for *Ae. vexans* and *Cx. tritaeniorhynchus*.[33] Competent mosquito vectors are present in RVF-free regions,[5,53–56] implicating the potential for the virus spread into these areas.

4.2 Virus Amplifiers

Domesticated ruminants are the primary species affected and likely the major amplifiers of the virus, but serological evidence suggests that a large number of African wildlife herbivorous and other species might also play a role in the RVF epidemiology.[57,58] Humans are highly susceptible to RVFV infection and develop a sufficient viremia to be a source of infection for mosquitoes and introduction of the disease into uninfected areas.

4.3 Transmission to Humans

For the reasons discussed above, the vast majority of human infections in sub-Saharan Africa result from direct or indirect contact with the blood or organs of infected animals. The virus can be transmitted to humans through the handling of animal tissues and body fluids during slaughtering or butchering, assisting with animal births, conducting veterinary procedures, or from the disposal of carcasses or fetuses. Therefore, certain occupational groups such as herders, farmers, slaughterhouse workers, and veterinarians are at higher risk of infection. The virus infects humans through inoculation, for example via a wound from an infected knife or through contact with broken skin, or through inhalation of aerosols produced during the slaughter of infected animals or obstetric procedures. The aerosol mode of transmission has also led to infection in laboratory workers. There is some evidence that humans may also become infected with RVFV by ingesting the unpasteurized or uncooked milk of infected animals. Human infections can result not only from the bites of infected vector competent mosquitoes, but also from bites by other blood-feeding insects such as midges, phlebotomids, stomoxids, and simulids, which might serve as mechanical transmitters of infection. To date, no human-to-human transmission of the virus has been documented. There are also no reports of RVFV transmission to healthcare workers and no evidence of outbreaks of RVF in urban areas.[46]

4.4 Inter-Epizootic Period and Virus Geographic Spread

RVFV isolates from one geographic area tend to cluster together, but genetic variants with distant origins are found within different genetic lineages, suggesting that the movement of infected livestock and the natural dispersal of mosquitoes allow the spread of the virus throughout continental Africa, Madagascar, and the Arabian Peninsula.[43] Mosquito bites are the principal infection mechanism of RVF in animals, but different dynamics of the contagious process during large epidemics suggests that other transmission mechanisms also play a role, including active vector dispersal, movements of infectious animals, passive vector dispersal, and/or multiple local emergences in endemic areas.[2,41] In countries where RVFV activity was not previously detected, outbreaks of the disease in animal and human populations result from the spread of a single lineage of the virus.[16,43] Recent molecular epidemiology studies in East Africa indicate ongoing RVFV activity and evolution during the IEP and highlight the importance of a cryptic enzootic transmission cycle that allows for the establishment of RVFV endemicity and to precipitate explosive outbreaks[59,60] when the herd immunity of livestock populations decreases to levels that are permissive for virus-wide spread.[44] RVFV transmission during IEP without noticeable outbreak or clinical cases has been reported in different species of African wildlife,[57,61,62] in cattle in Mayotte,[42] in sheep and goats in Mozambique,[63] and in humans in Tanzania,[64] Kenya,[65,66] and Gabon.[67] It has been postulated that during IEP, the virus persists in eggs of floodwater *Aedes* mosquito species or via low-level transmission between mosquitoes and vertebrates. The specific role domestic and wildlife animals play in the virus maintenance between outbreaks and virus amplification prior to noticeable outbreaks needs, however, to be described. Likewise, it is unknown why a low level of virus circulation during IEP does not result in clinical manifestations in livestock and humans. One explanation could be that sporadic clinical cases occurring during IEP are either underreported or misdiagnosed.

5. WHICH FACTORS ARE INVOLVED IN DISEASE PATHOGENESIS? WHAT ARE THE PATHOGENIC MECHANISMS?

In both animals and humans, the primary site of RVFV replication and the major site of tissue pathology is the liver. This finding has been consistently demonstrated by histopathological examination in natural cases and in experimentally infected animals.[36,68–71] The rapid onset of severe hepatic damage may explain many of the early clinical signs associated with severe RVF disease. Although RVFV is primarily hepatotropic, during severe infections the virus can be found in virtually all tissues and cell types.[28,46] Hepatic necrosis and increased liver enzymes are early markers for fatal RVF in humans; the latter are usually elevated seven- to eight-fold during the first 48 hours of illness in the majority of cases. In the mouse model, the pathogenesis of the virus in the liver and brain is mostly driven by chemokine and pro-inflammatory cytokine responses.[69] Reassortment events among RVFV strains have also been documented.[16,43,72] However, the impact of reassortment on RVFV replication, fitness, and host virulence remains to be investigated. A recent study by Morrill et al.[73] demonstrated that a single nucleotide heterogeneity at nucleotide 847 of the M segment (M847) affects RVFV virulence in mice. However, the effect of this single substitution on RVFV virulence in other species still needs to be determined. It is worth mentioning that recent results of molecular study by Grobbelaar et al.,[43] suggest that the natural history of RVFV, including its pathogenicity to humans, might be influenced by massive vaccination of ruminants in Africa with the live attenuated Smithburn neurotropic strain (SNS) of the virus, which is the only approved vaccine for veterinary use in Africa. The SNS vaccine strain is only partially attenuated and has been shown to be abortogenic and teratogenic.[74,75] RVFV isolate SA184/10 from a patient in South Africa, who was potentially exposed to co-infection with the SNS vaccine strain and wild virus from a needle injury while vaccinating sheep in 2010, was a reassortant grouping with the parent vaccine strain and was closely related to neurotropic and hepatotropic laboratory strains and isolates from countries that used the SNS vaccine on a large scale during a major outbreak of RVF.[43] The SNS vaccine virus was intensively passaged by intracranial inoculation of mice. Viruses attenuated through intracranial passage in mice may acquire new tissue tropism and pathogenic properties.[76,77] It has been recently demonstrated that aerosol exposure to RVFV causes earlier and more severe neuropathology in the murine model.[70]

6. WHAT ARE RVF CLINICAL MANIFESTATIONS?

Infection with RVFV results in several different disease syndromes in humans, none of which seems to predispose to other.[28,46] The majority of infections in humans are inapparent or associated with moderate to severe, non-fatal, flu-like febrile illness with headache, nausea, myalgia, and arthralgia joint pain. Some patients develop neck stiffness, sensitivity to light, loss of appetite, and vomiting. Less than 1% of human patients develop the hemorrhagic and/or encephalitic forms of the disease. The overall case fatality ratio is estimated to range from 0.5 to 2%, but it appears to be higher in recent outbreaks of the disease in East Africa and South Africa.[34,35,46] Human cases with jaundice, neurological disease, or hemorrhagic complications are at increased risk of fatality.[68,78,79] The onset of the meningoencephalitis form of the disease usually occurs 1 to 4 weeks after the disease onset. Clinical features include intense headache, loss of memory, hallucinations, confusion, disorientation, vertigo, convulsions, lethargy, and coma. Hemorrhagic symptoms of RVF appear 2 to 4 days after the onset of illness, and usually begin with evidence of severe liver impairment followed by bleeding manifestations (bleeding from venepuncture sites, petechiae, purpura, ecchymoses, gastrointestinal bleeding, bleeding from the nose or gums, menorrhagia). The case-fatality rate (CFR) for patients developing the hemorrhagic form of the disease can be as high as 50%. Death usually occurs 3 to 6 days after the onset of symptoms. In a minority of patients, the disease is complicated by the development of ocular lesions at the time of the initial illness or up to 4 weeks later. Estimates for the incidence of ocular complications range from less than 1% to 20% of human infections. The ocular disease usually presents as a loss of acuity of central vision, sometimes with development of scotomas. The lesions and the loss of visual acuity generally resolve over a period of months with variable residual scarring of the retina, but in instances of severe hemorrhage and detachment of the retina there may be permanent uni- or bilateral blindness.[46,80–82]

One of the most recent and detailed descriptions of the clinical features associated with moderate to severe RVF was reported from patients admitted to the Gizan regional referral hospital during the first outbreak of the disease in Saudi Arabia in 2000 (Table 6.1). Among the total of 165 consecutive patients who were treated and prospectively studied, the major clinical characteristics included hepatocellular failure (75.2% of patients), acute renal failure (41.2%), hemorrhagic manifestations (19.4%), retinitis (9.7%), and meningoencephalitis (4.2%). Of those patients, 56 died (33.9%), and there was no difference in CFR with regard to the sex of the patient. Hepatorenal failure, shock, and severe anemia were the major factors associated with patient death.[68]

TABLE 6.1 Major Clinical and Laboratory Findings in Moderate to Severe Cases of RVF ($n = 165$) Admitted to Gizan Regional Referral Hospital during an Outbreak of the Disease in Saudi Arabia in 2000. Modified from Al-Hazmi et al.[68]

Signs	Symptoms	Laboratory Results
Pallor (33.3)[a]	Nausea/vomiting (91.5)	Hemoglobin, 8 g/dL (19.4)
Dehydration (24.8)	Fever/chills (73.9)	Platelet count $<10^5$ platelets/L (51.8)
CNS abnormality (24.8)	Abdominal pain (46.1)	ALT[b] >3 ×[c] (73.8)
Hepatomegaly (21)	Diarrhea (43)	AST[d] >3 × (88.8)
Splenomegaly (20)	Headache (40.6)	GGT[e] >3 × (26.7)
Jaundice (10)	Myalgia (40)	LDH[f] >2 × (43.6)
Ecchymosis (8)	CNS symptoms (30.3)	CPK[g] >2 × (32.3)
Petechiae (5)	Oliguria/anuria (21.2)	Creatinine >150 μM (45.1)
Ascites (4)	Bleeding (17)	

[a] *(%) of patients;*
[b] *alanine aminotransferase;*
[c] *times higher the upper limit of normal ranges for laboratory values;*
[d] *aspartate aminotransferase;*
[e] *γ-glutamyl transferase;*
[f] *lactate dehydrogenase;*
[g] *creatine phosphokinase.*

7. HOW DO YOU DIAGNOSE?

The capacity for laboratory diagnosis of RVF is restricted to a limited number of reference laboratories worldwide. In the absence of hemorrhagic or specific organ manifestations, infections by VHF viruses are clinically difficult to recognize; consequently, definitive diagnosis depends largely on accurate laboratory tests. RVF may be suspected when there is a sudden outbreak of febrile illness with headache and myalgia in humans, in association with the occurrence of abortions in domestic ruminants and deaths of young animals. Cases of RVF are sometimes only recognized late after infection from the occurrence of ocular complications. Late recognition of RVF infections is especially a case during inter-epizootic periods when only very sporadic infections occur, and they are usually misdiagnosed. Hemorrhagic or encephalitic manifestations might also be indicative of RVF infection, and this is especially true in the rare instances when residents of RVF-free countries develop the illness following a visit to endemic areas.[46] Serum specimens are commonly used for RVF diagnosis in humans. Although viremia in infected individuals reaches high titers, it is of short duration, thus limiting the use of viral detection systems. Also most infected patients undergo subclinical or mild infections; therefore, antigen and nucleic acid detection assays should be run in parallel with antibody detecting techniques. Type-specific antibodies to RVFV are easily demonstrable shortly after exposure to the virus.[18]

Isolation of the virus is achieved in hamsters, infant or adult mice, and in various cell cultures.[18,28] Highly sensitive PCR assays for the detection and quantification of RVFV have been reported, including reverse transcriptase PCR (RT-PCR)[83–85] and real-time detection PCR (RTD-PCR) based on TaqMan probe technology.[86] More recently, the real-time reverse-transcription loop-mediated isothermal amplification assay (RT-LAMP) was developed and evaluated for the detection of a wide spectrum of RVFV isolates and clinical specimens. Apart from high analytical and diagnostic accuracy and speed of detection, another important practical advantage of the LAMP assay is that it utilizes simple and relatively inexpensive equipment, which renders it promising for use in resource-poor settings and as a portable device during RVF outbreaks in remote areas.[87] The recent findings in Kenyan patients[88] indicate that the quantitative real-time RT-PCR qRT-PCR can be used for rapid identification of patients with high viremia associated with poor prognosis, thereby enabling them to be targeted for special or intensive clinical management.

Diagnosis of recent infection is confirmed by demonstrating seroconversion or a four-fold or greater rise in titer of antibody in paired serum samples, or by detection of IgM antibody. The classical methods for the detection of antibodies to RVFV include hemagglutination inhibition, complement fixation, indirect immunofluorescence, and virus neutralization tests.[89] Disadvantages of these techniques include health risk to laboratory personnel[90,91] and restrictions for their use outside RVF endemic areas. Although regarded as a gold standard, the virus neutralization test is laborious, expensive, and requires 5–7 days for completion. It can be performed only when a standardized stock of live virus and tissue cultures are available. Consequently, it is rarely used, and then only in highly specialized reference laboratories.[18] Various ELISA formats were developed in recent years for specific detection of IgG and IgM, based on inactivated sucrose-acetone-extracted antigens derived from tissue culture or mouse brain.[92,93] However, the production of antigen for these assays also requires bio-containment facilities to limit the risk of exposure of laboratory personnel to infection. To address these problems, an indirect ELISA based on the recombinant nucleocapsid protein of RVFV has been recently developed for the detection of specific antibodies.[94,95] Viral antigen can be detected in blood and other tissues by a variety of immunological methods, including agar gel immunodiffusion and immunostaining assays. Histopathological examination of the liver reveals characteristic pathology.[28] Zaki et al.[96] reported immunofluorescence assay, which utilizes a pool of mouse IgG monoclonal conjugates reacting with a combination of virus-specific antigens. Although it was demonstrated to be highly reliable in detecting RVFV in patient sera, its use requires tissue culture amplification and handling of live virus. A sandwich ELISA based on safe procedure using specimens inactivated at 56°C for 1 h in the presence of 0.5% Tween-20 (v/v) before testing was recently reported for the detection of nucleocapsid protein of RVFV in various clinical specimens.[97]

8. HOW DO YOU DIFFERENTIATE THIS DISEASE FROM SIMILAR ENTITIES?

RVFV belongs to a group of VHF agents. These have worldwide distribution, but a specific pathogen is usually restricted to a known endemic region where its existence depends on the presence of natural reservoirs and/or competent arthropod vectors. As for most VHFs, the non-specific presentation of RVF makes it difficult to diagnose clinically. Therefore, the differential diagnosis concerns a broad array of conditions especially when first cases are encountered during a yet unrecognized outbreak. These include malaria, rickettsial infections, Q fever, typhoid fever, dysentery, plaque, brucellosis, leptospirosis, meningitis, other sepsis from bacterial infections, viral hepatitis, other VHFs, including Lassa fever, Crimean-Congo hemorrhagic fever, Marburg disease, Ebola fever, and the hemorrhagic fever with renal syndrome associated with hantavirus infections. Non-infectious causes of disseminated intravascular coagulopathy and acute leukemia should be also considered. Availability of laboratory results and epidemiological information usually helps to narrow the spectrum of differential diagnosis. The tentative cause can be assumed based on recent travel and exposure history (e.g., mosquito bite, contact with animals or animal products) in endemic regions. As for all VHFs, the RVF confirmatory diagnostic process has to consider all available laboratory results, clinical, pathological, and epidemiological data.

9. WHAT IS THE THERAPEUTIC APPROACH?

No specific treatment is currently available. Treatment is symptomatic and in more severe cases, general supportive therapy is provided. Fluid and blood pressure management guidelines for septic shock are recommended for VHFs due to common features in the pathogenesis of these two conditions.[98] Considering the high burden of malaria and tick-borne rickettsial disease in Africa, patients should be covered with broad spectrum antibacterial and antiparasitic therapy, until a diagnosis of RVF can be confirmed. Due to high viremia in severely ill patients, strict barrier and isolation nursing should be implemented, especially where intravenous transfusion of fresh frozen plasma, blood, and albumin is performed. Hemodialysis for patients with severe acute renal failure, and other intensive care and supportive measures such as mechanical ventilation, are provided as necessary.

10. WHAT ARE THE PREVENTIVE AND INFECTION CONTROL MEASURES?

An inactivated vaccine has been developed for human use,[99] but it is not licensed and has been used only experimentally to protect veterinary and laboratory personnel at high risk of exposure to RVF. Many candidate vaccines[24,100–104] are, however, under development. The most effective approach for protection of humans against RVF would be safe, efficacious, and affordable vaccination of animals in endemic regions. However, except for restricted animal use of the live attenuated Smithburn or inactivated vaccines in Africa,[105,106] no vaccines are licensed for veterinary applications elsewhere. Most recently, an animal vaccination program based on the use of a naturally occurring RVFV mutant (Clone-13) was introduced in South Africa.[107] Restricting or banning the movement of livestock may be effective in slowing the expansion of the virus from infected to uninfected areas. As outbreaks of RVF in animals usually precede human cases, an active animal health surveillance system to detect new cases is essential in providing early warning for public health authorities. Detection of animal cases at an early stage of an outbreak and sharing information about the cases between veterinary and public health sectors are crucial for implementation of control measures to minimize the severe consequences of RVF epidemics. Forecasting models and early warning systems for RVF using satellite images and weather/climate prediction data are useful for signaling increased risk of RVF outbreaks.[2]

During an outbreak of RVF, close contact with animals, particularly with their tissues and body fluids, either directly or indirectly, is the most significant risk factor for RVF virus infection. Therefore, in the absence of specific treatment and vaccine, raising awareness of the risk factors of RVF infection as well as the protective measures individuals can take to prevent mosquito bites is the only practical way to reduce human infection and deaths.[47]

Risk of becoming infected can be significantly reduced by avoiding exposure to blood or tissues of animals potentially infected with RVFV; this is of particular concern for persons working with animals in RVF-endemic areas. Veterinarians and others engaged in the livestock industry should be made aware of the potential health risks associated with exposure to zoonotic agents while carrying or handling tissues of sick animals, and precautions should be increased during RVF outbreaks. Public health messages should focus on reducing the risk of animal-to-human transmission as a result of unsafe animal husbandry and slaughtering practices. Gloves and other appropriate protective clothing should be worn and care taken when handling sick animals or their tissues or when slaughtering animals. In the epizootic regions, all animal products (blood, meat, and milk) should be cooked or pasteurized before eating. The importance of personal and community protection against mosquito bites through the use of impregnated mosquito nets, personal insect repellent, wearing light colored clothing, long-sleeved shirts and trousers, and avoiding outdoor activity, especially at peak mosquito biting times, should be strongly emphasized in affected areas.[46]

10.1 Isolation Precaution

Although no human-to-human transmission of RVF has been demonstrated, there is still a potential risk of transmission of the virus from infected patients to healthcare workers through contact with infected blood or tissues. Healthcare workers caring for patients with suspected or confirmed RVF should implement and thoroughly execute standard precautions. Samples taken from suspected human and animal cases of RVF for diagnosis should be handled by trained staff and processed in suitably equipped laboratories.

10.2 Post-Exposure

The antiviral drug ribavirin was shown to have a therapeutic efficacy against RVFV in mice,[108,109] various cell cultures, hamsters, and rhesus monkeys,[110] and it was suggested to be used even in benign RVF infections in humans to obviate the potentially serious complications. However, ribavirin did not prevent the late occurrence of encephalitis in patients in Saudi Arabia, and its use is now considered contraindicated.[46] A study by Reed at al.[70] showed that mice infected with RVFV by aerosol exposure and treated with ribavirin were not protected and developed severe neuropathology. Results of this study highlight the need for more candidate antivirals for RVF infection treatment and post-exposure prophylaxis, especially in the case of a potential aerosol exposure. A number of new antiviral drugs are under investigation.[111,112]

REFERENCES

1. Dar O, McIntyre S, Hogarth S, Heymann D. Rift Valley fever and a new paradigm of research and development for zoonotic disease control. *Emerg Infect Dis* 2013;**19**:189–93.
2. Paweska JT, Jansen van Vuren P. Rift Valley fever virus: a virus with potential for global emergence. In: Johnson N, editor. *The role of animals in emerging viral diseases*. Elsevier, Academic Press; 2013. p. 169–200.
3. Rich KM, Wanyoike F. An assessment of the regional and national socio-economic impacts of the 2007 Rift Valley fever outbreak in Kenya. *Am J Trop Med Hyg* 2010;**83**(Suppl. 2): S52–7.

4. Chevalier V. Relevance of Rift Valley fever to public health in the European Union. *Clin Microbiol Infect* 2013;**19**:705–8.
5. Chevalier V, Pepin M, Plee L, Lancelot R. Rift Valley fever—a threat for Europe? *EuroSurveillance* 2010;**15**:1950–6.
6. Mandel R, Flick R. Rift Valley fever virus an unrecognized emergin threat? *Human Vacc* 2010;**6**:597–601.
7. World Health Organization. International Health Regulations. <http://www.who.int/ihr/en/>; 2005.
8. Schmaljohn CS, Nichol ST. Bunyaviridae. In: Knipe DM, Howley PM, Griffin DE, Lamb RA, Martin MA, Roizman B, Straus SE, editors. *Fields virology*. 5th ed. Philadelphia, PA, USA: Lippincott, Williams & Wilkins; 2007. p. 1741–89.
9. Sherman MB, Freiberg AN, Holbrook MR, Watowich SJ. Single-particle cryo-electron microscopy of Rift Valley fever virus. *Virol* 2009;**387**:11–5.
10. Bouloy M, Weber F. Molecular biology of Rift Valley fever virus. *Open Virol J* 2010;**4**:8–14.
11. Gerrard SR, Nichol ST. Synthesis, proteolytic processing and complex formation of N-terminally nested precursor proteins of the Rift Valley fever virus glycoproteins. *Virol* 2007;**357**:124–33.
12. Piper CJ, Sorenson DR, Gerrard SR. Efficient celluar release of Rift Valley fever virus requires genomic RNA. *PLoS One* 2011;**6**:e18070.
13. Mandel RB, Koukuntla R, Mogler LJ, Carzoli AK, Freiberg AN, Holbrook MR, et al. A replication-incompetent Rift Valley fever vaccine:chimeric virus-like particles protect mice and rats agaist lethal challenge. *Virol* 2010;**397**:187–98.
14. Habjan M, Pichlmair A, Elliott RM, Overby AK, Glatter T, Gstaiger M, et al. NSs protein of Rift Valley fever virus induces the specific degradation of the double-stranded RNA-dependent protein kinase (PKR). *J Virol* 2009;**83**:4365–75.
15. Liu L, Celma CCP, Roy P. Rift Valley fever virus structural proteins: expression, characterization and assembly of recombinant proteins. *Virol J* 2008;**5**:82.
16. Bird BH, Khristova ML, Rollin PE, Ksiazek TG, Nichol ST, et al. Complete genome analysis of 33 ecologically and biologically diverse Rift Valley fever virus strains reveals widespread virus movement and low genetic diversity due to recent common ancestry. *J Virol* 2007;**81**:2805–6.
17. Jansen van Vuren P, Tiemessen CT, Paweska JT. Anti-nucleocapsid immune responses counteract pathogenic effects if Rift Valley fever virus infection in mice. *PLoS One* 2011;**6**:e2507.
18. Pepin M, Bouloy M, Bird BH, Kemp A, Paweska J. Rift Valley fever virus (Bunyaviridae: Phlebovirus): an update on pathogenesis, molecular epidemiology, vectors, diagnostics and prevention. *Vet Res* 2010;**41**:61.
19. Won S, Ikegami T, Peters CJ, Makino S. NSm protein of Rift Valley fever virus suppresses virus-induced apoptosis. *J Virol* 2007;**81**:13335–45.
20. Billecocq A, Spiegel M, Vialat P, Kohl A, Weber F, Bouloy M, et al. NSs protein of Rift Valley fever virus blocks interferon production by inhibiting host gene transcription. *J Virol* 2004;**78**:9798–806.
21. Ikegami T, Narayanan K, Won S, Kamitani W, Peters CJ, Makino S. Rift Valley fever virus NSs protein promotes post-transcriptional downregulation of protein kinase PKR and inhibits eIF2alpha phosphorylation. *PLoS Pathog* 2009;**5**:e1000287.
22. Kalveram B, Lihoradova O, Ikegami T. NSs protein of Rift Valley fever virus promotes postranslational downregulation of the TFIIH subunit p62. *J Virol* 2011;**85**:6234–43.

23. Le May N, Mansuroglu Z, Leger P, Josse T, Blot G, Billecocq A, et al. A SAP30 complex inhibits IFN-beta expression in Rift Valley fever virus infected cells. *PLoS Pathog* 2008;**4**:e13.
24. Bird BH, Maartens LH, Campbell S, Erasmus BJ, Erickson BR, Dodd KA, et al. Rift Valley fever virus lacking the NSs and NSm genes is safe, nonteratogenic, and confers protection from viremia, pyrexia, and abortion following challenge in adult and pregnant sheep. *J Virol* 2011;**85**:12901–9.
25. Crabtree MB, Kent Crockett RJ, Bird BH, Nichol ST, Erickson BR, et al. Infection and transmission of Rift Valley fever viruses lacking the NSs and/or NSm genes in mosquitoes: potential role for NSm in mosquito infection. *PLoS Negl Trop Dis* 2012;**6**:e1639.
26. Moutailler S, Roche B, Thiberge JM, Caro V, Rougeon F, Failloux AB. Host alternation is necessary to maintain the genome stability of Rift Valley fever virus. *PLoS Negl Trop Dis* 2011;**5**:e1156.
27. Daubney R, Hudson JR, Garnham PC. Enzootic hepatitis or Rift Valley fever. An undescribed virus disease of sheep, cattle and man from East Africa. *J Pathol Bacteriol* 1931;**34**:545–79.
28. Swanepoel R, Coetzer JA. Rift Valley fever. In: Coetzer JA, Tustin RC, editors. *Infectious diseases of livestock*. Southern Africa, Cape Town: Oxford University Press; 2004. p. 1037–70.
29. Morvan J, Saluzzo JF, Fontenille D, Rollin PE, Coulanges P. Rift Valley fever on the east coast of Madagascar. *Res Virol* 1991;**142**:475–82.
30. Meegan JM. Rift Valley fever in Egypt: an overview of the epizootics in 1977 and 1978. In: Swartz TA, Klinberg MA, Goldblum N, Papier CM, editors. *Contributions to epidemiology and biostatistics: Rift Valley fever*, 1981. Basel: S. Karger AG; 1981. p. 100–13.
31. Jouan A, Le Guenno B, Digoutte JP, Philippe B, Riou O, Adam F, et al. Epidemic in southern mauritania. *Ann Inst Pasteur Virol* 1988;**139**:307–8.
32. Ksiazek TG, Jouan A, Meegan JM, Le Guenno B, Wilson ML, Peters CJ, et al. Rift Valley fever among domestic animals in the recent West African outbreak. *Res Virol* 1989;**140**:67–77.
33. Jupp PG, Kemp A, Grobbelaar A, Leman P, Burt FJ, Alahmed AM, et al. The 2000 epidemic of Rift Valley fever in Saudi Arabia: mosquito vector studies. *Med Vet Entomol* 2002;**16**:245–52.
34. Mohamed M, Mosha F, Mghamba J, Zaki SR, Shieh W-J, Paweska J, et al. Epidemiologic and clinical aspects of a Rift Valley fever outbreak in humans in Tanzania, 2007. *Am J Trop Med Hyg* 2010;**83**(Suppl. 2):S22–7.
35. Nguku P, Sharif SK, Mutonga D, Amwayi S, Omolo J, Mohamed O, et al. An investigation of a major outbreak of Rift Valley fever in Kenya: 2006–2007. *Am J Trop Med Hyg* 2010;**83**(Suppl. 2):S5–13.
36. Shieh W-J, Paddock CD, Lederman E, Rao CY, Gould LH, Mohamed M, et al. Pathologic studies on suspect animal and human cases of Rift Valley fever from an outbreak in Eastern Africa, 2006–2007. *Am J Trop Med Hyg* 2010;**83**(Suppl. 2):S38–42.
37. Adriamandimby S, Randrianarivo-Solofoniaina A, Jeanmaire E, Ravolomanana L, Razafimanantsoa L, Rakotojoelinandrasana T, et al. Rift Valley fever during rainy seasons, Madagascar, 2008 and 2009. *Emerg Infect Dis* 2010;**16**:963–70.
38. Carroll SA, Reynes JM, Khristova ML, Andriamandimby SF, Rollin PE, Nichol ST. Genetic evidence for Rift Valley fever outbreaks in Madagascar resulting from virus introductions from the East African mainland rather than enzootic maintenance. *J Virol* 2011;**85**:6162–7.

39. Archer BN, Weyer J, Paweska J, Nkosi D, Leman P, San Tint K, et al. Outbreak of Rift Valley fever affecting veterinarians and farmers in South Africa, 2008. *S Afr Med J* 2011;**101**:263–6.
40. Archer BN, Thomas J, Weyer J, Cengimbo A, Essoya LD, Jacobs C, et al. Epidemiological investigations into outbreaks of Rift Valley fever in humans, South Africa, 2008–2011. *Emerg Infect Dis* 2013;**19**:1918–25.
41. Métras R, Porphyre T, Pfeiffer DU, Kemp A, Thomson P, Collins LM, et al. Exploratory space-time analyses of Rift Valley fever in South Africa in 2008–2011. *PLoS Negl Trop Dis* 2012;**6**:e1808.
42. Cêtre-Sossah C, Pédarrieu A, Guis H, Deferenz C, Bouloy M, Favre J, et al. Prevalence of Rift Valley fever among ruminants, Mayotte. *Emerg Infect Dis* 2012;**18**:972–5.
43. Grobbelaar AA, Weyer J, Leman PA, Kemp A, Paweska JT, Swanepoel R. Molecular epidemiology of Rift Valley fever virus. *Emerg Infect Dis* 2011;**17**:2270–6.
44. Murithi RM, Munyua P, Ithondeka PM, Macharia JM, Hightower A, Luman ET. Rift Valley fever in Kenya: history of epizootics and identification of vulnerable districts. *Epidemiol Infect* 2011;**139**:372–80.
45. LaBeaud AD, Ochiai Y, Peters CJ, Muchiri EM, King CH. Spectrum of Rift Valley fever virus transmission in Kenya: insight from three distinct regions. *J Trop Med Hyg* 2007;**76**:795–800.
46. Swanepoel R, Paweska JT. Rift Valley fever. In: Palmer SR, Soulsby L, Torgerson PR, Brown DWG, editors. *Oxford textbook of zoonoses: biology, clinical practise, and public health control*. Oxford University Press; 2011. p. 421–31.
47. Linthicum KG, Davies FG, Bailey CL, Kairo A. Mosquito species succession in a dambo in an East African forest. *Mosq News* 1983;**43**:464–70.
48. Linthicum KJ, Davies FG, Kairo A, Bailey CL. Rift Valley fever virus (family *Bunyaviridae*, genus *Phlebovirus*). Isolations from diptera collected during an inter-epizootic period in Kenya. *J Hyg* 1985;**95**:197–209.
49. Turell MJ, Linthicum KJ, Beaman JR. Transmission of Rift Valley fever virus by adult mosquitoes after ingestion of virus as larvae. *Am J Trop Med Hyg* 1990;**43**:677–80.
50. Fontenille D, Traore-Lamizana M, Diallo M, Thonnon J, Digoutte JP, Zeller HG. New vectors of Rift Valley fever in West Africa. *Emerg Infect Dis* 1998;**4**:289–93.
51. Zeller H, Fontenille D, Traorelamizana M, Thiongane Y, Digoutte JP. Enzootic activity of Rift Valley fever virus in Senegal. *Am J Trop Med Hyg* 1997;**56**:265–72.
52. Diallo M, Nabeth P, Ba K, Sall AA, Ba Y, Mondo M, et al. Mosquito vectors of the 1998–1999 outbreak of Rift Valley Fever and other arboviruses (Bagaza, Sanar, Wesselsbron and West Nile) in Mauritania and Senegal. *Med Vet Entemol* 2005;**19**:119–26.
53. Konrad SK, Miller SN. A temperature-limited assessment of the risk of Rift Valley fever transmission and establishmnet in the continental United States of America. *Geospat Health* 2012;**6**:161–70.
54. Martin V, Chevalier V, Ceccato P, Anyamba A, De Simone L, Lubroth J, et al. The impact of climate change on the epidemiology and control of Rift Valley fever. *Rev Sci Tech* 2008;**27**:413–26.
55. Turell MJ, Byrd BD, Harrison BA. Potential for populations of *Aedes j. japonicus* to transmit Rift Valley fever virus in the USA. *J Am Mosq Cont Ass* 2013;**29**:133–7.
56. Turell MJ, Dohm DJ, Mores CN, Terracina L, Wallette DL, Hribar LJ, et al. Potential for North American mosquitoes to transmit Rift Valley fever virus. *J Am Mosquito Control Ass* 2008;**24**:502–7.

57. Britch SC, Binepal YS, Ruder MG, Karithi HM, Linthicum KJ, Anyamba A, et al. Rift Valley fever risk map model and seroprevalence in selected wild ungulates and camels from Kenya. *PLoS One* 2013;**8**:e66626.
58. Olive MM, Goodman SM, Reynes JM. The role of wild mammals in the maintenance of Rift Valley fever virus. *J Wildl Dis* 2012;**48**:241–66.
59. Bird BH, Githinji JWK, Macharia JM, Kasiiti JL, Muriithi RM, Gacheru SG, et al. Multiple virus lineages sharing recent common ancestry were associated with a large Rift Valley fever outbreak among livestock in Kenya during 2006–2007. *J Virol* 2008;**82**:11152–66.
60. Nderitu L, Lee JS, Omolo J, Omulo S, O'Guinn ML, Hightower A, et al. Sequential Rift Valley fever outbreaks in Eastern Africa caused by multiple lineages of the virus. *J Infect Dis* 2011;**203**:655–65.
61. Evans A, Gakuya F, Paweska JT, Rostal M, Akoolo L, Jansen Van Vuren P, et al. Prevalence of antibodies against Rift Valley fever virus in Kenyan wildlife. *Epidemiol Infect* 2008;**136**:1261–9.
62. LaBeaud AD, Cross PC, Getz WM, Glinka A, King CH. Rift Valley fever virus infection in African buffalo (*Syncerus caffer*) herds in rural South Africa: evidence of interepidemic transmission. *Am J Trop Med Hyg* 2011;**84**:641–6.
63. Fafetine J, Neves L, Thompson PN, Paweska JT, Rutten VPMG, Coetzer JAW. Serological evidence of Rift Valley fever virus circulation in sheep and goats in Zambézia Province, Mozambique. *PLoS Negl Trop Dis* 2013;**7**:e2065.
64. Heinrich N, Saathoff E, Weller N, Clowes P, Kroidl I, Ntinginya E, et al. High seroprevalence of Rift Valley fever and evidence for endemic circulation in Mbeya region, Tanzania, in a cross-sectional study. *PLoS Negl Trop Dis* 2012;**6**:e1557.
65. LaBeaud AD, Muchiri EM, Ndzovu M, Mwanje MT, Muiruri S, Peters CJ, et al. Interepidemic Rift Valley fever virus seropositivity, Northeastern Kenya. *Emerg Infect Dis* 2009;**14**:1240–6.
66. LaBeaud AD, Muiruri S, Sutherland LJ, Dahir S, Gildengorin G, Morril J. Postepidemic analysis of Rift Valley fever virus transmission in northeastern Kenya: a village cohort study. *PLoS Negl Trop Dis* 2011;**5**:e1265.
67. Poourrut X, Nkoghe D, Souris M, Paupy C, Paweska J, Padilla C, et al. Rift Valley fever seroprevalence in human rural populations of Gabon. *PLoS Negl Trop Dis* 2010;**4**:e763.
68. Al-Hazmi M, Ayoola EA, Abdurahman M, Banzal S, Ashraf J, El-Bushra A, et al. Epidemic Rift Valley fever in Saudi Arabia; a clinical study of severe illness in humans. *Clin Infect Dis* 2003;**36**:245–52.
69. Gray KK, Worthu MN, Juelich TL, Agar SL, Pousssard A, Ragland D. Chemotactic and inflamatory responses in the liver and brain are associated with pathogenesis of Rift Valley fever virus infection in the mouse. *PLoS Negl Trop Dis* 2012;**6**:e1529.
70. Reed C, Lin K, Wilhelmsen C, Friedrich B, Nalca A, Keeney A, et al. Aerosol exposure to Rift Valley fever virus causes earlier and more severe neuropathology in the murine model, which has important implications for therapeutic development. *PLoS Neg Trop Dis* 2013;**7**: e2156.
71. Smith DR, Bird BH, Lewis B, Johnston SC, McCarthy S, Keeney A, et al. Development of a novel nonhuman primate model for Rift Valley fever. *J Virol* 2012;**86**:2109–10.
72. Sall AA, Zanotto PMDA, Sene OK, Zeller HG, Digoutte JP, Thiongane Y, et al. Genetic reassortment of Rift Valley fever virus in nature. *J Virol* 1999;**73**:8196–200.

73. Morril JC, Ikegami T, Yoshikawa-Iwata N, Lokugamage N, Won S, Teresaki K, et al. Rapid accumulation of virulent Rift Valley fever virus in mice from an attenuated virus carrying a single nucleotide substitution in the M RNA. *PLoS One* 2010;**5**:e9986.
74. Botros B, Omar A, Elian K, Mohamed G, Soliman A, Salib A, et al. Adverse response of non-indigenous cattle of European breeds to live attenuated Smithburn Rift Valley fever vaccine. *J Med Virol* 2006;**78**:787–91.
75. Kamal SA. Pathological studies on postvaccinal reactions of Rift Valley fever in goats. *Virol J* 2009;**6**:94.
76. Hayes EB. Is it time for a new yellow fever vaccine? *Vaccine* 2010;**28**:8073–6.
77. Swanepoel R, Erasmus BJ, Williams R, Taylor MB. Encephalitis and chorioretinitis associated with neurotropic African horsesickness virus infection in labolatory workers. Part III. Virological and serological investigations. *S Afr Med J* 1992;**81**:458–61.
78. Laughlin LW, Meegan JM, Strausbaugh LJ, Morens DM, Watten RH. Epidemic Rift Valley fever in Egypt: observations of the spectrum of human illness. *Trans R Soc Trop Med Hyg* 1979;**73**:630–3.
79. Madani TA, Al-Mazrou YY, Al-Jeffri MH, Mishkhas AA, Al-Rabeah AM, Turkistani AM. Rift Valley fever epidemic in Saudi Arabia: epidemiological, clinical, and laboratory characteristics. *Clin Infect Dis* 2003;**37**:1084–92.
80. Deutman AF, Klomp HJ. Rift Valley fever retinitis. *Am J Ophthalmol* 1981;**92**:38–42.
81. Siam AL, Meegan JM, Gharbawi KF. Rift Valley fever ocular manifestations; observations during the 1977 epidemic in Egypt. *Br J Ophthalmol* 1980;**64**:366–74.
82. Siam AL, Meegan JM. Ocular disease resulting from infection with Rift Valley fever virus. *Trans R Trop Med Hyg* 1980;**74**:539–41.
83. Garcia S, Crance JM, Billecocq A, Peinnequin A, Jouan A, Bouloy M, et al. Quantitative real-time PCR detection of Rift Valley fever virus and its application to evaluation of antiviral compounds. *J Clin Microbiol* 2001;**39**:4456–61.
84. Ibrahim MS, Turell MJ, Knauert FK, Lofts RS. Detection of Rift Valley fever virus in mosquitoes by RT-PCR. *Mol Cell Probes* 1997;49–53.
85. Sall AA, Macondo EA, Sene OK, Diagane M, Sylla R, Mondo M, et al. Use of reverse transcriptase PCR in early diagnosis of Rift Valley fever. *Clin Diag Lab Immunol* 2002;**9**:713–5.
86. Drosten C, Götting S, Schilling S, Asper M, Panning M, Schmitz H, et al. Rapid detection and quantification of RNA of Ebola and Marburg viruses, Lassa virus, Crimean-Congo hemorrhagic fever virus, Rift Valley fever virus, dengue virus, and yellow fever virus by real-time reverse transcription-PCR. *J Clin Microbiol* 2002;**40**:2323–30.
87. Le Roux CA, Kubo T, Grobbelaar AA, Jansen van Vuren P, Weyer J, Nel LH, et al. Development and evaluation of a real-time reverse transcription-loop-mediated isothermal amplification assay for rapid detection of Rift Valley fever virus in clinical specimens. *J Clin Micro* 2009;**47**:645–51.
88. Njenga KM, Paweska J, Wanjala R, Rao CY, Weiner M, Omballa V, et al. Using field quantitative real-time reverse transcription-PCR test to rapidly identify highly viremic Rift Valley fever cases. *J Clin Micro* 2009;**47**:1166–71.
89. Swanepoel R, Struthers JK, Erasmus MJ, Shepherd SP, McGillivray GM. Comparison of techniques for demonstrating antibodies to Rift Valley fever virus. *J Hyg* 1986;**97**:317–29.
90. Kitchen SF. Laboratory infections with the virus of Rift Valley fever. *Am J Trop Med Hyg* 1934;**14**:547–64.

91. Smithburn KC, Mahaffy AF, Haddow AJ, Kitchen SF, Smith JF. Rift Valley fever accidental infections among laboratory workers. *J Immunol* 1949;**62**:213–27.
92. Paweska JT, Burt FJ, Swanepoel R. Validation of enzyme-linked immunosorbent assay for the detection of IgG and IgM antibody to Rift Valley fever virus in humans. *J Virol Meth* 2005;**124**:173–81.
93. Paweska JT, Mortimer E, Leman PA, Swanepoel R. An inhibition enzyme linked immunosorbent assay for the detection of antibody to Rift Valley fever in humans, domestic and wild ruminants. *J Virol Meth* 2005;**127**:10–8.
94. Jansen van Vuren P, Potgieter AC, Paweska JT, Van Dijk AA. Preparation and evaluation of a recombinant Rift Valley fever virus N protein for the detection of IgG and IgM antibodies in humans and animals. *J Virol Meth* 2007;**140**:106–14.
95. Paweska JT, Jansen van Vuren P, Swanepoel R. Validation of an indirect ELISA based on a recombinant nucleocapsid protein of Rift Valley fever virus for the detection of IgG antibody in humans. *J Virol Meth* 2007;**146**:119–24.
96. Zaki A, Coudrier D, Yousef AI, Fakeeh M, Bouloy M, Billecocq A. Production of monoclonal antibodies against Rift Valley fever virus. Application for rapid diagnosis tests (virus detection and ELISA) in human sera. *J Virol Meth* 2006;**131**:34–40.
97. Jansen van Vuren P, Paweska JT. Laboratory safe detection of nucleocapsid protein of Rift Valley fever virus in human and animal specimens by a sandwich ELISA. *J Virol Meth* 2009;**157**:15–24.
98. Bausch DG, Moses LM, Boba A, Grant DS, Khan H. Lassa fever. In: Singh SH, Ruzek D, editors. *Viral hemorrhagic fevers*. Boca Raton FL, US: CRC Press Taylor & Francis Group; 2013. p. 261–86.
99. Rusnak JM, Gibbs P, Boudreau E, Clizbe DP, Pittman P. Immunogenicity and safety of an inactivated Rift Valley fever vaccine in a 19-year study. *Vaccine* 2011;**29**:3222–9.
100. Brennan B, Welch SR, McLees A, Elliott RM. Creation of a recombinant Rift Valley fever virus with a two-segmented genome. *J Virol* 2011;**85**:10310–8.
101. Dodd KA, Bird BH, Metcalfe MG, Nichol ST, Albarino CG. Single-dose immunization with virus replicon particles confers rapid robust protection against Rift Valley fever virus challenge. *J Virol* 2012;**86**:4204–12.
102. Kortekaas J, Oreshkova N, Cobos-Jiménez V, Vloet RPM, Potgieter CA, Moormann RJM. Creation of a nonspreading Rift Valley fever virus. *J Virol* 2011;**85**:12622–30.
103. Lihoradowa O, Ikegami T. Modifying the NSs gene to improve live-attenuated vaccine for Rift Valley fever. *Exp Rev Vac* 2012;**11**:1283–5.
104. Lorenzo G, Martin-Foglar R, Hevia E, Boshra H, Brun A. Protection against lethal Rift Valley fever virus (RVFV) infection in transgenic INFAR$^{-/-}$ mice induced by diffrent DNA vaccination regimes. *Vaccine* 2010;**28**:2937–44.
105. Kamal SA. Observations on Rift Valley fever virus and vaccines in Egypt. *Virol J* 2011;**8**:532.
106. Lagerqvist N, Moiane B, Bucht G, Fafetine J, Paweska JT, Lundkvist A, et al. Stability of a formalin-inactivated Rift Valley fever vaccine: evaluation of a vaccination campaign for cattle in Mozambique. *Vaccine* 2012;**30**:6534–40.
107. Von Teichman B, Engelbrecht A, Zulu G, Dungu B, Pardini A, Bouloy M. Safety and efficacy of of Rift Valley fever Smithburn and Clone 13 vaccines in calves. *Vaccine* 2011;**29**:5771–7.
108. Kende M, Alving CR, Rill WL, Swartz GM, Canonico PG. Enhanced efficacy of liposome-encapsulated ribavirin against Rift Valley fever virus infection in mice. *Antimicrob Agents Chemother* 1985;**27**:903–7.

109. Kende M, Lupton HW, Rill WL, Levy HB, Canonico PG. Enhanced therapeutic efficacy of poly (ICLC) and ribavirin combinations against Rift Valley fever virus infection in mice. *Antimicrob Agents Chemother* 1987;**31**:986–90.
110. Peters CJ, Reynolds JA, Slone TW, Jones DE, Stephen EL. Prophylaxis of Rift Valley fever with antiviral drugs, immune serum, an interferon inducer, and a macrophage activator. *Antiviral Res* 1986;**6**:285–97.
111. Narayanan A, Kehn-Hall K, Senina S, Hill L, van Duyne R, Guendel I, et al. Curcumin inhibits Rift Valley fever replication in human cells. *J Biol Chem* 2012;**40**:33198–214.
112. Scott T, Paweska JT, Arbuthnot P, Weinberg MS. Pathogenic effects of Rift Vally fever virus NSs gene are alleviated in cultured cells by expressed antiviral short hairpin RNAs. *Antivir Ther* 2012;**17**:643–56.

Chapter 7

Lujo Virus Hemorrhagic Fever

Janusz T. Paweska
Center for Emerging and Zoonotic Diseases, National Institute for Communicable Diseases of the National Health Laboratory Service, and School of Pathology, Faculty of Medical Sciences, University of Witwatersrand, South Africa

CASE PRESENTATION

A critically ill, 36-year-old female safari travel agent residing on a farm in Zambia was air evacuated from a hospital in Lusaka to the Morningside Medi-Clinic in Johannesburg, South Africa. She had been ill for 12 days and complained of severe headache, malaise, fever, severe chest pain, and sore throat, and then developed skin rash, swelling of the face, deterioration, and generalized tonic–clonic seizures. She was initially treated in Zambia for food poisoning and influenza, and then for suspected allergic reaction to antibiotic treatment (cephalosporin). When she arrived at the private hospital in Johannesburg, her papillary and corneal reflexes were absent, she had cerebral edema, acute respiratory distress syndrome, deteriorating renal function, thrombocytopenia, granulocytosis, and raised serum alanine and aspartate transaminase levels. The attending physician observed a classic sign of tick bite (eschar) on her right foot, which prompted treatment for rickettsiosis. Despite intensive care, including hemodialysis, she died 2 days later, with severe cerebral edema, generalized inflammatory capillary leak, and multiorgan failure. A male paramedic who cared for her during air evacuation developed similar severe illness and was evacuated from Zambia to the same hospital in Johannesburg as the index case, where he also died. The index case and a paramedic served as source of infection for three hospital staff members (two nurses and a cleaner) of which two were fatal. Due to the novel viral etiology of this outbreak, the agent (a new member of the family *Arenaviridae*), later named Lujo virus, was only identified as the specific cause several weeks after the first fatal cases of Lujo hemorrhagic fever emerged. The only patient who survived the outbreak was a nurse subjected to early treatment with ribavirin along with supportive therapy, including statin drugs and N-acetylcystein. Her treatment with oral form of the

Emerging Infectious Diseases. DOI: http://dx.doi.org/10.1016/B978-0-12-416975-3.00007-8

ribavirin started the day after hospitalization with a loading dose of 30 mg/kg (2 g) followed by 15 mg/kg every 6 hr (1 g) for 4 days and 7.5 mg/kg every 8 hr for 6 days. She had to be intubated and treatment was continued through nasogastric tube, but hepatitis and thrombocytopenia worsened and she became confused. It was realized that oral therapy might not be reaching the required inhibitory concentration to stop virus replication. A source of intravenous form of ribavirin was urgently sought, but could only be obtained after 1 week. On initiation of intravenous ribavirin administration there was an improvement and the patient was eventually discharged from the hospital 53 days after admission. Tracing and clinical monitoring of at least 368 contacts of the five known patients or fomites in Zambia and South Africa did not identify any additional clinical cases.

1. WHY THIS CASE WAS SIGNIFICANTLY IMPORTANT AS AN EMERGING INFECTION

The etiologic agents of viral hemorrhagic fevers (VHFs) are taxonomically diverse RNA viruses, including members of the families *Arenaviridae*, *Bunyaviridae*, *Flaviviridae*, and *Filoviridae*, with distinct natural histories, transmission cycles, and modes of human exposure. Importation of these viruses from endemic into non-endemic countries carries severe implications, including tendency to cause public panic, disruption of social life, commerce, widespread anxiety about extreme virulence, mysterious origins, fear of wider spread, travel-associated and healthcare-acquired infections. Management of VHF cases represents a challenging task for diagnostic laboratories, medical and public health systems and requires highly specialized control and prevention measures.[1–3] While significant progress has been made over the last decade in research on the molecular biology, epidemiology, and pathogenesis of VHFs, there are currently no licensed vaccines or chemotherapeutics available to combat most of VHFs.[4] There is an international concern about the potential use of VHF viruses as bioweapons. Due to these reasons they are classified as Biosafety Level 4 (BSL-4) pathogens, and must be handled in maximum biocontainment facilities to prevent human and animal exposures, and intentional misuse. At least six of known arenaviruses are capable of causing some of the most lethal VHFs in humans in Africa and South America (Table 7.1). Until recently, Lassa virus (LASV) was the only known arenavirus to cause VHF in humans in West Africa.[5] The importation of a previously unrecognized arenavirus to South Africa via air medical evacuation of an ill patient from Zambia resulted in highly fatal nosocomial outbreak in Johannesburg caused by a new member of the family *Arenaviridae*, named Lujo virus (LUJV) in recognition of its geographic origin and geographic location of the first recognized outbreak (Lusaka, Zambia, and Johannesburg, South Africa).[6,7]

TABLE 7.1 Old and New World Arenaviruses Causing Viral Hemorrhagic Fevers in Humans

Virus Name and Abbreviation	Year of First Reported Outbreak	Disease	Natural Host	Transmission	World Distribution
Lassa (LASV)[OW]	1969	Lassa fever	*Mastomys* sp.	Urine, saliva	West Africa
Lujo (LUJV)[OW]	2008	Lujo HF[a]	Unknown	Unknown	Sub-Saharan Africa
Guantarito (GTOV)[NW]	1990	Venezuelan HF	*Zygodontomys brevicauda*	Urine, saliva	Venezuela
Junin (JUNV)[NW]	1958	Argentine HF	*Calomys musculinus*	Urine, saliva	Argentina
Machupo (MACV)[NW]	1963	Bolivian HF	*Calomys callosus*	Urine, saliva	Bolivia
Sabiá (SABV)[NW]	1990	Brazilian HF	Unidentified rodents	Unknown	Brazil

[a] *Hemorrhagic fever.*
[OW] Old World arenavirus,
[NW] New World arenavirus.

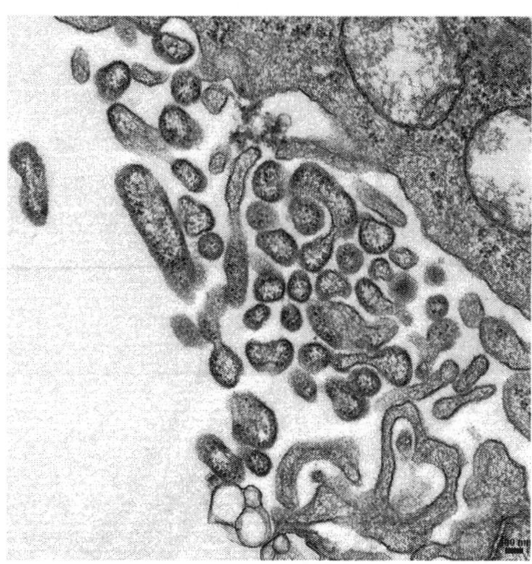

FIGURE 7.1 Electron micrograph of LUJV. Sandy appearance of electron-dense particles representing host ribosomes incorporated into virions. *Micrograph courtesy of S. Zaki and C. Goldsmith, Center for Disease Control, Atlanta, Georgia, USA.*

2. WHAT IS THE CAUSATIVE AGENT?

The *Arenaviridae* family is a large and genetically diverse group of over 30 viruses broadly divided into two major complexes or serogroups based on serological cross-reactions, genetic, and geographic relationships, namely the New World (NW) or Tacaribe complex and the Old World (OW) or Lassa-Lymphocytic Choriomeningitis complex.[8–12] The name for the family comes from the Latin word for sand (*arena*) because of the grainy appearance of electron-dense particles given by host ribosomes incorporated into virions (Figure 7.1). Virions are pleomorphic, ranging in size from 60 to 300 nm, and have a lipid envelop. Arenaviruses have a bi-segmented, single-stranded RNA genome consisting of a small (S) and a large (L) segment, which together total about 11 kb, each encoding for two proteins in ambisense coding strategy.[13] The S segment encodes for the nucleocapsid protein (NP) and the glycoprotein precursor, subsequently cleaved into the envelope proteins GP1 and GP2. The GP1 associates with cellular receptors. The GP2 is a transmembrane protein that mediates fusion of the viral and cellular membranes after internalization of the virus into cytoplasmic endosomes. The L segment encodes for the viral RNA-dependent RNA polymerase (L protein) and a zinc-binding matrix protein (Z protein), which is required for budding of virions, but also antagonizes host cell interferon responses by various mechanisms.[14–17] Genome sequencing and phylogenetic analysis of LUJV[6,7] showed it to be genetically unique, and very distinct from previously characterized arenaviruses. The LUJV G1 glycoprotein sequence is highly diverse and almost equidistant from that of other OW and NW arenaviruses, indicating a potential distinctive receptor tropism.

3. WHAT IS THE FREQUENCY OF THE DISEASE?

To date only five cases of Lujo hemorrhagic fever (LHF) have been recognized and laboratory confirmed following a nosocomial outbreak in South Africa.[7] The index case of this outbreak was airlifted from Zambia to South Africa in 2008 for medical treatment and subsequently served as a source of infection to four healthcare workers (HCWs). Tracing and clinical monitoring of more than 368 contacts of the LHF patients or fomites in Zambia and South Africa, including 94 staff members of the Morningside Medi-Clinic in Johannesburg, which was the epicenter of the outbreak, did not identify any additional cases.[7,18] However, LHF contacts in Zambia and South Africa were not serologically tested to assess the possibility of less severe infections. The distribution or prevalence of the virus in human and rodent populations in sub-Saharan Africa is unknown. Similarly, ecology, epidemiology, including host range, natural transmission cycle, distribution, and the mode of LUJV transmission from a reservoir host to humans remain a mystery. The geographic distribution of each arenavirus is largely determined by the range of its natural vertebrate reservoir hosts, mostly rodents. LCMV is the only arenavirus to have a worldwide distribution due to its association with the ubiquitous *Mus musculus*. Other arenaviruses are distributed either in Americas or in Africa.[8,11] The unique phylogenetic position of LUJV prompts various hypotheses regarding the likely reservoir host, including murinae or non-murinae hosts. To date, only one arenavirus has been linked with non-rodent species, namely, the Tacaribe virus having a fruit bat reservoir.[8] The natural reservoirs of arenaviruses in Africa are rodents of the family *Muridae*, especially the multimammate rat (*Mastomys natalensis*). The *M. natalensis* is abundantly distributed in the region where the index case of the LUJV associated-outbreak lived and worked, but pathogen discovery expedition in the area following the outbreak could not find the virus in *M. natalensis* although the number of rodents captured and tested was rather limited.[19]

4. HOW IS THE VIRUS TRANSMITTED?

Rodent-to-human arenavirus transmission occurs through contact with virus-contaminated rodent excreta, commonly via inhalation of dust or aerosolized materials, or vomites soiled with rodent feces or urine, or ingestion of contaminated food.[20,21] Rodent consumption has been postulated as possible risk behavior.[22] Person-to-person transmission occurs via direct contact with infected blood, urine, or pharyngeal secretions.[23] Severe outbreaks reported in healthcare facilities were often a result of inadequate standard precautions.[24,25] Poor hygiene in households increases risk of rodent infestation and consequently contracting Lassa fever.[26] Transmission of arenaviruses may also occur during transplantation of infected organs, as it has been documented for LCM[27] and Dandenong[28] viruses. Although, the primary source and the route of infection of the first recognized event of LUJV transmission to a man have not been established, there was evidence of rodent activity on the index case farm.[7] Out of the five LHF cases, four were nosocomial in origin, which might indicate that LUJV is easily transmissible from person to person. However, the first LHF cases in hospitals in Johannesburg were initially managed without strict barrier nursing practices and LUJV transmission was confirmed in a limited number of HCWs. Moreover, secondary and tertiary transmission took place only in a single private hospital among those HCWs who had very close contact with LHF patients, including exposure to blood and bodily fluids in enclosed settings such as flight medical evacuation, or involved in high risk procedures, such as insertion of intravascular catheters, dialysis, and endotracheal intubation.[7,18,29] Although no specific exposures were recorded during the outbreak among attending HCWs, contact with infected patients or fomites might be inadvertent especially during highly charged outbreaks, and specific exposure to infected secretions or tissues might be unnoted or not recalled as being significant.

5. WHICH FACTORS ARE INVOLVED IN DISEASE PATHOGENESIS? WHAT ARE THE PATHOGENIC MECHANISMS?

Knowledge of the pathogenesis of LUJV is very limited. Based on the five cases recognized to date, the clinical disease associated with LUJV infection is remarkably similar to infection with LASV despite distinct genetic relationship between the two viruses. However, the case fatality rate (CFR) in hospitalized patients with Lassa fever (LF) ranges from 20 to 50%, but typically is much less than 30%,[25] whereas CFR associated with LHF was 80%,[7] suggesting that LUJV is more virulent that LASV. As with all VHFs, microvascular instability, impaired homeostasis, and multi-organ failure seem to be also the hallmarks of LHF. Severe or fatal VHF cases are not only due to uncontrolled viral replication and destruction of host cells by viral lysis, but also because of immunopathogenic effects resulting from altered host immune responses, including uncontrolled inflammatory responses.[30]

Studies on the role of inflammatory mediators in the pathogenesis of LF indicate that low levels of interleukin-8 and interferon-inducible protein 10 in serum are associated with fatal infections.[31] Studies on LCMV have implicated the nucleoprotein of arenaviruses as the pathogenesis marker as a result of its ability to counteract type I IFN, specifically inhibiting the interferon regulatory factor 3 (IRF3).[14] To date, two different cell surface molecules have been implicated as cellular receptors for arenaviruses. OW arenaviruses (LASV, certain strains of LCMV) and NW clade C viruses use α-dystroglycan (α-DG) to enter cells.[32] Viruses with a high binding affinity for α-DG replicate preferentially in the white pulp of the spleen and infect large numbers of lymphocytes. The ability of these cells to act as antigen presenting cells results in impairment of host immune responses. Although immunosuppression may be important for the establishment of persistent infection in rodent hosts, in which the infection is subclinical, in humans it may lead to serious illness. Immunosuppressive strains of LCMV and highly pathogenic LASV have been shown to bind to α-DG with very high affinity, which is crucial for their ability to infect dendritic and Schwann cells to cause a generalized immunosuppression.[33] The pathogenic NW clade B arenaviruses Junin virus (JUNV), Machupo virus (MACV), Guanarito virus (GTOV), and Sabi virus (SABV) use both human transferrin receptor 1 (TfR1) and the TfR1 orthologues of their reservoir rodent species. This molecule has a number of properties that favor their replication and disease pathology in humans.[34,35] Full genome sequencing of LUJV revealed that it lacks the sequences to support the use of α-DG and TfR1 receptors.[6]

Recent *in vivo* experiments further demonstrated the unique virulence properties of LUJV compared to NW and OW arenaviruses. Both JUNV and LASV are highly virulent in newborn and weaning mice whereas LUJV was non-lethal in mice regardless of the route of inoculation or viral dose. In contrast, LUJV caused more severe, rapidly progressive, and uniformly lethal hemorrhagic disease in strain 13/N guinea pigs compared to LASV.[36] Moreover, previous studies of JUNV, GTOV, MACV, and LASV using rodents and other small animals failed to demonstrate consistent signs of HF.[37] LUJV infection of guinea pigs resulted in severe infarction, fibrin deposition, and hemorrhage in multiple organs, suggesting DIC. All animals developed leucopenia, lymphopenia, thrombocytopenia, coagulopathy and elevated transaminase levels. Despite very high viral load in these animals 5 days post infection (dpi), pro-inflammatory cytokine/chemokine genes (e.g., IL-1b, RANTES) were down-regulated early during infection while as early as 5 dpi mediators of macrophage and neutrophil activation and inflammation (e.g., IL-8, MCP-1) were significantly up-regulated. The IL-10 was also detected in some organs 9 dpi, suggesting that LUJV limits and controls pro-inflammatory host responses in order to allow its replication.[36] Results of reverse-genetic studies suggest that LUJV has unique molecular motifs, which may further augment the apparently enhanced virulence of LUJV by increasing viral replication or interfering with host immunoregulatory responses.[38]

6. WHAT ARE LHF CLINICAL MANIFESTATIONS?

Clinical presentation and histological lesions in the course of infection with LUJV[7,18] are similar to those of other arenavirus VHFs.[20,25,39–44] The putative incubation periods for LHF patients ranged from 7 to 13 days, and the periods of illness for the four fatal patients ranged from 10 to 13 days. All patients initially sought treatment for non-specific febrile illness with fever, headache, and myalgia, followed by the development of diarrhea and pharyngitis.[7] Terminal features included severe respiratory distress, neurological signs, and circulatory collapse. Severe bleeding or hemorrhage was not a prominent clinical feature.

Clinical laboratory findings for LHF patients included thrombocytopenia ($20-104 \times 10^9$cells/L), and increased aspirate aminotransferase (AST) with maximum AST values recorded in the four patients who died ranging from 549 IU/L to 2486 IU/L, compared with 240 IU/L in the one survivor who was treated with ribavirin. Three patients had leukocyte counts in the normal range, and two had leukopenia on admission, whereas leukocytosis developed in four patients during the illness. Hepatocyte necrosis and skin vasculitis compatible with VHF histopathological lesions were observed in tissue biopsies taken from two fatal cases. Major symptoms and signs in patients infected with LUJV are summarized in Table 7.2.

TABLE 7.2 Major Signs and Symptoms in LHF Patients ($n = 5$)

Signs and Symptoms	Frequency (%)
Fever	100
Headache	100
Myalgia	100
Sore throat	100
Diarrhea	100
Respiratory distress	100
Neurological signs	100
Morbilliform rash	60
Neck and facial swelling	60
Chest pain	40
Petechial rash	20
Bleeding	20
Thrombocytopenia	100
Increased aspirate aminotransferase	80
Leukocytosis	80

7. HOW DO YOU DIAGNOSE?

Early diagnosis of VHFs is crucial for the timely implementation of infection control measures and to prevent further spread of the disease. The unique genomic sequence and high antigenic diversity compared to other OW and NW arenaviruses greatly complicated initial diagnosis of LHF by molecular, antigen, and serologic detection techniques during the first emergence of this disease in 2008.[7] Since then, however, no commercial assays have been made available for diagnosis of LHF and those developed in-house are performed in limited and highly specialized laboratories. Common diagnostic techniques for VHFs include cell culture, serologic assays, including enzyme-linked immunosorbent assay (ELISA) or immunofluorescent antibody assay (IFA), antigen detection assays, immunohistochemical staining of formalin-fixed tissues, and the reverse transcriptase polymerase chain reaction (RT-PCR). A very challenging aspect of the routine applications of these techniques is general lack of diagnostic validation data due to the very rare occurrence of some of these diseases and limited availability of clinical materials. The development of molecular and serological diagnostic tools highly specific for LUJV is not only needed for differential laboratory diagnosis of VHFs but also for surveillance programs aimed at discovery of the reservoir host(s) and mapping areas at risk of LUJV virus infection.

LUJV was isolated in Vero cell culture during a period of 2 to 13 days of illness. Of the total of seven blood and two liver samples taken from five LHF patients all were positive by virus isolation. The liver samples were also positive by the RT-PCR, but the assay yielded false-negative results in four out of the seven blood samples.[7] In four fatal cases infected with LUJV, IgM and IgG could not be demonstrated during a period of 10–13 days from disease onset to death. This corresponds to observations made in fatal LF cases.[5,45] In one patient who survived, the IgM was first detected on day 20, with peak on day 30, followed by decreasing but still detectable levels 11 months after onset of the disease. The IgG was first detectable on day 37 with still increasing levels by day 333 post-onset.[18] The long persistence of IgM in an LHF patient corresponds to recent findings with LF and some of the South American arenavirus HFs. Possible explanations for the sustained IgM response include inhibition of antibody class switching, presence of IgM^+ memory B cells or decreased T-helper cell ($CD4^+$) function.[46] LUJV glycoprotein 1 (GP1) moderately cross-reacts with immune serum to LASV, weakly with Mobala and LCM immune sera, but does not cross-react with sera against Junin, Mopeia, and Ippy viruses.[18]

8. HOW DO YOU DIFFERENTIATE THIS DISEASE FROM SIMILAR ENTITIES?

VHFs are found worldwide, but the distribution of a specific virus is restricted by the distribution of its natural reservoir or arthropod vector. Several members of the family *Arenaviridae* can cause severe hemorrhagic fevers in humans (Table 7.1), thus representing a serious public health problem in endemic areas of Africa and South America.[36,47,48] The non-specific presentation of LHF and other VHFs makes them very difficult to diagnose clinically, especially early in the course of infection. The differential diagnosis includes a broad array of parasitic, bacterial, viral and non-infectious causes, including malaria, rickettsial infections, Q fever, typhoid fever, dysentery, plaque, tularemia, brucellosis, leptospirosis, other sepsis from bacterial infections, viral hepatitis, other viral hemorrhagic fevers, non-infectious causes of disseminated intravascular coagulopathy, and acute leukemia. Although it is important to maintain a broad differential diagnosis, in practice this is a very challenging task due to biosafety requirements, limitations of all identification methods at a single diagnostic laboratory, and general lack of a simple method for detection of all VHF causative agents. Differential diagnosis may be narrowed as the disease progresses or as laboratory results and epidemiological information become available. For a patient presenting with clinical symptoms of VHF, the likely causative agent can be tentatively determined based on recent travel and potential exposure history in endemic regions. Therefore, for accurate diagnosis it is necessary to combine all available laboratory results, clinical, pathological, and epidemiological data.

9. WHAT IS THE THERAPEUTIC APPROACH?

It remains unknown if the survival of the single patient during the LHF outbreak in 2008 can be attributed to a milder infection or individual lower susceptibility allowing for the development of humoral and cellular immune responses, or if it was the result of ribavirin and supportive treatment, including statin drugs and N-acetylcystein,[5] or combination of all. Statin drugs appear to have immunomodulatory, anti-inflammatory, antimicrobial, and vasculature-stabilizing properties whereas N-acetylcystein acts as the antioxidant and free radical scavenger. Ribavirin suppresses pro-inflammatory cytokines and Th-2 cytokines, involved in humoral immunity, while maintaining normal Th-1 cytokine expression that favors cellular immunity.[49] Whether humoral responses to either of the LUJV antigens correlate with protection against disease is also unknown. The danger of arenaviruses for human health, the increased emergence of new viral species in recent years, and the lack of effective tools for their control or prevention, makes the search for novel antiviral compounds an urgent effort. Unfortunately, therapeutic options for all VHFs are still very limited, thus early differential diagnosis has implications for containment and clinical management.[50] Fluid and blood pressure management guidelines for septic shock are recommended for VHFs due to common features in the pathogenesis of these two conditions.[5] Considering the high burden of malaria and tick-borne rickettsial disease in Africa, patients should be immediately covered with broad spectrum anti-bacterial and/or anti-parasitic therapy, until a diagnosis of LHF can be confirmed.

A number of vaccine candidates have been developed for VHFs, but only yellow fever vaccine is widely available for public use. Early treatment with immune plasma was effective in Junin and LASV infections.[51] Ribavirin appears to be helpful if given early in the course of Crimean-Congo hemorrhagic fever, or hemorrhagic fever with renal syndrome, and is recommended in post-exposure prophylaxis and early treatment of arenavirus infections.[1] It has been shown to be an effective treatment for LF, especially when started within the first 6 days of illness. Licensed vaccine is available in Argentina for Junin virus,[52] but there is no vaccine for LF and most of the VHFs caused by arenaviruses. However, several candidates are under development with successful trials in primates.[53] Recent identification of HLA-restricted $CD8^+$ T cell epitopes that are either cross-reactive or species specific has a promise in the development of a multivalent vaccine strategy against arenaviruses pathogenic for humans.[54]

10. WHAT ARE THE PREVENTIVE AND INFECTION CONTROL MEASURES?

It must be emphasized that outbreaks of most VHFs are unpredictable in nature, rare, and usually their recognition is delayed. In this context management of VHFs requires a highly organized, specialized, and well-prepared public health response system in order to control these dangerous diseases.[1,2,55] A suspected case of VHF has to be immediately reported to the infectious control professionals and respective state health departments, and specific barrier infectious control measures should be implemented immediately. Infection control of LHF relies on classic public health principles of identification and isolation of cases and monitoring of their contacts. Constant vigilance and attention to standard precautions, although basic, should be strictly applied to the daily, routine medical practice and not only in hospitals and clinics but also during medical air evacuation procedures. The International Health Regulations are designed to protect against the international spread of these dangerous infections.[56] One of the major weaknesses recognized during the 2008 LHF outbreak in Johannesburg was the lack of prior consultation by air evacuation services with South African Port Control officers employed by the South African Department of Health and timely consultations of these officials with their own local medical experts. Although airport health officials checked the flight that brought the index case from Lusaka to Johannesburg, the list of symptoms concerned appears to cover more common diseases and is one of the reasons why the index case has not been recognized as a potential high consequence communicable threat.[18] It is also unacceptable that the attending doctor at Morningside Medi-Clinic in Johannesburg was not given any documentation related to the patients' disease history.[57] International transfer of patients must follow strict international health regulations to minimize the possibility of rapid spread of dangerous infectious diseases by air transport.

10.1 Isolation Precautions

The early non-specific clinical presentation of LHF presents a challenge to case identification and timely isolation of patients. The low and spatially restricted level of LUJV secondary spread during the 2008 nosocomial outbreak in South Africa[18] indicates that aerosol transmission is not highly efficient and infection with the virus requires very close, unprotected skin or mucosal contact with either patient's excreta, vomitus, blood, or direct, unprotected exposure to contaminated fomites (e.g., bedding, utensils, bedpans, hospital equipment) or might result from accidentally acquired infection, e.g., by a needle stick. The low secondary attack rate of LUJV, as for most VHF pathogens, affords a measure of reassurance even when cases are unrecognized, providing strict barrier nursing is maintained by HCWs. Clinically, LHF-suspected patients should be regarded as infectious and thus kept under VHF isolation precautions. Placement of a patient in a negative pressure isolation room, if affordable, would be beneficial, but hermetically sealed isolation chambers are not required. It has been demonstrated that HCWs in a poor-resourced hospital in West Africa using simple barrier nursing methods had no higher risk for infection with LASF than the local population.[58] Specific VHF barrier nursing precautions include the use of surgical masks, double gloves, gowns, protective aprons, face shields, and shoe covers. Positive-pressure respirators or N95 should be used when performing procedures that might generate aerosols, such as endotracheal intubation. Modern hospitals have all the resources required to protect the HCWs from exposure to VHF agents, but public panic associated with their emergence may result in reluctance among HCWs to care for patients. Due to the high severity of LHF, intensive care units (ICU) have to be utilized for patients' management and isolation. The ICU physicians, nurses, and all ancillary staff need to be adequately trained in barrier nursing procedures and be able to state their concerns before caring for VHF patients. Training should be offered periodically so staff can retain their knowledge and necessary skills. The designated hospitals should have procedures in place to verify that ICUs are adequately maintained and decontaminated after occupancy by VHF patients.

10.2 Post-exposure Management

Post-exposure management strategy, including procedures for accidental exposure to blood and bodily fluids, should be in place and reviewed when new knowledge on LHF treatment and prevention is available to enhance HCW safety. Although no firm conclusion can be drawn about the prophylactic value of ribavirin during the 2008 LHF outbreak, this antiviral drug is recommended for post-exposure LF prophylaxis. Absorption of oral ribavirin from the gut may pose a barrier given the gastrointestinal symptoms in LHF. Until the time more data are available on the efficacy of oral ribavirin treatment as recommended for LF,[5] post-exposure LHF prophylaxis should be given IV.

REFERENCES

1. Bannister B. Viral haemorrhagic fevers imported into non-endemic countries: risk assessment and management. *British Med Bull* 2010;**95** 193–25.
2. Borio L, Inglesby T, Peters CJ, Schmaljohon AL, Hughers JM, Jahrling PB, et al. Hemorrhagic fever viruses as biological weapons: medical and public health management. *J Am Med Ass* 2002;**287**:2391–405.
3. Keeton C. South African doctors move quickly to contain new virus. *Bull WHO* 2008;**86**:912–3.
4. Boulton ML, Wells EV. Major strategies for prevention and control of viral hemorrhagic fevers. In: Singh S, Ruzek D, editors. *Viral hemorrhagic fevers*. Boca Raton FL, US: CRC Press Taylor & Francis Group; 2013. p. 159–76.
5. Bausch DG, Moses LM, Boba A, Grant DS, Khan H. Lassa fever. In: Singh SH, Ruzek D, editors. *Viral hemorrhagic fevers*. Boca Raton FL, US: CRC Press Taylor & Francis Group; 2013. p. 261–86.
6. Briese T, Paweska JT, McMullan LK, Hutchison SK, Street C, Palacios G, et al. Genetic detection and characterization of Lujo virus, a new hemorrhagic fever-associated arenavirus from southern Africa. *PLoS Pathog* 2009;**5**:e1000455.
7. Paweska JT, Jansen van Vuren P, Wyer J. Lujo virus hemorrhagic fever. In: Singh SH, Ruzek D, editors. *Viral hemorrhagic fevers*. Boca Raton FL, US: CRC Press Taylor & Francis Group; 2013. p. 287–303.
8. Bowen MD, Peters CJ, Nichol ST. Phylogenetic analysis of the arenaviridae: patterns of virus evolution and evidence for cospeciation between arenaviruses and their rodent hosts. *Mol Phylogenet Evol* 1997;**8**:301–16.
9. Charrel RN, de Lamballerie X. Arenaviruses other than Lassa virus. *Antiviral Res* 2003;**57**:89–100.
10. Charrel RN, de Lamballerie X, Emonet S. Phylogeny of the genus Arenavirus. *Curr Opin Microbiol* 2008;**11**:362–8.
11. Gonzalez JP, Emonet S, de Lamballerie X, Charrel R. Arenaviruses. *Curr Top Microbiol Immunol* 2007;**315**:253–88.
12. Moncayo AC, Hice CL, Watts DM, Travassos de Rosa AP, Guzman H, Russell KL, et al. Allpahuayo virus: a newly recognized arenavirus (arenaviridae) from arboreal rice rats (oecomys bicolor and oecomys paricola) in northeastern Peru. *Virology* 2001;**284**:277–86.
13. Buchmeier M, de la Torre JC, Peters CJ. Arenaviridae: the viruses and their replication. In: Knipe DM, Howley PM, editors. *Fields Virology*. 5th ed. Philadelphia: Lippencott Williams & Wilkins; 2007. p. 1791–827.
14. Borrow PL, Martinez-Sobrido JC, de la Torre L. Inhibition of the type I interferon antiviral response during Arenavirus infection. *Viruses* 2010;**2**:2443–80.
15. Kranzusch PJ, Whelan SP. Arenavirus Z protein controls viral RNA synthesis by locking a polymerase-promoter complex. *Proc Natl Acad Sci USA* 2001;**108**:19743–8.

16. Schlie K, Maisa A, Freiberg F, Groseth A, Strecker T, Garten W. Viral protein determinants of Lassa virus entry and release from polarized epithelial cells. *J Virol* 2010;**84**:3178–88.
17. York J, Nunberg JH. A novel zinc-binding domain is essential for formation of the functional Junin virus envelope glycoprotein complex. *J Virol* 2007;**81**:13385–91.
18. Paweska JT, Sewlall NH, Ksiazek TG, Blumberg LH, Hale MJ, Lipkin WI, et al. Nosocomial outbreak of novel arenavirus infection, Southern Africa. *Emerg Infect Dis* 2009;**15**:1598–602.
19. Ishii A, Thomas Y, Moonga L, Nakamura I, Ohnuma A, Hang'omebe B, et al. Novel arenavirus, Zambia. *Emerg Infect Dis.* 2011;**17**:1921–4.
20. McCormick JB, King IJ, Webb PA, Johnson KM, O'Sullivan R, Smith ES, et al. A case-control study of the clinical diagnosis and course of Lassa fever. *J Infect Dis* 1987;**155**:445–55.
21. Stephenson EH, Larson EW, Dominik JW. Effect of environmental factors on aerosol-induced Lassa virus infection. *J Med Virol* 1984;**14**:295–303.
22. Ter Meulen J, Lukashevich I, Sidibe K, Inapogui A, Marx M, Dorlemann A, et al. Hunting of peridomestic rodents and consumption of their meat as possible risk factors for rodent-to-human transmission of Lassa virus in the Republic of Guinea. *Am J Trop Med Hyg* 1996;**55**:661–6.
23. Kerneis S, Koivogui L, Magassouba NF, Koulemou K, Lewis R, Aplogan A, et al. Prevalence and risk factors of Lassa seropositivity in inhabitants of the forest region of Guinea; a cross-sectional study. *PLoS Negl Trop Dis* 2009;**3**:e548.
24. Bajani MD, Tomori O, Rollin PE, Harry TO, Bukbuk ND, Wilson L, et al. A survey for antibodies to Lassa virus among health workers in Nigeria. *Trans R Sco Trop Med Hyg* 1997;**91**:379–81.
25. Frame JD, Baldwin JM, Gocke GD, Troup JM. Lassa fever, a new virus disease of man from West Africa. I. Clinical description and pathological findings. *Am J Trop Med Hyg* 1970;**19**:670–6.
26. Bonner PC, Schmidt WP, Belmain SR, Oshin B, Borchert M. Poor housing quality increases risk of rodent infestation and Lassa fever in refugee camps of Sierra Leone. *Am J Trop Med Hyg* 2007;**77**:169–75.
27. Fisher SA, Graham MB, Kuehnert MJ, Kotton CN, Srinivasan A, Marty FM, et al. Transmission of lymphocytic choriomeningitis virus by organ transplantation. *N Engl J Med* 2006;**354**:2235–49.
28. Palacios G, Druce J, Du L, Tran T, Birch C, Briese T, et al. A new arenavirus in a cluster of fatal transplant-associated diseases. *N Engl J Med* 2008;**358**:991–8.
29. Sewlall N. Arenavirus outbreak with nosocomial transmission: infection control and the lessons learned. *S Afr Anaesth Analg* 2011;**17**:54–6.
30. Huzell LM, Cann JA, Lackemeyer M, Wahl Jensen V, Jahrling PB, et al. General disease pathology in filoviral and arenaviral infections. In: Singh SH, Ruzek D, editors. *Viral hemorrhagic fevers*, 2013. Boca Raton FL, US: CRC Press Taylor & Francis Group; 2013. p. 261–86.
31. Mahanty S, Bausch DG, Thomas RL, Goba S, Bah A, Peters CJ, et al. Low levels of interleukin-8 and interferon-inducible protein-10 in serum are associated with fatal infections in acute Lassa fever. *J Infect Dis* 2001;**183**:1713–21.
32. Spiropoulou CF, Kunz S, Rollin PE, Campbell KP, Oldstone MB. New World arenaviruses clade C, but not clade A and B viruses, utilizes α-dystroglycan as its major receptor. *J Virol* 2002;**76**:5140–6.

33. Smelt SC, Borrow P, Kunz S, Cao W, Tishon A, Lewicki H, et al. Differences in affinity of binding of lymphocytic choriomeningitis virus strains to the cellular receptor α-dystroglycan correlate with viral tropism and disease kinetics. *J Virol* 2001;**75**:448–57.
34. Flanagan ML, Oldenburg J, Reignier T, Holt N, Hamilton GA, Martin VK, et al. New World clade B arenaviruses can use transferrin receptor 1 (TfR1)-dependent and -independent entry pathways and glycoproteins from human pathogenic strains are associated with the use of TfR1. *J Virol* 2008;**82**:938–48.
35. Radoshitzky SR, Kuhn JH, Spiropoulou CF, Albarino CG, Nguyen DP, Salazar-Bravo J, et al. Receptor determinants of zoonotic transmission of New World hemorrhagic fever arenaviruses. *PNAS* 2008;**105**:2664–9.
36. Bird BH, Dodd KA, Erickson BR, Albariño CG, Chakrabarti AK, McMullan LK, et al. Severe hemorrhagic fever in strain 13/N guinea pigs infected with Lujo virus. *PLoS Negl Trop Dis* 2012;**6**:e1801.
37. Goven BB, Hollbrook MR. Animal models of highly pathogenic RNA viral infections: hemorrhagic fever viruses. *Antiviral Res* 2008;**78**:79–90.
38. Bergeron É, Chakrabarti AK, Bird BH, Dodd KA, McMullan LK, Spiropoulou CF, et al. Reverse genetics recovery of Lujo virus and role of virus RNA. *J Virol* 2012;**86**:10759–65.
39. De Manzione N, Salas RA, Paredes H, Godoy O, Rojas L, Araoz F, et al. Venezuelan hemorrhagic fever: clinical and epidemiological studies of 165 cases. *Clinical Infect Dis* 1998;**26**:308–13.
40. Elsner B, Schwarz E, Mando OGJ, Maiztegui J, Vilches A. Pathology of 12 cases of Argentine hemorrhagic fever. *Am J Trop Med Hyg* 1973;**22**:229–36.
41. Maiztegui JI. Clinical and epidemiological patterns of Argentine haemorrhagic fever. *Bull WHO* 1975;**52**:567–75.
42. Monat TP, Casals J. Diagnosis of Lassa fever and the isolation and management of patients. *Bull WHO* 1975;**52**:707–15.
43. Salas R, Pacheco ME, Ramos B, Taibo ME, Jaimes E, Vasquez C, et al. Venezuelan heamorrhagic fever. *Lancet* 1991;**338**:1033–6.
44. Walker DH, McCormick JB, Johnson KM, Webb PA, Komba-Kono G, Elliott LH, et al. Pathologic and virologic study of fatal Lassa fever in man. *Am Ass Pathol* 1982;**107**:349–56.
45. Bausch DG, Hadi CM, Khan SH, Lertora JJ. Review on the literature and proposed guidelines for the uses of oral ribavirin as postexposure prophylaxis for Lassa fever. *Clin Infect Dis* 2010;**51**:1435–41.
46. Branco LM, Grove JN, Boisen ML, Shaffer JG, Goba A, Fullah M, et al. Emerging trends in Lassa fever: redefining the role of immunoglobulin M and inflammation in diagnosing acute infection. *Virology J* 2011;**8**:478.
47. Aguilar PV, Camargo W, Vargas J, Guevera C, Roca Y, Felices V, et al. Reemergence of Bolivian hemorrhagic fever, 2007–2008. *Emerg Infect Dis* 2009;**15**:1526–8.
48. Tesh RB. Viral haemorrhagic fevers of South America. *Bimédica* 2002;**22**:287–95.
49. Ning Q, Brown D, Parodo J, Cattral M, Gorczynski R, Cole E, et al. Ribavirin inhibits viral-induced macrophage production of TNF, IL-1, the procoagulant fgl2 prothrombinase and preserves Th1 cytokine production but inhibits Th2 cytokine response. *J Immunol* 1998;**160**:3487–93.
50. Gowen BB, Bray M. Progress in the experimental therapy of severe arenaviral infections. *Future Microbiol* 2011;**6**:1429–41.
51. Enria DA, Briggiler AM, Sánchez Z. Treatment of Argentine hemorrhagic fever. *Antiviral Res* 2008;**78**:132–9.

52. Maiztegui JI, McKee Jr KT, Barrera-Oro JG, Harrison LH, Gibbs PH, Feuillade MR, et al. Protective efficacy of a live attenuated vaccine against Argentine hemorrhagic fever. *J Infect Dis* 1998;**177**:277–83.
53. Geisbert TW, Jones S, Fritz EA, Schurtleft AC, Geisbert JB, Libscher R, et al. Development of a new vaccine for the prevention of Lassa fever. *PLoS Med* 2005;**2**:e183.
54. Kotturi MF, Botten J, Sidney HH, Bui L, Giancola M, Maybeno J, et al. A multivalent and cross-protective vaccine strategy against arenaviruses associated with human disease. *PLoS Pathog* 2009;**5**:e1000695.
55. Franz DR, Jahrling PB, Friedlander AM, McClain DJ, Hoover DL, Bryne WR, et al. Clinical recognition and management of patients exposed to biological warfare agents. *J Am Med Ass* 1997;**278**:399–411.
56. World Health Organisation. International Health Regulations. <http://www.who.int/ihr/en/>; 2005.
57. Bateman C. Arenavirus deaths—emergency air services tighten up. *S Afr Med J* 2008;**98**:910–4.
58. Helmick CG, Webb PA, Scribner CL, Krebs JW, McCormick JB. No evidence for increased risk of Lassa fever infection in hospital staff. *Lancet* 1986;**328**:1202–5.

Chapter 8

Toscana Virus Infection

Remi N. Charrel

Aix Marseille Université, IRD French Institute of Research for Development, EHESP French School of Public Health, EPV UMR_D 190 "Emergence des Pathologies Virales", & IHU Méditerranée Infection, APHM Public Hospitals of Marseille 13385, Marseille, France

CASE PRESENTATION

At the beginning of summer, a 17-year-old male, originating from and living in France, without noticeable medical background, visited his medical practitioner for brutal and recent onset characterized by fever, severe headache, and vomiting. After examination, his GP referred the patient to the emergency ward of the public hospital for suspected aseptic meningitis. Upon admission, he presented with the same manifestations in a febrile context (39°C) in spite of self-medication with paracetamol. Examination revealed stiff neck in the absence of photophobia or signs suggesting encephalitis. Laboratory tests showed an increased white blood cell count (WBC) at 11,700/mm^3 (80% neutrophils) and a normal C-reactive protein. The cerebrospinal fluid (CSF) contained 840 leukocytes/mm^3 (70% lymphocytes) with normal glucose and protein levels. All molecular tests performed on CSF that targeted enteroviruses, herpes simplex virus types 1 and 2, *Neisseria meningitidis*, and *Streptococcus pneumoniae* were negative. The patient reported no recent history of travel. Treatment was started with amoxicillin and acyclovir, and the patient was transferred into the Infectious Diseases department. The day after, he was no longer febrile, but still complained of severe headache and vomiting. Routine analysis of the CSF including direct microscopic examination, bacterial culture, and conventional PCR assays for 16 S and 18 S were negative. Real-time RT-PCR for enterovirus, West Nile virus, and Toscana virus were performed. The real-time RT-PCR for Toscana virus was positive (cycle threshold value: 38). Despite low viral load in the CSF, the virus strain was isolated in Vero cells and further characterized by complete genome sequencing.[1] Serology using indirect immunofluorescence showed the presence of IgM and IgG in the acute serum. Clinical manifestations rapidly resolved and the patient was discharged the next day. Three days after the diagnosis was established, a field campaign was organized to collect sandflies in the

Emerging Infectious Diseases. DOI: http://dx.doi.org/10.1016/B978-0-12-416975-3.00008-X

vicinity close to the patient's home. Four CDC miniature light traps were placed in the garden around the house, where the patient used to spend time at nightfall. A total of 17 *Phlebotomus perniciosus* were trapped but were negative for Toscana virus RNA through real-time RT-PCR. However, the trapping of sandflies in the vicinity close to the patient's home confirmed that the patient may have been exposed at home to sandflies, known as the vector of Toscana virus.

(This is a published case report.[1])

1. WHY THIS CASE WAS SIGNIFICANTLY IMPORTANT AS AN EMERGING INFECTION

Neurological infections caused by Toscana virus, a sandfly-borne phlebovirus, are common during the summer in regions bordering the Mediterranean Sea (Figure 8.1). Studies collating series of Toscana virus cases demonstrate that Toscana virus is the third cause of aseptic meningitis and meningoencephalitis in the regions of Spain, France, and Italy where the virus circulates in sandflies. Seroprevalence studies performed in southern Europe indicate that a significant proportion of the exposed population (5 to >50%, depending on the studies and on the regions) possess antibodies that react with Toscana virus. According to these data, Toscana virus is the most prevalent arthropod-borne virus in Europe far ahead of tick-borne encephalitis virus, West Nile virus, or dengue virus. As an indicator of medical and scientific attention, a search using the keyword "Toscana virus" retrieved 204 related

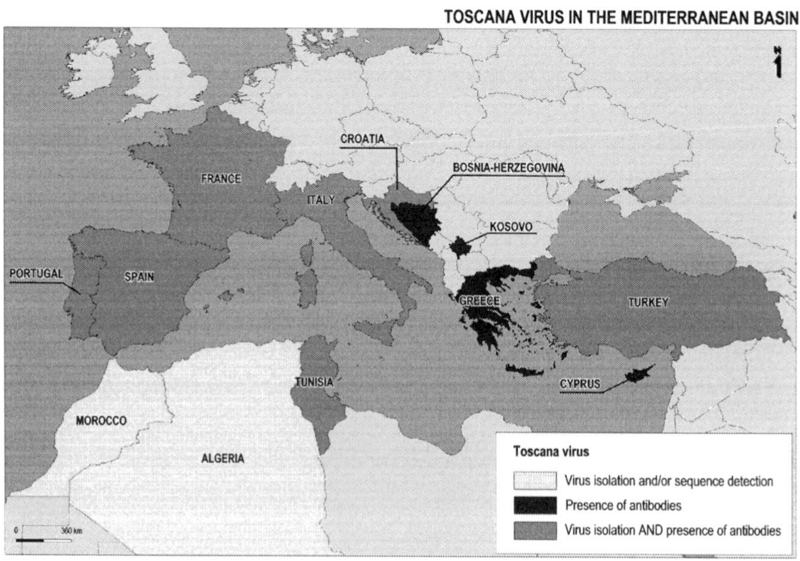

FIGURE 8.1 Toscana virus in the Mediterranean Sea.

articles in PubMed, which correspond to fewer than five articles per year since virus discovery. There is a great need to draw attention to Toscana virus. Because it was discovered in the early 1970s, Toscana virus should be considered as a neglected pathogen rather than an emerging pathogen. As a countermeasure, availability of affordable commercial kits for molecular diagnosis and serology should be aimed for in order to enable dissemination of the diagnostic capacities outside of the few reference centers.

2. WHAT IS THE CAUSATIVE AGENT?

Toscana virus (Toscana virus; *Bunyaviridae*; *Phlebovirus*) is an enveloped, single-strand, negative-sense RNA virus. First isolated in Italy in 1971, Toscana virus is an arthropod-borne virus transmitted by phlebotomine sandflies (*Phlebotomus* sp. of the subgenus *Laroussius*). According to the International Classification on the Taxonomy of Viruses, Toscana virus belongs to the species *Sandfly fever Naples virus*, which includes also several other viruses such as Naples virus and Theran virus; in addition, there are several recently discovered phleboviruses that should be included in this species together with Toscana virus; this is not the role of ICTV since this organism classifies virus down to the level of species, not below the species. Obviously, these newly discovered viruses (Massilia, Granada, Punique) have to be considered as variant genotypes of the species *Sandfly fever Naples virus*. There is a very limited knowledge on the capacity of these variants' genotypes to infect humans and to cause disease in humans.

Previous phylogenetic analysis of Toscana virus first identified two geographically associated lineages: the lineage A endemic in Italy (including the Sardinia and the Sicily) and the lineage B endemic in Spain.[2] New sequences of Toscana virus isolates originating from other countries confirmed this dichotomy: Turkish and Tunisian Toscana virus are more closely related to Italian strains than to the Spanish ones. In contrast, Moroccan Toscana virus is more closely related to Spanish strains than to the Italian ones.[3] This dichotomy could be linked to the existence of different vectors (*P. perniciosus*, *P. perfiliewi*, other species of *Phlebotomus*) and/or to the existence of two geographically restricted races of *P. perniciosus*, the main vector of Toscana virus.[2] Interestingly, Toscana virus strains that circulate in France were reported to belong to either of these two lineages.[4] Recently, a new lineage of Toscana virus was described in Croatia suggesting a larger genetic diversity than initially believed, and underlining the need to increase the number of Toscana virus sequences and their geographic diversity.[5]

3. WHAT IS THE FREQUENCY OF THE DISEASE?

Because of the arthropod-borne nature of Toscana virus, at-risk areas depend on the distribution of the sandfly species that are capable of transmitting the virus.

Sandfly distribution is considerably influenced by climatic changes and environmental modifications,[6] and recent data suggest that there is tendency to northbound expansion into Austria, Germany, and Switzerland.[7]

Human cases have been laboratory documented using direct diagnosis (RT-PCR or virus isolation) in Italy, France, Spain, Cyprus, Greece, Portugal, and Croatia.[5,8,9] During the last decade, direct and indirect evidence was increasingly reported from the aforementioned regions, but also from regions where Toscana virus was unrecognized such as Morocco and Tunisia in northern Africa, Turkey, and the Mediterranean islands (Corsica, Malta, Elba, Sardinia, and Sicily).[10]

Recent studies also indicate that Toscana virus is also widely distributed in northern Africa and in Turkey. Since seroprevalence rates frequently reach 5% and can raise up to >50% depending on the region, Toscana virus is one of the most prevalent arboviruses in Europe.

Interestingly, there are no data from south-eastern Asian countries such as Taiwan, Hong Kong, and Malaysia, and no reports from Australia and New Zealand. Whether or not this accurately reflects the absence of Toscana virus in these regions remains to be investigated, since this could be falsely reassuring due to the lack of specific studies conducted in these regions.

To summarize the extent of recent knowledge about Toscana virus, direct detection was reported in Portugal, Spain, France, Italy, Croatia, Turkey, Morocco, and Tunisia; and indirect (serology-based) evidence was published in Portugal, Spain, France, Italy, Croatia, Greece, Bosnia-Herzegovina, Kosovo, Cyprus, Turkey, and Tunisia.

4. HOW IS THE VIRUS TRANSMITTED?

Toscana virus is transmitted to humans and other vertebrates when infected female sandflies take a blood meal. At present, there are no data to assess that either humans or other vertebrates are acting as reservoir for Toscana virus; it is commonly accepted that they are dead-end hosts, and thus do not play a significant role in nature in the virus life cycle. Considering the hypothesis, it seems reasonable to assume that the primary reservoir host is the sandfly in which the virus replicates. Detection of Toscana virus from female sandflies but also from males indicates that the virus is not exclusively acquired through blood meal, and that alternative modes of transmission between sandflies are to be considered. Transovarial and venereal transmission of Toscana virus were demonstrated at the vector level through experiments conducted in insectariums with colonized sandflies.[11–16] Currently, maintenance and transmission of Toscana virus in nature appears to depend on the presence of appropriate vector species and their abundance in a local environment. Studies aiming at a better characterization of the determinants driving the ecology of Toscana virus together with environmental parameters within a defined geographic area must be encouraged.

This knowledge is pivotal for virus control and could be beneficial both for public health and for predicting their potential for emergence in new regions.

5. WHICH FACTORS ARE INVOLVED IN DISEASE PATHOGENESIS? WHAT ARE THE PATHOGENIC MECHANISMS?

Various clinical forms of Toscana virus infection are described. They vary from mild febrile illness to peripheral or central neurological manifestations that can be very severe (encephalitis) although fatalities are exceptional. Non-symptomatic infections are suspected but there is no laboratory-documented case, and if they exist their frequency is unknown. The reasons for such a wide array of clinical features are unexplored.

6. WHAT ARE THE CLINICAL MANIFESTATIONS?

After a short incubation period (2–7 days), the onset is brutal with non-specific signs of viral febrile illness associated or not with central nervous system manifestations (aseptic meningitis or meningoencephalitis).[17,18] Neuroinvasive infections usually begin with headache, fever, nausea, vomiting, and myalgia. Physical examination may show neck rigidity, Kernig sign, and in some cases consciousness, tremors, paresis, and nystagmus. In most cases, the cerebrospinal fluid contains more than 5–10 cells/mL with normal levels of glucose and proteins. Leucopenia or leucocytosis can be observed. The outcome is usually favorable without reported sequelae. Although the vast majority of cases have a favorable outcome, some severe cases of Toscana virus infections were described.[8,9,18–20] Recent studies have reported a larger extent for neurological symptoms than initially identified. Other neurological manifestations such as deafness,[21,22] persistent personality alterations,[23] fasciitis and myositis,[24] and speech disorders and paresis[25] were reported.

7. HOW DO YOU DIAGNOSE?

Toscana virus infection can be diagnosed using virus isolation, serological tests, and molecular tests (RT-PCR). These methods are often combined in clinical microbiology laboratories. In the absence of direct evidence of Toscana virus (RT-PCR or virus isolation), demonstration of a seroconversion or presence of Toscana virus IgM in the CSF is needed to assess a confirmed case. To avoid misclassification of suspect Toscana virus cases, the case definition of the National Notifiable Diseases Surveillance System of the US Center for Disease Control and Prevention should be used.[26] As demonstrated from this case, virus isolation should be attempted whenever biological material is available since it is the gold standard technique. It is important to underline that commonly used serological tests to detect IgG and IgM such as immunofluorescence (IFA) and ELISA have a poor capacity to discriminate between Toscana virus and other related virus belonging to the same serocomplex (Naples, Massilia, Granada, Punique, Theran). Because of antigenic closeliness, cross-reactivity must be anticipated and ELISA/IFA results must be confirmed by neutralization tests. The increasing number of recently discovered sandfly-borne phleboviruses will render confirmation using neutralization assays more and more indispensable. There are several molecular tests that can be used for direct detection of the genome of Toscana virus in CSF, in serums (and also in sandflies).[4,27] It is important to underline that there is a striking need for standardized diagnostic kits (commercially available at affordable prices) to include the detection of Toscana virus in syndromes such as "fever of unknown origin" and "CNS infections," and to organize large-scale epidemiologic studies in at-risk regions.

8. HOW DO YOU DIFFERENTIATE THIS DISEASE FROM SIMILAR ENTITIES?

The few studies reporting series of cases are in agreement for denying the existence of specific characters of infection due to Toscana virus.[20,28,29] For instance, clinical and paraclinical signs are similar for aseptic meningitis due to enteroviruses and to Toscana virus. Only the seasonal aspect may be differential since Toscana virus cases can occur only when sandflies are circulating, which may drastically vary from one region to another, but can be roughly estimated between March and November with a clear peak in July and August. Other than neuroinvasive infections, summer fevers are generally not laboratory documented since these patients are not hospitalized and thus escape from subsequent specific virological investigations. Research projects focused on general medicine practices might be organized to study the incidence of this virus in febrile illness.

9. WHAT IS THE THERAPEUTIC APPROACH?

There are no approved drugs for treatment of Toscana virus infections. The majority of *in vitro* studies were done using Sicilian virus that is very distantly related to Toscana virus.[30,31]

None of the volunteers who were treated with ribavirin were infected when they were experimentally infected with Sicilian virus.[32] The pyrazine derivatives T-705 and T-1106 showed *in vitro* activity against Naples virus with a lower toxicity than ribavirin.[33,34]

10. WHAT ARE THE PREVENTIVE AND INFECTION CONTROL MEASURES?

There is no licensed vaccine for Toscana virus. Repellents, insecticides, and impregnated bed nets may be efficient at the individual level, but rarely used because sandflies are commonly unnoticed.[35–38] Insecticide spraying significantly decreases the incidence of sandfly-borne diseases only if the spraying is continuous, whereas sporadic campaigns are considered to be ineffective. Alternatives such as insecticide-impregnated curtains,[39] insecticide-impregnated dog collars,[40] insecticide-treated sugar baits, pheromone dispenser baits,[41,42] and cultivation of noxious plants against sandflies[43] have been recently developed and are being studied.

REFERENCES

1. Nougairede A, Bichaud L, Thiberville SD, et al. Isolation of Toscana virus from the cerebrospinal fluid of a man with meningitis in Marseille, France, 2010. *Vector Borne Zoonotic Dis* 2013;**13**:685–8.
2. Sanbonmatsu-Gámez S, Pérez-Ruiz M, Collao X, et al. Toscana virus in Spain. *Emerg Infect Dis* 2005;**11**:1701–7.
3. Ergünay K, Saygan MB, Aydoğan S, et al. Sandfly fever virus activity in central/northern anatolia, Turkey: first report of Toscana virus infections. *Clin Microbiol Infect* 2011;**17**:575–81.
4. Charrel RN, Izri A, Temmam S, et al. Cocirculation of 2 genotypes of Toscana virus, southeastern France. *Emerg Infect Dis* 2007;**13**:465–8.
5. Punda-Polić V, Mohar B, Duh D, et al. Evidence of an autochthonous Toscana virus strain in Croatia. *J Clin Virol* 2012;**55**:4–7.
6. Weaver SC, Reisen WK. Present and future arboviral threats. *Antiviral Res* 2010;**85**:328–45.
7. Fischer D, Moeller P, Thomas SM, Naucke TJ, Beierkuhnlein C. Combining climatic projections and dispersal ability: a method for estimating the responses of sandfly vector species to climate change. *PLoS Negl Trop Dis* 2011;**5**:e1407.
8. Depaquit J, Grandadam M, Fouque F, Andry PE, Peyrefitte C. Arthropod-borne viruses transmitted by Phlebotomine sandflies in Europe: a review. *Euro Surveill* 2010;**15**:19507.
9. Charrel RN, Gallian P, Navarro-Mari JM, et al. Emergence of Toscana virus in Europe. *Emerg Infect Dis* 2005;**11**:1657–63.
10. Alkan C, Bichaud L, de Lamballerie X, Alten B, Gould EA, Charrel RN. Sandfly-borne phleboviruses of Eurasia and Africa: epidemiology, genetic diversity, geographic range, control measures. *Antiviral Res* 2013;**100**:54–74.
11. Ciufolini MG, Maroli M, Guandalini E, Marchi A, Verani P. Experimental studies on the maintenance of Toscana and Arbia viruses (Bunyaviridae: Phlebovirus). *Am J Trop Med Hyg* 1989;**40**:669–75.
12. Ciufolini MG, Maroli M, Verani P. Laboratory reared sandflies (Diptera: Psychodidae) and studies on phleboviruses. *Parassitologia* 1991;**33**(Suppl):137–42.
13. Ciufolini MG, Maroli M, Verani P. Growth of two phleboviruses after experimental infection of their suspected sand fly vector, Phlebotomus perniciosus (Diptera: Psychodidae). *Am J Trop Med Hyg* 1985;**34**:174–9.
14. Maroli M, Ciufolini MG, Verani P. Vertical transmission of Toscana virus in the sandfly, Phlebotomus perniciosus, via the second gonotrophic cycle. *Med Vet Entomol* 1993;**7**:283–6.
15. Tesh RB, Modi GB. Maintenance of Toscana virus in Phlebotomus perniciosus by vertical transmission. *Am J Trop Med Hyg* 1987;**36**:189–93.

16. Tesh RB, Lubroth J, Guzman H. Simulation of arbovirus overwintering: survival of Toscana virus (Bunyaviridae: Phlebovirus) in its natural sand fly vector Phlebotomus perniciosus. *Am J Trop Med Hyg* 1992;**47**:574–81.
17. Dionisio D, Valassina M, Ciufolini MG, et al. Encephalitis without meningitis due to sandfly fever virus serotype toscana. *Clin Infect Dis* 2001;**32**:1241–3.
18. Baldelli F, Ciufolini MG, Francisci D, et al. Unusual presentation of life-threatening Toscana virus meningoencephalitis. *Clin Infect Dis* 2004;**38**:515–20.
19. Cusi MG, Savellini GG, Zanelli G. Toscana virus epidemiology: from Italy to beyond. *Open Virol J* 2010;**4**:22–8.
20. Vocale C, Bartoletti M, Rossini G, et al. Toscana virus infections in northern Italy: laboratory and clinical evaluation. *Vector Borne Zoonotic Dis* 2012;**12**:526–9.
21. Martinez-Garcia FA, Moreno-Docon A, Segovia-Hernandez M, Fernandez-Barreiro A. Deafness as a sequela of Toscana virus meningitis. *Med Clin (Barc)* 2008;**130**:639.
22 Paul C, Schwarz TF, Meyer CG, Jager G. Neurological symptoms after an infection by the sandfly fever virus. *Dtsch Med Wochenschr* 1995;**120**:1468–72.
23. Serata D, Rapinesi C, Del Casale A, et al. Personality changes after Toscana virus encephalitis in a 49-year-old man: a case report. *Int J Neurosci* 2011;**121**:165–9.
24. Doudier B, Ninove L, Million M, de Lamballerie X, Charrel RN, Brouqui P. Unusual Toscana virus encephalitis in southern France. *Med Mal Infect* 2011;**41**:50–1.
25. Sanbonmatsu-Gamez S, Perez-Ruiz M, Palop-Borras B, Navarro-Mari JM. Unusual manifestation of Toscana virus infection, Spain. *Emerg Infect Dis* 2009;**15**:347–8.
26. Anonymous. *Case definitions: nationally notifiable conditions infectious and non-infectious case*. Atlanta, GA: Centers for Disease Control and Prevention; 2012.
27. Cusi MG, Savellini GG. Diagnostic tools for Toscana virus infection. *Expert Rev Anti Infect Ther* 2011;**9**:799–805.
28. Jaijakul S, Arias CA, Hossain M, Arduino RC, Wootton SH, Hasbun R. Toscana meningoencephalitis: a comparison to other viral central nervous system infections. *J Clin Virol* 2012;**55**:204–8.
29. de Ory F, Avellón A, Echevarría JE, et al. Viral infections of the central nervous system in Spain: a prospective study. *J Med Virol* 2013;**85**:554–62.
30. Kirsi JJ, North JA, McKernan PA, et al. Broad-spectrum antiviral activity of 2-beta-D-ribofuranosylselenazole-4-carboxamide, a new antiviral agent. *Antimicrob Agents Chemother* 1983;**24**:353–61.
31. Crance JM, Gratier D, Guimet J, Jouan A. Inhibition of sandfly fever Sicilian virus (Phlebovirus) replication in vitro by antiviral compounds. *Res Virol* 1997;**148**:353–65.
32. Huggins JW. Prospects for treatment of viral hemorrhagic fevers with ribavirin, a broad-spectrum antiviral drug. *Rev Infect Dis* 1989;**11**(Suppl. 4):S750–61.
33. Gowen BB, Wong MH, Jung KH, et al. In vitro and in vivo activities of T-705 against arenavirus and bunyavirus infections. *Antimicrob Agents Chemother* 2007;**51**:3168–76.
34. Gowen BB, Wong MH, Jung KH, Smee DF, Morrey JD, Furuta Y. Efficacy of favipiravir (T-705) and T-1106 pyrazine derivatives in phlebovirus disease models. *Antiviral Res* 2010;**86**:121–7.
35. Alten B, Caglar SS, Kaynas S, Simsekand FM. Evaluation of protective efficacy of K-OTAB impregnated bednets for cutaneous leishmaniasis control in Southeast AnatoliaTurkey. *J Vector Ecol* 2003;**28**:53–64.
36. Elnaiem DA, Elnahas AM, Aboud MA. Protective efficacy of lambdacyhalothrin-impregnated bednets against Phlebotomus orientalis, the vector of visceral leishmaniasis in Sudan. *Med Vet Entomol* 1999;**13**:310–4.

37. Jalouk L, Al Ahmed M, Gradoni L, Maroli M. Insecticide-treated bednets to prevent anthroponotic cutaneous leishmaniasis in aleppo governate, Syria: results from two trials. *Trans R Soc Trop Med Hyg* 2007;**101**:360–7.
38. Faiman R, Cuño R, Warburg A. Control of phlebotomine sand flies with vertical fine-mesh nets. *J Med Entomol* 2009;**46**:820–31.
39. Maroli M, Majori G. Permethrin-impregnated curtains against phlebotomine sand flies (Diptera: Psychodidae): laboratory and field studies. *Parassitologia* 1991;**33**:399–404.
40. Killick-Kendrick R, Killick-Kendrick M, Focheux C. Protection of dogs from the bites of phlebotomine sand flies by deltamethrin collars for the control of canine leishmaniasis. *Med Vet Entomol* 1997;**11**:105–11.
41. Bray DP, Alves GB, Dorval ME, Brazil RP, Hamilton JG. Synthetic sex pheromone attracts the leishmaniasis vector Lutzomyia longipalpis to experimental chicken sheds treated with insecticide. *Parasit Vectors* 2010;**3**:16.
42. Bray DP, Bandi KK, Brazil RP, Oliveira AG, Hamilton JG. Synthetic sex pheromone attracts the leishmaniasis vector Lutzomyia longipalpis (Diptera: Psychodidae) to traps in the field. *J Med Entomol* 2009;**46**:428–34.
43. Schlein Y, Jacobson RL, Müller GC. Sand fly feeding on noxious plants: a potential method for the control of leishmaniasis. *Am J Trop Med Hyg* 2001;**65**:300–3.

Chapter 9

Ebola Virus Disease

Pierre Formenty
Emerging and Epidemic Zoonotic Diseases Team (CED/EZD), World Health Organization

CASE PRESENTATION

In the Taï National Park, Côte d'Ivoire, the behavior of a community of free-living chimpanzees has been studied since 1979[1]. In early November 1994, several decomposed corpses of chimpanzees were found and one chimpanzee that had died less than 12 hours previously was dissected on November 16 by three research workers. There were signs of hemorrhage and non-clotting blood. Eight days later, one of the researchers, a 34-year-old woman, became ill.

CLINICAL COURSE

On November 24 (day 1), around 6 pm, the patient started shivering with fever (Figure 9.1). She took a curative dose of halofantrine for suspected malaria. On the third day, as there was no notable improvement, the patient was transported by car to Abidjan (600 km) and admitted to a clinic. Despite persistent chills, headache, and myalgia her general condition was satisfactory. Physical examination of the abdomen, heart, lung, throat, and tongue was normal.

On day 4, she was treated with intravenous quinine (1.6 g daily) for suspected malaria. Her temperature remained around 40°C and quinine was discontinued due to progressive deafness.

On day 5 the patient developed diarrhea (seven stools/day) without blood traces, then nausea, vomiting, and anorexia. A non-itching rash developed on her left shoulder, spread to her back, and finally became generalized. She also suffered from central nervous disorders such as temporary loss of memory, anxiety, confusion, and irritability. Urinary output failed from day 5 to day 7. Repeated blood examinations did not reveal parasites, and blood cultures remained negative. She was rehydrated with Ringer-lactate solutions and antibiotics were given, initially pefloxacin per os, then amoxicillin

Emerging Infectious Diseases. DOI: http://dx.doi.org/10.1016/B978-0-12-416975-3.00009-1

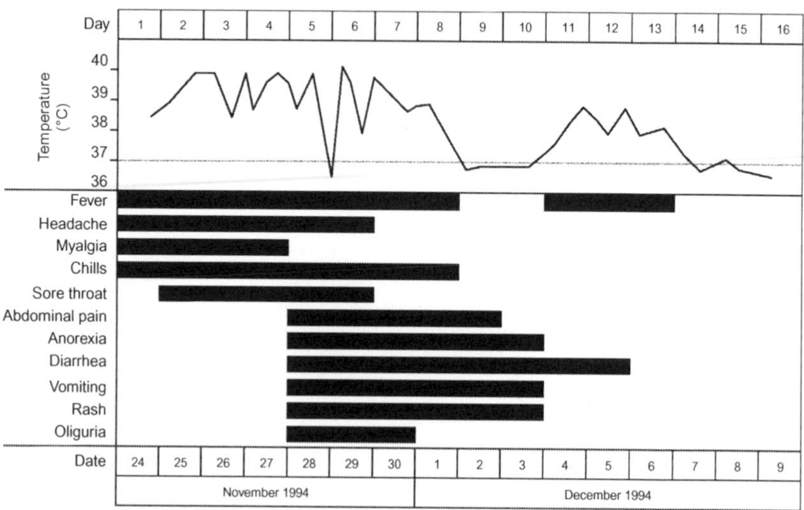

FIGURE 9.1 Clinical course of the disease in a patient with Taï Forest virus disease.

clavulanate iv. A chest X-ray was normal. A diagnosis of Gram-negative sepsis was evoked.

On day 6 there was no improvement in the patient's condition and she was repatriated to Switzerland on day 7 in a Swiss Air Ambulance jet. Due to geographical closeness with Liberia, Lassa fever was considered. During the flight, the patient wore a mask and the physician and nurse wore surgical masks, gloves, and gowns.

On day 8, she was admitted to the University Hospital of Basel. She was transferred to a single double-door isolation room with negative pressure. All healthcare workers wore high-quality gloves, gowns, and dust-and-mist masks (3 M). Because her illness started a week after the autopsy of a non-human primate, infection control practice did comply with the CDC guidelines.[2] On admission, she was tired but awake. Physical examination revealed tender spleen and liver on palpation. An ultrasound scan of the abdomen was normal. The differential diagnosis included a viral or bacterial infection of unknown origin, dengue fever, rickettsial disease, hantavirus infection, leptospirosis, typhoid fever, and malaria. A form of hemorrhagic fever (Lassa fever or Ebola) was considered as unlikely. She was treated with ciprofloxacin iv and doxycycline iv for suspected Gram-negative sepsis, leptospirosis, or rickettsial disease.

On day 9 she became afebrile for 2 days; therefore, clothes were disposed of as usual in the double-bag technique and used dishes were returned to the kitchen. On day 11, strict barrier isolation was reduced to body substance isolation practice. The same day, fever recurred. Diarrhea changed to constipation and she started to eat normally.

On day 15, she was discharged from the hospital having lost 6 kg (10% of her initial weight). She only fully recovered after 6 weeks. A month after onset, the patient's hair became dry, lost its elasticity, and began falling out in large quantities. Hair loss lasted about 3 months.

LABORATORY INVESTIGATIONS

Persistent thrombocytopenia and an early lymphopenia followed by neutrophilic hyperleucocytosis were observed. The coagulation profile and platelets count evoked an intravascular coagulation. The most notable biochemical finding was ASAT activity three-fold higher than ALAT activity. Gamma glutamyltransferase and alkaline phosphatases were normal and slightly elevated, respectively. Creatinine kinase was slightly above normal on day 8, but was normal thereafter. Lactate dehydrogenase was 20 times normal on day 7. Amylase levels increased from day 11 to 14. As of day 15, all values progressively stabilized.

Tests for parasites in blood and stool, blood cultures, and cultures of urine were negative. Urine samples were examined on days 5, 10, and 12: hematuria was observed on day 5 and proteinuria on day 10. Electrolyte levels (sodium, potassium, calcium, phosphate) and lipid levels (triglycerides and cholesterol) were investigated from day 7 to 14 and were normal.

While in the Swiss hospital, serological tests for dengue, hantavirus, hepatitis B, C, and E, leptospirosis, *Rickettsia mooseri* and *R. conori*, and brucellosis were negative. Epstein–Barr and cytomegalovirus were IgG positive but IgM negative. Due to the lack of bleeding, diagnosis of hemorrhagic fever (Lassa or Ebola) had not been required.

VIROLOGICAL INVESTIGATION

The first days of December 1994, an epidemiological investigation was undertaken to elucidate the cause of mortalities within the community of chimpanzees under study.[3] The survey indicated a highly lethal epidemic with hemorrhagic syndrome. Etiology was unknown and differential diagnosis included a viral infection of unknown origin, anthrax, dengue, and the African hemorrhagic fevers. Thinking that the patient could have been contaminated during the autopsy, we asked the Ivoirian clinic for a patient serum in order to conduct a proper investigation. The patient serum was sent to the Pasteur Institute on December 14. We asked for different diagnostics and, notably, anthrax, dengue, Congo-Crimean hemorrhagic fever, Rift Valley fever, yellow fever, Chikungunya, Lassa, Ebola, Marburg, and Hantaviruses. Viral isolation was attempted on patient serum because it had been taken during the febrile phase of illness on November 26 (day 3).

A novel strain of the Ebola virus named "Côte d'Ivoire" was isolated;[4] the virus has been recently renamed Taï Forest virus (TAFV). Antibodies and antigen titers in sera were determined by indirect fluorescent antibody (IFA)[5] test and enzyme-linked immunosorbent assay (ELISA).[6,7]

(This is a published case report, *The Journal of Infectious Diseases* 1999.[4])

1. BRIEF JUSTIFICATION ON WHY THIS CASE WAS IDENTIFIED AS EMERGENT

In November 1994, when the case described above was reported from Côte d'Ivoire,[8] it was the first case of Ebola virus disease (EVD) reported since the case of Tandala in 1979 in the Democratic Republic of Congo. After 15 years of epidemiological silence, Ebola was re-emerging in Africa and for the first time in West Africa. This case was also the first documented human infection associated with naturally infected non-human primates in Africa.

2. WHAT IS THE CAUSATIVE AGENT?

Genus *Ebolavirus* is one of the three members of the *Filoviridae* family (filovirus) together with genus *Marburgvirus* and genus *Cuevavirus.* All are enveloped, single-strand, negative-sense RNA viruses. Genus *Ebolavirus* is comprised of five distinct species: species *Bundibugyo ebolavirus*, virus: Bundibugyo virus (BDBV); species *Zaire ebolavirus*, virus: Ebola virus (EBOV); species *Reston ebolavirus*, virus: Reston virus (RESTV); species *Sudan ebolavirus*, virus: Sudan virus (SUDV); and species *Taï Forest ebolavirus*, virus: Taï Forest virus (TAFV).[9]

BDBV, EBOV, and SUDV have been associated with large Ebola virus disease (EVD) outbreaks in Africa, whereas RESTV and TAFV have not. EVD is a febrile hemorrhagic illness, which causes death in 25–90% of all clinically ill cases.

TAFV has thus far caused only one human infection in Africa (described in section 1) and the patient survived. The patient's mild form of the disease may be due to the virus strain and/or to the mode of contamination and/or to the biological response of the patient. Considering the high mortality rate observed among chimpanzees (25%) during the TAFV event in 1994,[3] TAFV should be considered as potentially highly pathogenic in humans with the potential to cause severe EVD outbreaks.

The RESTV found in the Philippines can infect humans but no illness or death in humans has been reported to date. Among workers in contact with monkeys or pigs infected with Ebola Reston, several human infections have been documented and were clinically asymptomatic.[10] RESTV appears to be less capable of causing disease in humans than the other Ebola species. However, the evidence available relates only to healthy adult males. It would be premature to conclude the health effects of the virus on all population groups, such as immunocompromised persons, persons with underlying medical conditions, pregnant women, and children. More studies of RESTV are needed before definitive conclusions can be made about the pathogenicity and virulence of this virus in humans. In 2009, a group of experts consulted by WHO conclude that RESTV should be considered potentially pathogenic for humans.[11]

3. WHAT IS THE FREQUENCY OF THE DISEASE?

Ebola first appeared in 1976 in two simultaneous outbreaks, in Nzara, Sudan, and in Yambuku, Democratic Republic of Congo (DRC).[12,13] From 1976 to December 2012 a total of 23 outbreaks or isolated cases of Ebola have been reported; during these events a total of 2388 Ebola cases including 1590 deaths were reported.[14]

Since its discovery in 1976, Ebola virus disease (EVD) has mostly occurred in sub-Saharan Africa. Sudan (1976, 1979, 2004), Democratic Republic of Congo (DRC) (1976, 1977, 1995, 2007, 2008), Gabon (1994, 1996, 2001, 2002), Uganda (2000, 2007, 2011, 2012), and Republic of the Congo (2001, 2002, 2003, 2005) have reported EVD epidemics. Côte d'Ivoire (1994) reported one case of Ebola Taï Forest virus, a female research worker who was infected when performing an autopsy on an infected chimpanzee. No secondary transmission occurred and the patient survived the infection. In addition, serosurveys conducted in human or animal populations reported serological evidence of Ebola infection in the following African countries: Guinea, Sierra Leone, Liberia, Nigeria, Cameroon, Chad, Central African Republic, Ethiopia, and Madagascar.

Studies have reported serological evidence of Ebola virus infection in orangutans in Indonesia;[15] in several bat species in China,[16] and in *Rousettus leschenaultia* fruit bats, Bangladesh.[17] In addition, there has been the identification of genetically distinct filovirus—provisionally named Lloviu virus—in dead insectivorous bats in caves in Spain.[18] Reston virus infection was confirmed by reverse transcriptase polymerase chain reaction (RT-PCR) on samples collected from domestic pigs from three farms in Shanghai, China, during February to September 2011.[19]

As more and more studies are finding evidence of filovirus infection in bats from Africa to Asia, the geographic distribution of Ebola and Marburg virus should be considered to overlap with the range of fruit bats of the Pteropodidae family (Figure 9.2).

4. WHAT ARE THE TRANSMISSION ROUTES?

4.1 Reported Routes of Transmission

Ebola is first introduced into the human population through close contact with the blood, secretions, organs, or other bodily fluids of infected animals (Figure 9.3). After analysis of 23 Ebola outbreaks from 1976 to 2012, it appears that, in Africa, primary human infection (index case) was documented

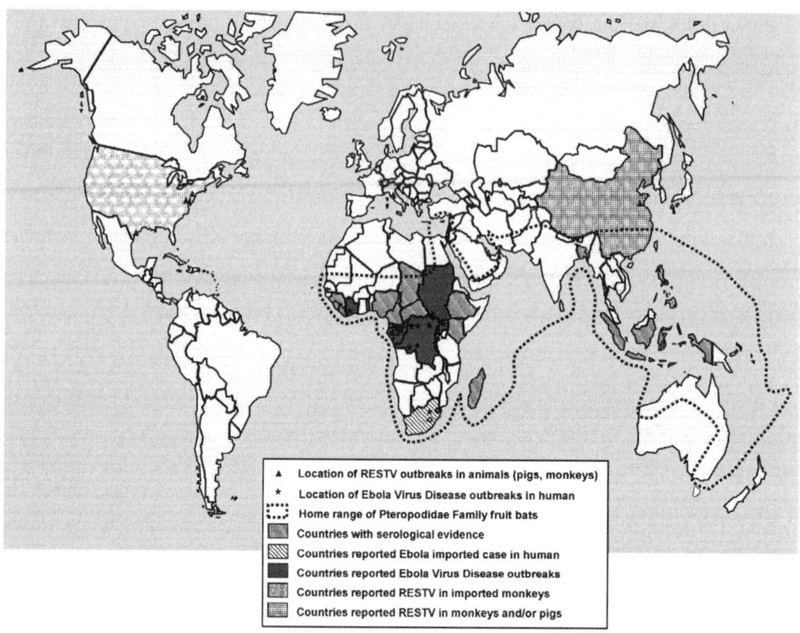

FIGURE 9.2 Geographic distribution of Ebola virus disease outbreaks.

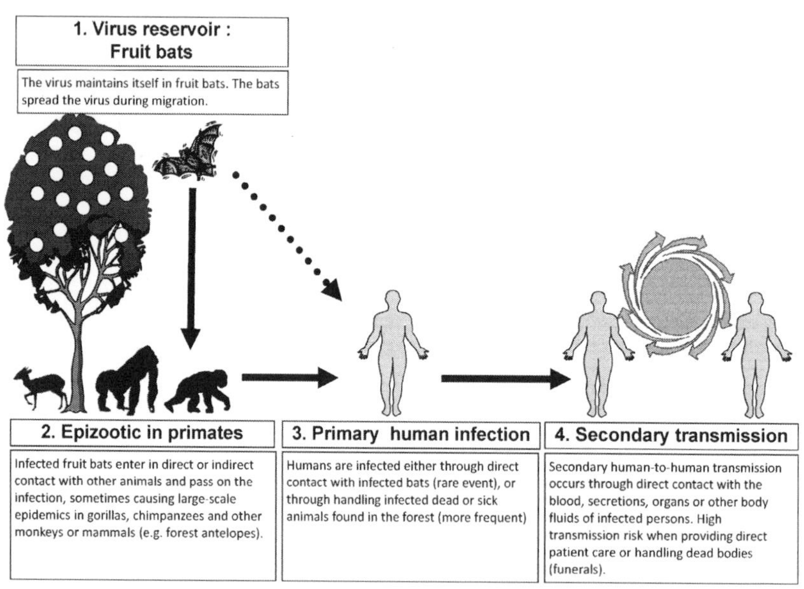

FIGURE 9.3 Hypothesis of Ebola virus transmission at the human–animal interface.

through the handling of infected chimpanzees (6 events), gorillas (4), monkeys (4), forest antelopes (3), fruit bats (1), and porcupines (1) found dead or ill in the rainforest.[20] For some EVD events there were several introductions in the human population.[21] Of note, it was not possible to identify the source of primary human infection for 13 events out of 23. In Africa, fruit bats, particularly species of the genera *Hypsignathus monstrosus*, *Epomops franqueti*, and *Myonycteris torquata*, are considered possible natural hosts for Ebola virus.[22]

Secondary human-to-human transmission spreads in the community resulting from close contact with the blood, secretions, organs, or other bodily fluids of infected people. Burial ceremonies where mourners have direct contact with the body of the deceased person can also play a role in the transmission of Ebola. Transmission via infected semen can occur up to 7 weeks after clinical recovery.

Healthcare workers have frequently been infected while treating Ebola patients. This has occurred through close contact without the use of correct infection control precautions and adequate barrier nursing procedures. For example, healthcare workers not wearing gloves and/or masks and/or goggles and who are exposed to direct contact with infected patients' blood are at risk.

Among workers in contact with monkeys or pigs infected with RESTV, several human infections have been documented and were reported to be clinically asymptomatic.

4.2 Possible Other Routes of Transmission

Reston virus (RESTV) infection discovery in pigs in the Philippines in 2008 marked the first time this virus was found in pigs.[23,24] An intensive collaborative effort of animal health and public health authorities, at national, regional, and international levels, was necessary to assess the risk of RESTV in animal and humans. Possible routes of infection from pigs to humans include: transmission when handling infected pigs in farms, during slaughtering activities and preparation of meat, food chain, xenotransplantation, vaccines, cosmetics, etc.

Experimental infection demonstrated that pigs replicated Zaire ebolavirus (ZEBOV) to high titers mainly in the respiratory tract, and thus identified an unexpected site of virus amplification and shedding linked to possible virus transmission.[25] Additional experimental infection studies showed ZEBOV transmission from pigs to cynomolgus macaques without direct contact.[26]

Given results of recent experimental inoculations, pigs should be considered as a potential amplifier host during Ebola outbreaks in Africa and Asia. Although no other domestic animals have yet been confirmed as having an association with filovirus outbreaks, as a precautionary measure they should be considered as potential amplifier hosts until proven otherwise.

New modes of infection and transmission are possible for Ebola at the human–animal interface and need to be investigated during future outbreaks. If confirmed, this will impact prevention and control measures during Ebola outbreaks.

5. WHAT ARE THE CLINICAL MANIFESTATIONS?

The clinical manifestations of EVD were described during the initial outbreaks in DRC and Sudan in 1976; more details were reported for a large number of patients after the 1995 Kikwit epidemic, and shorter descriptions have appeared for subsequent outbreaks.[27,28] It seems that different species of Ebola virus might cause somewhat different clinical manifestations.

Generally, Ebola virus disease is a severe acute viral illness often characterized by the sudden onset of fever followed by a 2–3-day initial period with non-specific symptoms: fever, severe headache, muscle pain, intense weakness, sore throat, and sometimes conjunctival injection. This is followed by a 2–4-day deteriorating period with severe sore throat, chest and abdominal pain, maculopapular skin rash on the trunk and shoulders, diarrhea, vomiting, impaired kidney and liver function, and in some cases both internal and external bleeding. During this phase laboratory findings show low counts of white blood cells and platelets as well as elevated liver enzymes.

For fatal cases, the 2–4-day terminal period is characterized by hemorrhage, hiccups, somnolence, delirium, and coma. Bleeding is manifested as maculopapular skin rash, petechiae, ecchymosis, uncontrolled bleeding from venepuncture sites, and postmortem evidence of visceral hemorrhagic effusions. Abortion is a common consequence of infection, and infants born from mothers dying of Ebola infection are fatally infected. Death in shock usually occurs 6–9 days after onset of clinical disease.

During the EVD initial period, blood sampling shows that all sick individuals have circulating viral antigen. However, recent reports of EBOV and SUDV have shown that fatal illness is associated with high and increasing amounts of virus in the bloodstream. Conversely, patients who survive infection show a decrease in circulating virus with clinical improvement around day 7–10. In most cases, this improvement coincides with the appearance of Ebola virus-specific antibodies.[29,30]

Although Ebola virus disappears quickly from the bloodstream of recovering patients, it was isolated from seminal fluid of a few patients up to the 61st day after onset of illness in a laboratory acquired case.

The EVD incubation period (interval from infection to onset of symptoms) varies between 2 and 21 days (mean 4–10 days). During EVD outbreaks, the case-fatality rate has varied from outbreak to outbreak between 25 and 90%.

6. HOW DO YOU DIAGNOSE?

Ebola virus infections can only be diagnosed definitively in the laboratory by a number of different tests:[31,32]

- ELISA
- Antigen detection tests
- Serum neutralization test

- RT-PCR assay
- Virus isolation by cell culture

Tests on samples from patients are an extreme biohazard risk and should only be conducted under maximum biological containment conditions.

7. HOW DO YOU DIFFERENTIATE THIS DISEASE FROM SIMILAR ENTITIES?

EVD differential diagnoses include malaria, typhoid fever, shigellosis, cholera, leptospirosis, plague, rickettsiosis, relapsing fever, meningitis, hepatitis, and other viral hemorrhagic fevers (VHFs).

The discovery of TAFV in November 1994 (see section 1) emphasizes the difficulty of diagnosing new tropical diseases. In tropical Africa, many cases present malaria-like syndromes and the majority are caused by malaria. The others cases can be bacterial (i.e., thypoid fever, typhus, leptospirosis) or viral. If malaria tests are negative and antibiotic treatment and blood cultures unsuccessful, then an arboviral or hemorrhagic fever infection has to be considered. More than 100 arboviruses can cause human infections, but the number of tests required can be reduced according to the geographical origin of the infection and the clinical signs. Furthermore, few viruses present a risk of mortality and the need for isolation measures: Lassa, Ebola, Marburg, Rift Valley fever, and Crimean-Congo hemorrhagic fever in Africa.

In addition, infection by Ebola viruses is not always associated with hemorrhages. The absence of bleeding is insufficient to reject this etiology. Less than 40% of Ebola infections present with gum bleeding, petechia, hematemesis, or melena. Beside non-specific but constant symptoms, high fever, myalgia, headache, and nausea, the most predictive signs for Ebola and Marburg infection is abdominal pain often accompanied by diarrhea.

In conclusion, the laboratory diagnostic remains essential to confirm EVD cases.

8. WHAT IS THE THERAPEUTIC APPROACH?

There is no specific treatment or vaccine for Ebola. Severely ill patients must be given symptomatic treatment and intensive care. The therapeutic approach is based on:

- Palliative care: rehydration is essential (oral or others depending on circumstances), maintenance of electrolyte balance (for example, with a potassium supplement), kidney and liver function support
- Symptomatic treatment: painkillers, anti-emetic against vomiting, anxiolytics to combat anxiety, antibiotics, antimalarial remedies
- Intensive care: use of oxygen

- In the event of severe bleeding and if intravenous therapy is an option: transfusion of blood or previously tested blood components (red blood cells, platelet concentrates, fresh frozen plasma)
- Use of equipment to monitor biochemical and blood values of patients to maintain the electrolyte balance
- The use of products containing salicylates (i.e., acetylsalicylic acid/aspirin) or other non-steroidal anti-inflammatory drugs (NSAIDS) is prohibited
- At the current state of knowledge, serotherapy is not recommended for the treatment of Ebola or Marburg
- Outcome of the laboratory diagnostic test for Ebola (important for classifying suspected cases in the field, for prognosis when the presence of antibodies is detected, and for discharge of patients)

Several candidate vaccines are being developed,[33] but it will be another few years until outbreak response teams working in the field will have vaccines available. Similarly, several candidate drugs show promise but their safety and efficacy in humans is not yet known. The new vaccines and therapies against Ebola and Marburg (i.e., antivirals, vaccines) have not yet been released for large-scale use in the field. A strategy involving post-exposure immunization and/or treatment with recombinant vaccines and/or monoclonal antibodies (e.g., after contact with an Ebola case or exposure to the virus) may be proposed soon.[34,35] But the research protocols, insofar as pre-established protocols exist, need to be submitted to the ethical research committees of the countries concerned prior to their implementation.

9. WHAT ARE THE PREVENTIVE AND INFECTION CONTROL MEASURES?

9.1 Preventive Measures

9.1.1 Controlling Ebola in Domestic Animals

There is no animal vaccine available against RESTV. Routine cleaning and disinfection of pig or monkey farms (with sodium hypochlorite or other detergents) is expected to be effective in inactivating the virus. If an RESTV outbreak is suspected, the premises should be quarantined immediately. Restricting or banning the movement of animals from infected farms to other areas can reduce the spread of the disease.

For the Philippines, as RESTV outbreaks in pigs and monkeys have preceded human infections, the establishment of an active animal health surveillance system to detect new cases is essential in providing early warning for veterinary and human public health authorities.

Experimental inoculations in pigs with different Ebola viruses have been reported and show that pigs are susceptible to filovirus infection and shed the virus. Therefore, pigs should be considered as a potential amplifier host during Ebola outbreaks in Africa and Asia. Although no other domestic animals have yet been confirmed as having an association with EVD outbreaks, as a precautionary measure they should be considered as potential amplifier hosts until proven otherwise. Precautionary measures are needed in pig farms in Africa to avoid pigs becoming infected through contact with fruit bats. Such infection could potentially amplify the virus and cause or contribute to Ebola outbreaks.

9.1.2 Reducing the Risk of Ebola Infection in People

In the absence of effective treatment and a human vaccine, raising awareness of the risk factors of Ebola infection and the protective measures individuals can take is the only way to reduce human infection and death.[36–38]

In Africa, during EVD outbreaks, educational public health messages for risk reduction should focus on several factors:

- Reducing the risk of wildlife-to-human transmission from contact with infected fruit bats or monkeys/apes and the consumption of their raw meat. Animals should be handled with gloves and other appropriate protective clothing. Their products (blood and meat) should be thoroughly cooked before consumption.
- Reducing the risk of human-to-human transmission in the community arising from direct or close contact with infected patients, particularly with their bodily fluids. Close physical contact with Ebola patients should be avoided. Gloves and appropriate personal protective equipment should be worn when taking care of ill patients at home. Regular hand washing is required after visiting sick relatives in hospital, as well as after taking care of ill patients at home.
- Communities affected by Ebola should inform the population about the nature of the disease and about outbreak containment measures, including burial of the deceased. People who have died from Ebola should be promptly and safely buried.
- Precautionary measures are needed in Africa to avoid pig farms infected through contact with fruit bats amplifying the virus and causing EHF outbreaks.

For RESTV, educational public health messages should focus on reducing the risk of pig-to-human transmission as a result of unsafe animal husbandry and slaughtering practices, and unsafe consumption of fresh blood, raw milk, or animal tissue. Gloves and other appropriate protective clothing should be worn when handling sick animals or their tissues or when slaughtering animals. In the regions where RESTV has been reported/detected in pigs, all animal products (blood, meat, and milk) should be thoroughly cooked before eating.

9.2 Infection Control Measures

Human-to-human transmission of the Ebola virus is primarily associated with direct contact with blood and bodily fluids. Transmission to healthcare workers has been reported when appropriate infection control measures have not been observed. During Ebola outbreaks, only strict compliance with biosafety guidelines (appropriate laboratory practices, infection control precautions, barrier nursing procedures, use of personal protective equipment by healthcare workers handling patients, disinfection of contaminated objects and areas, safe burials, etc.) can prevent the epidemic from spreading and reduce the number of victims.

Healthcare workers caring for patients with suspected or confirmed Ebola virus should apply infection control precautions to avoid any exposure to the patient's blood and bodily fluids and/or direct unprotected contact with the possibly contaminated environment. Therefore, the provision of healthcare for suspected or confirmed Ebola patients requires specific control measures and the reinforcement of standard precautions, particularly basic hand hygiene, the use of personal protective equipment, safe injections practices, and safe burial practices.[39,40]

Laboratory workers are also at risk. Samples taken from suspected human and animal Ebola cases for diagnosis should be handled by trained staff and processed in suitably equipped laboratories.

REFERENCES

1. Boesch C, Boesch H. Hunting behavior of wild chimpanzees in the Taï National Park. *Am J Phys Anthropol* 1989;**78**:547–73.
2. Centers for Disease Control. Management of patient with suspected viral hemorrhagic fever. *Morbility and Mortality Weekly Report* 1988;(Supplement, 27/s–3):1–16.
3. Formenty P, Boesch C, Wyers M, Steiner C, Donati F, Dind F, et al. Ebola outbreak among wild chimpanzees living in a rain forest of Côte-d'Ivoire. *J Infect Dis* 1999;**179**(Suppl. 1): S120–6.
4. Formenty P, Hatz C, Le Guenno B, Stoll A, Rogenmoser Ph, Widmer AF. Human infection due to Ebola virus, subtype Côte-d'Ivoire: clinical and biologic presentation. *J Infect Dis* 1999;**179**(Suppl. 1):S48–53.
5. Johnson KM, Elliot LH, Heymann DL. Preparation of polyvalent viral immunofluorescent intracellular antigens and use in human serosurveys. *J Clin Microbiol* 1981;**14**:527–9.
6. Kziazek TG. Laboratory diagnostic of filovirus infections in nonhuman primates. *Lab Anim* 1991;**30**:34–46.
7. Kziazek T, Rollin P, Jahrling P, Johnson E, Dalgard D, Peters CJ. Enzyme immunosorbent assay for Ebola virus antigens in tissues of infected primates. *J Clin Microb* 1992;**30**: 947–50.
8. Le Guenno B, Formenty P, Wyers M, Gounon P, Walker F, Boesch C. Isolation and partial characterization of a new strain of Ebola virus. *Lancet* 1995;**1**:664–6.

9. Kuhn JH, Bao Y, Bavari S, Becker S, Bradfute S, Brister JR, et al. Virus nomenclature below the species level: a standardized nomenclature for natural variants of viruses assigned to the family Filoviridae. *Arch Virol* 2013;**158**:301–11.
10. Miranda ME, Miranda NL. Reston ebolavirus in humans and animals in the Philippines: a review. *J Infect Dis* 2011;**204**(Suppl. 3):S757–60.
11. World Health Organization. WHO experts consultation on Ebola Reston pathogenicity in humans. (WHO/HSE/EPR/2009.2). 2009, Geneva Switzerland. 25 pages.
12. WHO/International Study Team. Ebola haemorrhagic fever in Sudan, 1976. *Bull World Health Organ* 1978;**56**:247–70.
13. WHO/International Study Team. Ebola haemorrhagic fever in Zaire, 1976. *Bull World Health Organ* 1978;**56**:271–93.
14. World Health Organization. Ebola haemorrhagic fever, Fact sheet No. 103. August 2012 (http://www.who.int/mediacentre/factsheets/fs103/en/index.html).
15. Nidom CA, Nakayama E, Nidom RV, Alamudi MY, Daulay S, Dharmayanti IN, et al. Serological evidence of Ebola virus infection in Indonesian orangutans. *PLoS One* 2012;**7**: e40740.
16. Yuan J, Zhang Y, Li J, Zhang Y, Wang LF, Shi Z. Serological evidence of ebolavirus infection in bats, China. *Virol J* 2012;**9**:236.
17. Olival KJ, Islam A, Yu M, Anthony SJ, Epstein JH, Khan SA, et al. Ebola virus antibodies in fruit bats, Bangladesh. *Emerg Infect Dis* 2013;**19**:270–3.
18. Negredo A, Palacios G, Vázquez-Morón S, González F, Dopazo H, et al. Discovery of an Ebolavirus-like filovirus in Europe. *PLoS Pathog* 2011;**7**:e1002304.
19. Pan Y, Zhang W, Cui L, Hua X, Wang M, Zeng Q. Reston virus in domestic pigs in China. *Arch Virol* 2012. Available from: http://dx.doi.org/10.1007/s00705-012-1477-6. [Epub ahead of print].
20. Leroy EM, Epelboin A, Mondonge V, Pourrut X, Gonzalez JP, Muyembe-Tamfum JJ, et al. Human Ebola outbreak resulting from direct exposure to fruit bats in Luebo, Democratic Republic of Congo, 2007. *Vector-Borne and Zoonotic Dis* 2009;**9**: 723–8.
21. Leroy EM, Formenty P, Rouquet P, Souquière S, Kilbourne A, Froment J-M, et al. Multiple Ebola virus transmission events and rapid decline of Central African wildlife. *Science* 2004;**303**:387–90.
22. Leroy EM, Kumulungui B, Pourrut X, Rouquet P, Hassanin A, Yaba P, et al. Fruit bats as reservoirs of Ebola virus. *Nature* 2005;**438**:575–6.
23. Barrette RW, Metwally SA, Rowland JM, Xu L, Zaki SR, Nichol ST, et al. Discovery of swine as a host for the Reston ebolavirus. *Science* 2009;**325**:204–6.
24. World Health Organization. Ebola Reston in pigs and humans, Philippines. *Wkly Epidemiol Rec* 2009;**84**:49–50.
25. Kobinger GP, Leung A, Neufeld J, Richardson JS, Falzarano D, Smith G, et al. Replication, pathogenicity, shedding, and transmission of Zaire ebolavirus in pigs. *J Infect Dis* 2011;**204**:200–8.
26. Weingartl HM, Embury-Hyatt C, Nfon C, Leung A, Smith G, Kobinger G. Transmission of Ebola virus from pigs to non-human primates. *Sci Rep* 2012;**2**:811.
27. Bwaka MA, Bonnet MJ, Calain P, Colebunders R, De Roo A, Guimard Y, et al. Ebola hemorrhagic fever in Kikwit, Democratic Republic of the Congo: clinical observations in 103 patients. *J Infect Dis* 1999;**179**(Suppl. 1):S1–7.
28. Kortepeter MG, Bausch DG, Bray M. Basic clinical and laboratory features of filoviral hemorrhagic fever. *J Infect Dis* 2011;**204**(Suppl. 3):S810–6.

29. Towner JS, Rollin PE, Bausch DG, et al. Rapid diagnosis of Ebola hemorrhagic fever by reverse transcription-PCR in an outbreak setting and assessment of patient viral load as a predictor of outcome. *J Virol* 2004;**78**:4330–41.
30. Sanchez A, Lukwiya M, Bausch D, Mahanty S, Sanchez AJ, Wagoner KD, et al. Analysis of human peripheral blood samples from fatal and nonfatal cases of Ebola (Sudan) hemorrhagic fever: cellular responses, virus load, and nitric oxide levels. *J Virol* 2004;**78**: 10370–7.
31. Ksiazek TG, Rollin PE, Williams AJ, Bressler DS, Martin ML, Swanepoel R, et al. Clinical virology of Ebola Hemorrhagic Fever (EHF): virus, virus antigen, and IgG and IgM antibody findings among EHF patients in Kikwit, Democratic Republic of the Congo, 1995. *J Infect Dis* 1999;**179**(Suppl. 1):S177–87.
32. Grolla A, Lucht A, Dick D, Strong JE, Feldmann H. Laboratory diagnosis of Ebola and Marburg hemorrhagic fever. *Bull Soc Pathol Exot* 2005;**98**:205–9.
33. Falzarano D, Geisbert TW, Feldmann H. Progress in filovirus vaccine development: evaluating the potential for clinical use. *Expert Rev Vaccines* 2011;**10**:63–77.
34. Qiu X, Wong G, Fernando L, Audet J, Bello A, Strong J, et al. mAbs and Ad-Vectored IFN-α therapy rescue Ebola-infected nonhuman primates when administered after the detection of viremia and symptoms. *Sci Transl Med* 2013;**16**:207ra143. Available from: http://dx.doi.org/10.1126/scitranslmed.3006605.
35. Pettitt J, Zeitlin L, Kim do H, Working C, Johnson JC, Bohorov O, et al. Therapeutic intervention of Ebola virus infection in rhesus macaques with the MB-003 monoclonal antibody cocktail. *Sci Transl Med* 2013;**5**:199ra113. Available from: http://dx.doi.org/10.1126/scitranslmed.3006608.
36. World Health Organization. Communication for Behavioural Impact (COMBI). A toolkit for behavioural and social communication in outbreak response. 2012, Geneva Switzerland. 126 pages.
37. Epelboin A, Formenty P, Anoko J, Allarangar Y. Humanisation and informed consent for people and populations during responses to VHF in central Africa (2003–2008). In: Biquet JM, editor. *Humanitarian Stakes No. 1, Infection control measures and individual rights: an ethical dilemma for medical staff.* Geneva (Switzerland): MSF; 2008. p. 25–37.
38. Hewlett BS, Epelboin A, Hewlett BL, Formenty P. Medical anthropology and Ebola in Congo: cultural models and humanistic care. *Bull Soc Pathol Exot* 2005;**98**:230–6.
39. World Health Organization. Standard precautions in health care—aide-memoire. 2007, Geneva Switzerland. 7 pages.
40. World Health Organization. Interim Infection Control Recommendations for Care of Patients with Suspected or Confirmed Filovirus (Ebola, Marburg) Haemorrhagic Fever. 2008, Geneva Switzerland. 7 pages.

Chapter 10

Crimean-Congo Hemorrhagic Fever

Önder Ergönül
School of Medicine, Koç University, Istanbul, Turkey

CASE PRESENTATION

A seventeen-year-old female dealing with livestock husbandry from a village in the northeastern part of central Anatolia was admitted to Ankara Numune Education and Research Hospital. Her complaints started 5 days ago, which were fever, malaise, headache, myalgia, nausea, vomiting, diarrhea, epistaxis, and gastrointestinal and vaginal bleeding.

On her admission, she had fever of 39°C, epistaxis, conjunctival injection, and irritation of the abdomen with palpation. The white blood cell (WBC) count was 1110/mm^3, hemoglobin was 9.3 g/L, and platelet count was 10,600/mm^3. At the second day of hospitalization, her WBC dropped to 513/mm^3, hemoglobin level to 5.3 g/l, and platelet count (PLT) to 8910/mm^3. Large bruises and ecchymoses at the anticubital fossae were detected. The level of AST elevated to 3195 IU, and ALT to 1443 IU. The maximum level of lactate dehydrogenase (LDH) was 8190 IU, creatinine phosphokinase (CPK) 1427 IU, and amylase 458 IU. Prothrombin time (PT) was 22.9 seconds, activated partial prothrombin time (aPTT) was 65 seconds, international normalized ratio (INR) 1.23, and fibrinogen 2.03 g/l. Since neutropenic fever was considered as the possible diagnosis, cefepime 2 g bid and amikacine 1 g a day was started on the day of admission, empirically.

The serological tests for differential diagnosis such as lyme, leptospirosis, brucelosis, Q fever, toxoplasmosis, rubella, cytomegalovirus, herpes, and hepatitis A, B, and C were performed. Routine cultures of urine, feces, and blood were done.

At the third day of hospitalization, viral hemorrhagic fever was considered in etiology, and oral ribavirin was given. The dose of ribavirin was 4 g qd for 4 days, and 2.4 g qd for 6 days. Throughout her stay, the patient was given 40 units of fresh frozen plasma, and 16 units of thrombocyte. On the

Emerging Infectious Diseases. DOI: http://dx.doi.org/10.1016/B978-0-12-416975-3.00010-8

fourth day of ribavirin, WBC count was 4520/mm^3, and PLT count was 98,000/mm^3. At the eighth day of hospital stay, she had a sudden abdominal pain. By ultrasound, the appendix could not be visualized; however, lymphadenopathies at the right inguinal region and fluid between the muscles were noted. The patient was referred to the general surgery department, and explorative laparotomy was performed with a suspicion of acute appendicitis. Before the operation, her WBC count was 9210/mm^3, PLT count 173,000/mm^3, and INR 1.03. The operation team was informed about the potential risk of transmission and the role of protection against a potential viral hemorrhagic fever. In her operation, the hemorrhage between the internal, external oblique, and rectus abdominus muscles, intramural hematoma, and ischemia at the cecum were noted; the appendix was normal. The patient was discharged at the end of 11 days with total cure. The patient's serum samples and the operation team had been sent to the Pasteur Institute in Lyon, and the CCHF IgM level of the patient was reported to be positive a few weeks after the operation. None of the members of the operation team had been infected; IgM and IgG levels of the operation team were negative.

(This is a published case report, *Journal of Infection*, 2005.[1])

1. WHY THIS CASE WAS SIGNIFICANTLY IMPORTANT AS AN EMERGING INFECTION

Crimean-Congo hemorrhagic fever (CCHF) is a fatal viral infection described in parts of Africa, Asia, Eastern Europe, and the Middle East and causes severe diseases in humans, with the reported mortality rate of 3–30%.[2]

2. WHAT IS THE CAUSATIVE AGENT?

The disease is caused by Crimean-Congo hemorrhagic fever virus. The virus is an enveloped, segmented, negative-strand RNA virus, and a member of the *Nairovirus* genus of the *Bunyaviridae* family. The virus has a spherical shape approximately 100 nm in diameter. The envelope contains two glycoprotein spikes (GN and GC) 8–10 nm in length and encloses three negative-strand RNAs, the large (L), medium (M), and small (S) RNAs that are associated with protein to form nucleocapsids (Figure 10.1). When viewed by negative stain electron microscopy, CCHF virions are distinct from other viruses within the *Bunyaviridae* family, as they possess very small morphologic surface units with no central holes arranged in no obvious order.[3] There are at least eight genetically distinct clades based on the S segment of the genome, and only one of these clades, AP92, was previously known as causing no symptomatic disease. A study from Turkey reported a mild symptomatic child infected by a clade very closely related to the AP92 strain.[4]

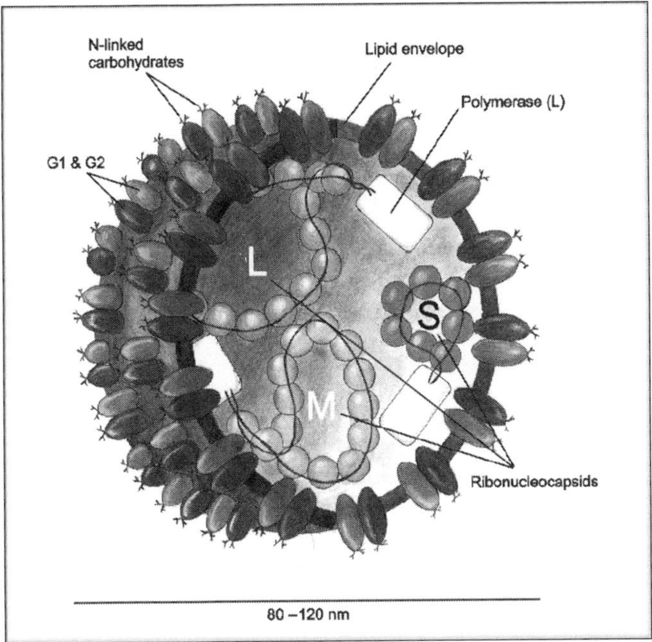

FIGURE 10.1 The structure of Crimean-Congo hemorrhagic fever virus.

3. WHAT IS THE FREQUENCY OF THE DISEASE?

The geographic range of CCHFV is known to be the most extensive one among the tickborne viruses related to human health, and the second most widespread of all medically important arboviruses after dengue viruses.[2] Crimean hemorrhagic fever (CHF) was first described as a clinical entity in 1944–1945, when about 200 Soviet military personnel were infected while assisting peasants in the devastated Crimea after the Nazi invasion. By the year 2000, new outbreaks had been reported from Pakistan, Iran, Senegal, Albania, Kosovo, Bulgaria, Turkey, Greece, Kenya, Mauritania, and recently from India (Figure 10.2). The serologic evidence for CCHFV was documented from Egypt, Portugal, Hungary, France, and Benin, although no human case had been reported.[5]

4. HOW IS THE VIRUS TRANSMITTED?

(See Case Presentation.)

CCHFV circulates in the nature in an enzootic tick–vertebrate–tick cycle. Humans have been infected with CCHFV after contact with livestock and other animals, and the virus causes disease among animals.[6] Antibody surveys among livestock in endemic areas have shown high prevalence

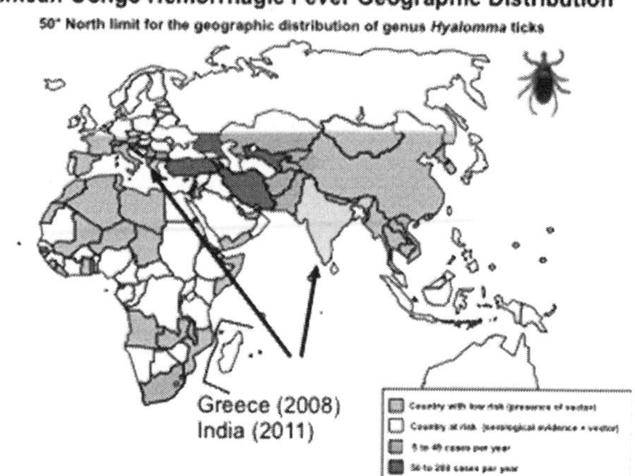

FIGURE 10.2 Distribution of CCHF in the world.

among both cattle and sheep. CCHF viral infection has been demonstrated more commonly among smaller wildlife species such as hares and hedgehogs that act as hosts for the immature stages of the tick vectors.[7] Birds may play a role in the transportation of CCHFV-infected ticks between countries.

Humans become infected through the bites of ticks, by contact with a patient with CCHF during the acute phase of infection, or by contact with blood or tissues from viremic livestock.[2,7,8] The cases were distributed among the actively working population[7] that were eventually exposed to the ticks. The great majority of the affected cases deal with agriculture and/or husbandry. Healthcare workers (HCWs) are the second most affected groups in the literature. The gender distribution differs between countries, according to the participation of women in agricultural work. In Turkey, the male to female ratio was reported to be one to one[9] (Figure 10.3).

5. WHICH FACTORS ARE INVOLVED IN DISEASE PATHOGENESIS? WHAT ARE THE PATHOGENIC MECHANISMS?

After inoculation, the virus first replicates in dendritic cells and other local tissues, with subsequent migration to regional lymph nodes and then dissemination through the lymph and blood monocytes to a broad range of tissues and organs, including the liver, spleen, and lymph nodes. Migration of tissue macrophages results in secondary infection of permissive parenchymal cells. Although lymphocytes remain free of infection, they may be destroyed in massive numbers over the course of illness through apoptosis, as seen in other forms of septic shock. The synthesis of cell surface tissue factor

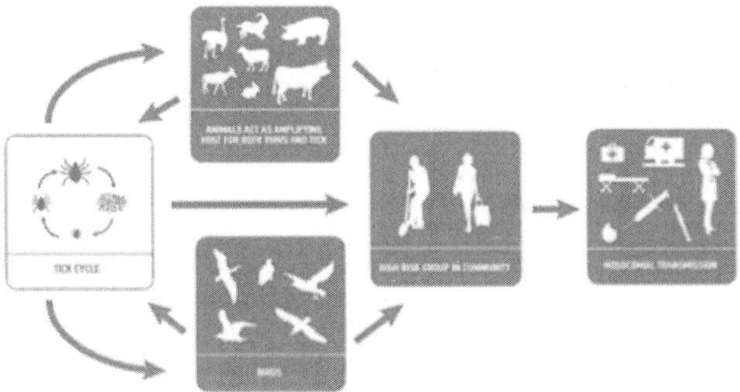

FIGURE 10.3 Transmission routes of CCHF virus.

triggers the extrinsic coagulation pathway. Impaired hemostasis may entail endothelial cell, platelet, and/or coagulation factor dysfunction. Disseminated intravascular coagulopathy (DIC) is frequently noted in Crimean-Congo hemorrhagic fever virus infections. Reduced levels of coagulation factors may be secondary to hepatic dysfunction and/or disseminated intravascular coagulation. In addition, CCHFV may lead to a hemorrhagic diathesis through direct damage of platelets and endothelial cells and/or indirectly through immunological and inflammatory pathways.[10–12] These changes appear to be largely the consequence of the release of cytokines, chemokines, and other pro-inflammatory mediators from virus-infected monocytes and macrophages.[13,14] Tissue damage may be mediated through direct necrosis of infected cells or indirectly through apoptosis of immune cells. The hepatocytes were reported to be particularly effected in CCHFV infection.[15]

6. WHAT ARE THE CLINICAL MANIFESTATIONS?

(See Case Presentation.)

The incubation period for CCHFV ranges from 1 to 9 days. Patients initially exhibit a non-specific prodrome, which typically lasts less than 1 week. Symptoms typically include high fever, headache, malaise, arthralgias, myalgias, nausea, abdominal pain, and rarely diarrhea.[13] Early signs typically include fever, hypotension, conjunctivitis, and cutaneous flushing or a skin rash. Later, patients may develop signs of progressive hemorrhagic diathesis, such as petechiae, mucous membrane and conjunctival hemorrhage; hematuria; hematemesis; and melena. Disseminated intravascular coagulation and circulatory shock may ensue. Death is typically preceded by hemorrhagic diathesis, shock, and multi-organ system failure 1 to 2 weeks following onset of symptoms. The disease was reported to be milder among children[16] (Figure 10.4).

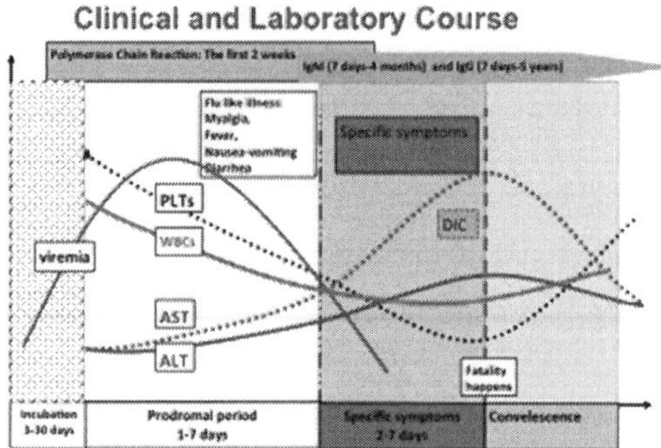

FIGURE 10.4 The clinical and laboratory course of CCHF infection.

Laboratory abnormalities usually include leukopenia, thrombocytopenia, and elevated liver enzymes. Anemia is not usually seen at the early phase of the disease, but may develop late in the disease course. Coagulation abnormalities may include prolonged bleeding time, prothrombin time, and activated partial thromboplastin time; elevated fibrin degradation products; and decreased fibrinogen (Table 10.1).

7. HOW DO YOU DIAGNOSE?

(See Case Presentation.)

The diagnosis is performed by the detection of the viral RNA genome and/or the antigen, and the detection of specific IgM antibodies in human serum or blood (Table 10.2). Antigen detection (by ELISA—enzyme-linked immunosorbent assay) and reverse transcriptase polymerase chain reaction (RT-PCR) are the most useful diagnostic techniques in the acute clinical setting. Leukopenia, particularly neutropenia, thrombocytopenia, high levels of liver enzymes alanine aminotransferase (ASL) and aspartate aminotransferase (AST), and lactate dehydrogenase (LDH) are regularly reported in patients.

7.1 Viral Isolation

The most definitive test is viral culture; however, the time to diagnosis, which is 2–10 days for the virus to grow, is too long for management of acute cases. Moreover, the need for high containment facilities renders this technique more of a confirmatory test and research tool. CCHFV has been isolated most frequently by intracranial inoculation of newborn suckling mice.[17]

TABLE 10.1 The Laboratory Characteristics of CCHFV Infection

Test	Findings and Comments
Complete blood count	
White blood cell count	Moderate or severe leukopenia, sometimes leucosytosis
Platelet count	Mild to severe decrease
Hemoglobin and/or hematocrit	Could be decreased later in disease course
Liver enzymes	Increased, usually AST > ALT
Coagulation studies (INR, PT, PTT, fibrinogen, fibrin split products, platelets, D-dimer)	Hemophagocytosis and DIC are common
Lactate dehydrogenase	A level greater than 4 mmol/L (36 mg/dL) may indicate persistent hypo-perfusion and sepsis
Creatinin phosphokinase	Elevated
Blood urea nitrogen and creatinine	Renal failure may occur late in disease course, but proteinuria may occur

7.2 Molecular Detection: RT-PCR

RT-PCR is a very sensitive, rapid test, although careful attention must be paid to the potential for false-positive results. One-step real-time RT-PCR assays using primers to the same nucleoprotein gene have been developed but the development of these assays has been hampered by the high diversity of the genome sequence.[18–20]

7.3 Indirect Serological Diagnosis

The serological diagnosis of VHF infection is based on the detection of specific IgM and IgG antibodies induced by the immune response principally to the nucleoprotein, which is recognized as the predominant antigen. Seroconversion with detection of CCHFV IgM antibodies or a ≥four-fold increase in antibody titer between two successive blood samples is evidence of a recent infection. The serological diagnosis is valid after several days post onset of the disease; nevertheless, the antibody response is rarely observed in fatal cases.[21] ELISA is the most common technique for CCHFV antibody detection with a sensitivity of more than 90%. ELISA was reported to be more sensitive than indirect fluorescent antibody (IFA) test.[22] Usually,

TABLE 10.2 Summary for the Case Management for CCHFV Infection

Evaluation of the cases
Clinical symptoms
Early symptoms (first days at the end of incubation): myalgia, fever, diarrhea
Late symptoms (3–10 days after incubation): bleeding from various sites
Patient history
Referral from endemic area
Outdoor activities (picnic, tracking, etc.) in endemic area
History of tick bite or contact with bodily fluids of the infected people
Dealing with husbandry in endemic area
Laboratory tests: findings compatible with hemophagocytosis (low thrombocyte and white blood cell count, elevated AST, ALT, LDH, CPK)
Preventive measures
a. Isolate the patient b. Inform and educate HCWs and caregivers of the patients c. Assess the risk for transmission and use the barrier precautions accordingly
Investigations for confirmation
Serum for PCR and ELISA
IgM positivity or PCR positive confirms diagnosis, IgG positivity cannot
Sera for differential diagnosis
Decision making for therapy
Start ribavirin for early cases
Do not neglect other causes of clinical picture. Doxycycline or equivalent should be considered for the diseases in differential diagnosis list
Hematological support
Fresh frozen plasma to improve the homeostasis
Thrombocyte solutions
Respiratory support
Follow-up
No recurrence was reported. Therefore, there is no definition of follow-up
Post-exposure prophylaxis
HCWs, or other individuals who are exposed to the virus, should be assessed for the level of the risk. Individuals in high-risk groups should receive ribavirin, whereas individuals in low risk groups should be followed up with complete blood counts and biochemical tests for 14 days

IgM and IgG antibodies are detected 4–5 days post onset of symptoms. The IgM titer is maximal 2 to 3 weeks after onset of the disease, and the IgM antibodies generally disappear within 4 months. The IgG antibodies remain detectable for several years.[17]

8. HOW DO YOU DIFFERENTIATE THIS DISEASE FROM SIMILAR ENTITIES?

(See Case Presentation.)

CCHFV has an extended geographical distribution including Africa, southern Europe, the Middle East, Russia, India, and China. Other viral etiologies have to be considered according to the origin of the patient and the risks of potential exposure. These would include Alkhurma and Rift Valley fever in the Middle East; Omsk hemorrhagic fever in Russia; Kyasanur Forest disease in India; hantaviruses in Europe and Asia; Lassa, Ebola, Marburg, Rift Valley fever, and yellow fever in Africa; and dengue in various locations.[23] In tropical and subtropical countries, malaria is the most important alternative diagnosis to be excluded in cases of suspected VHF. The differential diagnosis list should include hepatitis viruses, influenza, *Neisseria meningitidis*, leptospirosis, borreliosis, typhoid, rickettsiosis, and Q fever (*Coxiella burnetii*) staphylococcal or Gram-negative sepsis, toxic shock syndrome, salmonellosis and shigellosis, psittacosis, trypanosomiasis, septicemic plague, rubella, measles, and hemorrhagic smallpox[2,17,24] (Table 10.3).

Non-infectious processes associated with bleeding diathesis, which should be included in the differential diagnosis, include idiopathic or thrombotic thrombocytopenic purpura, HELP syndrome among pregnant women, hemolytic uremic syndrome, acute leukemia, vitamin B12 deficiency, and collagen-vascular diseases.[2,25]

9. WHAT IS THE THERAPEUTIC APPROACH?

(See Case Presentation.)

Supportive therapy is an essential part of case management. Thrombocyte solutions and fresh frozen plasma are given by monitoring the bleeding status of patients. Ribavirin is the only antiviral drug that has been used to treat viral hemorrhagic fever syndromes, including CCHF and Lassa fever.[26–28] Viruses in the *Bunyaviridae* family are generally sensitive to ribavirin.[29] Ribavirin was shown to be effective against CCHFV *in vitro*.[30–32] In suckling mice, ribavirin treatment reduced CCHF virus growth in the liver; significantly decreased, but did not prevent, viremia; and significantly reduced mortality and extended the geometric mean time to death.[30] In clinical practice, ribavirin was found to be beneficial, especially at the earlier phase of the infection.[33–36] Ribavirin is placed on the WHO essential medicines list (15th Model List of Essential Medicines, March 2007) to be used against

TABLE 10.3 Differential Diagnosis List for Viral Hemorrhagic Fevers

Disease Categories	Differentials
Infections	
Brucellosis	Pancytopenia, Wright agglutination
Q fever	Serology (ELISA or IFAT)
Rickettsia	Weil–Felix test
Ehrlichiosis	Serology (ELISA)
Hanta	Pulmonary or renal involvement, serology, PCR
Leptospira	Agglutination
Salmonella	Widal test
Non-infectious reasons	
Vitamin B12 deficiency	Pancytopenia, and B12 level in serum
Febrile neutropenia	Underlying disease
HELLP syndrome	Geographic location
Drug side effects	
Metamizole	History

CCHFV infection. In a recent review, ethical concerns about conducting a randomized controlled trial of ribavirin in the treatment of Crimean-Congo hemorrhagic fever were detailed.[37] Despite the need for more evidence about the impact of ribavirin in treatment, the authors described why it was not ethical to conduct a randomized controlled trial on such a fatal disease with only one antiviral alternative.[5,37] A recent observational study reported the beneficial effect of ribavirin in reduction of case-fatality rate among moderately ill patients, whereas steroids were found to be beneficial especially among more severe patients. The same study demonstrated that grouping of the patients by severity scoring was necessary for case management and also for the observational drug assessment studies.[38]

Bulgarian investigators suggested that immunotherapy treatment of seven patients with severe CCHF via passive simultaneous transfer of two different specific immunoglobulin preparations, CCHF-bulin (for intramuscular use) and CCHF-venin (for intravenous use), prepared from the plasma of CCHF survivor donors boosted with one dose of CCHF vaccine, resulted in quick recovery of all patients,[39] and they suggested that the intravenous preparation be used for treatment of all cases of CCHF.[40] A further study included 22 severe patients from Turkey; prompt administration of CCHFV hyperimmunoglobulin was suggested as an alternative treatment approach, especially for high-risk individuals.[41] Further studies with larger sample size and more detailed design are necessary.

10. WHAT ARE THE PREVENTIVE AND INFECTION CONTROL MEASURES?

(See Case Presentation.)

Bunyaviruses are highly infectious after direct contact with infected blood and bodily secretions. A suspected case of CCHF must be immediately reported to the infection control professional and to the local or state health department. The infection control professional should notify the clinical laboratory as well as other clinicians and public health authorities.

10.1 Isolation Precautions

Direct contact with infected blood and bodily fluids has accounted for the majority of person-to-person transmission. Therefore, it was recommend that in the case of any patient with suspected or documented CCHFV infection, specific barrier precautions should be implemented immediately. Airborne transmission was suspected in one Lassa fever outbreak in 1969,[42] but there has never been any documented case of airborne transmission of that virus to humans. A prospective serological study from Sierra Leone suggested that the hospital staff who cared for Lassa fever patients using simple barrier nursing methods have no higher risk of infection than the local population.[43] A similar study from Turkey reported the lack of airborne transmission of CCHF to the HCWs after a CCHF epidemic.[44] Usually, the standard precautions including health hygiene, using gloves, gowns, face shields, and masks are sufficient protection. However, in the case of a procedure which may generate an aerosol, HCWs should consider wearing an N95 or FFP2 respirator (European Norm (EN) 61010-1).[45]

An integrated strategy for the control of accidental exposure to blood and bodily fluids is the critical step while providing protection among HCWs.[45] Sharps containers are the foremost safety equipment, which should be available at all times to all units. The use of safety engineered devices should also be considered in order to decrease the risk of needlestick injuries.[45]

10.2 Post-exposure Management

Post-exposure management systems are an integral part of an effort to enhance HCW safety. In CCHF[33] and Lassa fever,[28] use of oral ribavirin as post-exposure prophylaxis was well described as an effective and beneficial drug. Ribavirin prophylaxis is generally well tolerated, potentially useful and should therefore be recommended for HCWs who are at high risk of

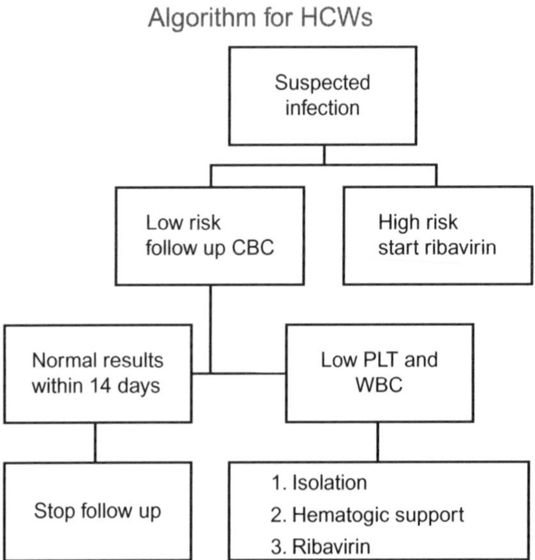

FIGURE 10.5 Algorithm for post-exposure management.

exposures such as percutaneous injuries or splash of contaminated blood or bodily fluid to the face or mucosal surfaces of the HCWs[45,46] (Figure 10.5).

In conclusion, many recent developments were noted in epidemiology, pathogenesis, and early detection; however, there is neither alternative antiviral therapy nor a vaccine in the pipeline, yet.

REFERENCES

1. Celikbas A, Ergonul O, Dokuzoguz B, Eren S, Baykam N, Polat-Duzgun A. Crimean Congo hemorrhagic fever infection simulating acute appendicitis. *J Infect* 2005;**50**:363–5.
2. Ergonul O. Crimean-Congo haemorrhagic fever. *Lancet Infect Dis* 2006;**6**:203–14.
3. Whitehouse CA. Crimean-Congo hemorrhagic fever. *Antiviral Res* 2004;**64**:145–60.
4. Midilli K, Gargili A, Ergonul O, et al. The first clinical case due to AP92 like strain of Crimean-Congo hemorrhagic fever virus and a field survey. *BMC Infect Dis* 2009;**9**:90.
5. Ergonul O. Crimean-Congo hemorrhagic fever virus: new outbreaks, new discoveries. *Curr Opin Virol* 2012;**2**:215–20.
6. Shepherd AJ, Swanepoel R, Shepherd SP, McGillivray GM, Searle LA. Antibody to Crimean-Congo hemorrhagic fever virus in wild mammals from southern Africa. *Am J Trop Med Hyg* 1987;**36**:133–42.
7. Hoogstraal H. The epidemiology of tick-borne Crimean-Congo hemorrhagic fever in Asia, Europe, and Africa. *J Med Entomol* 1979;**15**:307–417.
8. Watts DM, Ksiasek TG, Linthicum KJ, Hoogstraal H. Crimean-Congo hemorrhagic fever. In: Monath TP, editor. *The Arboviruses: Epidemiology and Ecology*. Boca Raton, FL, USA: CRC; 1988.

9. Yilmaz GR, Buzgan T, Irmak H, et al. The epidemiology of Crimean-Congo hemorrhagic fever in Turkey, 2002–2007. *Int J Infect Dis* 2009;**13**:380–6.
10. Chen JP, Cosgriff TM. Hemorrhagic fever virus-induced changes in hemostasis and vascular biology. *Blood Coagul Fibrinolysis* 2000;**11**:461–83.
11. Peters CJ, Zaki SR. Role of the endothelium in viral hemorrhagic fevers. *Crit Care Med* 2002;**30**:S268–73.
12. Geisbert TW, Jahrling PB. Exotic emerging viral diseases: progress and challenges. *Nat Med* 2004;**10**:S110–21.
13. Bray M. Comparative pathogenesis of Crimean Congo hemorrhagic fever and Ebola hemorrhagic fever. In: Ergonul O, Whitehouse CA, editors. *Crimean Congo Hemorrhagic Fever: A Global Perspective*. Dordrecht (NL): Springer; 2007. p. 221–31.
14. Ergonul O, Tuncbilek S, Baykam N, Celikbas A, Dokuzoguz B. Evaluation of serum levels of interleukin (IL)-6, IL-10, and tumor necrosis factor-alpha in patients with Crimean-Congo hemorrhagic fever. *J Infect Dis* 2006;**193**:941–4.
15. Rodrigues R, Paranhos-Baccala G, Vernet G, Peyrefitte CN. Crimean-Congo hemorrhagic fever virus-infected hepatocytes induce ER-stress and apoptosis crosstalk. *PLoS One* 2012;**7**:e29712.
16. Tezer H, Sucakli IA, Sayli TR, et al. Crimean-Congo hemorrhagic fever in children. *J Clin Virol* 2010;**48**:184–6.
17. Zeller H. Laboratory diagnosis of Crimean Congo hemorrhagic fever. In: Ergonul O, Whitehouse CA, editors. *Crimean Congo Hemorrhagic Fever: A Global Perspective*. Dordrecht (NL): Springer; 2007. p. 233–43.
18. Drosten C, Kummerer BM, Schmitz H, Gunther S. Molecular diagnostics of viral hemorrhagic fevers. *Antiviral Res* 2003;**57**:61–87.
19. Drosten C, Gottig S, Schilling S, et al. Rapid detection and quantification of RNA of Ebola and Marburg viruses, Lassa virus, Crimean-Congo hemorrhagic fever virus, Rift Valley fever virus, dengue virus, and yellow fever virus by real-time reverse transcription-PCR. *J Clin Microbiol* 2002;**40**:2323–30.
20. Atkinson B, Chamberlain J, Logue CH, et al. Development of a real-time RT-PCR assay for the detection of Crimean-Congo hemorrhagic fever virus. *Vector Borne Zoonotic Dis* 2012;**12**:786–93.
21. Ergonul O, Celikbas A, Baykam N, Eren S, Dokuzoguz B. Analysis of risk-factors among patients with Crimean-Congo haemorrhagic fever virus infection: severity criteria revisited. *Clin Microbiol Infect* 2006;**12**:551–4.
22. Burt FJ, Swanepoel R, Braack LE. Enzyme-linked immunosorbent assays for the detection of antibody to Crimean-Congo haemorrhagic fever virus in the sera of livestock and wild vertebrates. *Epidemiol Infect* 1993;**111**:547–57.
23. Ergonul O. Clinical and pathologic features of Crimean Congo hemorrhagic fever. In: Ergonul O, Whitehouse CA, editors. *Crimean Congo Hemorrhagic Fever: A Global Perspective*. Dordrecht (NL): Springer; 2007. p. 207–20.
24. Borio L, Inglesby T, Peters CJ, et al. Hemorrhagic fever viruses as biological weapons: medical and public health management. *JAMA* 2002;**287**:2391–405.
25. Ergonul O, Celikbas A, Yildirim U, et al. Pregnancy and Crimean-Congo haemorrhagic fever. *Clin Microbiol Infect* 2010;**16**:647–50.
26. Ergonul O. Treatment of Crimean Congo hemorrhagic fever. In: Ergonul O, Whitehouse CA, editors. *Crimean Congo Hemorrhagic Fever: A Global Perspective*. Dordrecht (NL): Springer; 2007. p. 245–60.

27. McCormick JB, King IJ, Webb PA, et al. Lassa fever. Effective therapy with ribavirin. *N Engl J Med* 1986;**314**:20–6.
28. Bausch DG, Hadi CM, Khan SH, Lertora JJ. Review of the literature and proposed guidelines for the use of oral ribavirin as postexposure prophylaxis for Lassa fever. *Clin Infect Dis* 2010;**51**:1435–41.
29. Sidwell RW, Smee DF. Viruses of the Bunya- and Togaviridae families: potential as bioterrorism agents and means of control. *Antiviral Res* 2003;**57**:101–11.
30. Tignor GH, Hanham CA. Ribavirin efficacy in an in vivo model of Crimean-Congo hemorrhagic fever virus (CCHF) infection. *Antiviral Res* 1993;**22**:309–25.
31. Watts DM, Ussery MA, Nash D, Peters CJ. Inhibition of Crimean-Congo hemorrhagic fever viral infectivity yields in vitro by ribavirin. *Am J Trop Med Hyg* 1989;**41**:581–5.
32. Paragas J, Whitehouse CA, Endy TP, Bray M. A simple assay for determining antiviral activity against Crimean-Congo hemorrhagic fever virus. *Antiviral Res* 2004;**62**:21–5.
33. Ergonul O. Treatment of Crimean-Congo hemorrhagic fever. *Antiviral Res* 2008;**78**:125–31.
34. Tasdelen Fisgin N, Ergonul O, Doganci L, Tulek N. The role of ribavirin in the therapy of Crimean-Congo hemorrhagic fever: early use is promising. *Eur J Clin Microbiol Infect Dis* 2009.
35. Izadi S, Salehi M. Evaluation of the efficacy of ribavirin therapy on survival of Crimean–Congo hemorrhagic fever patients: a case-control study. *Jpn J Infect Dis* 2009;**62**:11–5.
36. Ozbey SB. Impact of early ribavirin use on fatality of CCHF. *Klimik J* 2010;**23**:6–10.
37. Arda B, Aciduman A, Johnston JC. A randomised controlled trial of ribavirin in Crimean Congo haemorrhagic fever: ethical considerations. *J Med Ethics* 2012;**38**:117–20.
38. Dokuzoguz B, Celikbas AK, Gok SE, Baykam N, Eroglu MN, Ergonul O. Severity Scoring index for Crimean-Congo hemorrhagic fever and the impact of ribavirin and corticosteroids on fatality. *Clin Infect Dis* 2013;**57**:1270–4.
39. Vassilenko SM, Vassilev TL, Bozadjiev LG, Bineva IL, Kazarov GZ. Specific intravenous immunoglobulin for Crimean-Congo haemorrhagic fever. *Lancet* 1990;**335**:791–2.
40. Dimitrov DS. Antibodies to CCHFV for prophylaxis and treatment. In: Ergonul O, Whitehouse CA, editors. *Crimean-Congo Hemorrhagic Fever: A Global Perspective*. Dordrecht (NL): Springer; 2007. p. 261–9.
41. Kubar A, Haciomeroglu M, Ozkul A, et al. Prompt administration of Crimean-Congo hemorrhagic fever (CCHF) virus hyperimmunoglobulin in patients diagnosed with CCHF and viral load monitorization by reverse transcriptase-PCR. *Jpn J Infect Dis* 2011;**64**:439–43.
42. Carey DE, Kemp GE, White HA, et al. Lassa fever. Epidemiological aspects of the 1970 epidemic, Jos, Nigeria. *Trans R Soc Trop Med Hyg* 1972;**66**:402–8.
43. Helmick CG, Webb PA, Scribner CL, Krebs JW, McCormick JB. No evidence for increased risk of Lassa fever infection in hospital staff. *Lancet* 1986;**2**:1202–5.
44. Ergonul O, Zeller HG, Celikbas A, Dokuzoguz B. The lack of Crimean-Congo hemorrhagic fever virus antibodies in healthcare workers in an endemic region. *Int J Infect Dis* 2007;**11**:48–51.
45. Tarantola A, Ergonul O, Tattevin P. Estimates and prevention of Crimean Congo hemorrhagic fever risks for health care workers. In: Ergonul O, Whitehouse CA, editors. *Crimean-Congo Hemorrhagic Fever: A Global Perspective*. Dordrecht (NL): Springer; 2007. p. 281–94.
46. Celikbas A., Dokuzoguz B., Baykam N., et al. Crimean-Congo hemorrhagic fever infection among healthcare workers in Turkey. Emerg Infect Dis; in press.

Chapter 11

Phlebotomus Fever—Sandfly Fever

Koray Ergunay

Hacettepe University Faculty of Medicine, Department of Medical Microbiology, Virology Unit, Ankara, Turkey

CASE PRESENTATION

During mid-August, a 27-year-old male was admitted to the emergency ward with high fever, chills, severe headache, joint pain, watery diarrhea, nausea and, vomiting, which had started the day before. The initial complete physical examination demonstrated a fever of 38.9°C, generalized muscle tenderness, and multiple skin lesions suggesting insect bites on the upper left limb. Neurological examination was normal without meningeal signs. Medical history revealed no previous disease of significance but an exposure to mosquitoes during his stay at his cousin's cottage 5 days ago. He had vomited three times during the last 24 hours. No apparent risk for infectious gastroenteritis could be identified. Laboratory evaluation demonstrated decreased leukocyte count ($3.8 \times 10^3/\mu L$) with relative lymphocytosis, decreased platelet count ($1.32 \times 10^5/\mu L$), elevated alanine aminotransferase (ALT, 101 U/L), aspartate aminotransferase (AST, 128 U/L), gamma glutamyl transpeptidase (GGT, 107 U/L), creatinine phosphokinase (CPK, 428 U/L), and lactate dehydrogenase (LDH, 354 U/L) levels. Hemoglobulin, C-reactive protein, total protein, blood urea nitrogen (BUN), albumin, creatinine, prothrombin time (PT), activated partial thromboplastin time (aPTT), international normalized ratio (INR), and the chest X-ray were within normal limits. The patient was transferred to the infectious diseases department with the preliminary diagnosis of undifferentiated viral febrile condition, and blood and stool samples were submitted for microbiological analyses. Symptomatic treatment with intravenous rehydration and anti-pyretics was initiated.

The patient responded favorably to the treatment and fever subsided within 48 hours. Platelet and leukocyte counts returned to normal limits in 36–48 hours while hepatic enzymes remained over the threshold for another

Emerging Infectious Diseases. DOI: http://dx.doi.org/10.1016/B978-0-12-416975-3.00011-X

24–48 hours. No leukocyte, erythrocyte, or parasite eggs were detected in stool microscopy. Blood and stool cultures were evaluated as negative for pathogenic bacteria. Serological assays ruled out acute hepatitis by hepatitis A–E viruses. Due to the history and observation of insect bites, West Nile virus and phlebovirus assays were also performed in blood samples obtained during admission. While West Nile virus as well as Toscana and sandfly fever Sicilian virus immunoglobulins could not be detected, a pan-phlebovirus polymerase chain reaction demonstrated reactive results. Sequencing of the amplicons revealed complete homology to the sandfly fever Sicilian virus variant, sandfly fever Turkey virus, and diagnosis of sandfly fever was established. A serum sample collected on the fifth day was reactive for sandfly fever Sicilian virus IgM antibodies via commercial assays. The patient was discharged with recovery on the sixth day of admission with normalized biochemical and hematologic parameters. During the follow-up visit on the fourth week after discharge, the patient complained about weakness, fatigue, and lack of concentration, which had lasted about 10 days. A convalescent serum obtained during the visit demonstrated an IgG seroconversion.

1. WHY THIS CASE WAS SIGNIFICANTLY IMPORTANT AS AN EMERGING INFECTION

Sandfly fever, also known as phlebotomus, pappataci, or 3-day fever, is an arthropod-borne febrile disease transmitted by phlebotomine sandflies. Sandfly fever is prevalent in many regions including the countries in the Mediterranean Basin, northern Africa, the Middle East, and parts of Central and Southern Asia where it constitutes a significant health problem affecting non-immune persons.[1] Although the disease is not frequently associated with mortality and residual sequelae, it is a highly incapacitating and debilitating condition that can significantly affect the individual and the population.[2] In endemic regions, epidemics, especially occurring during summer months, have been noted. Sandfly fever is also of concern as a travel-related infection.[3–6] Historically, the disease affected a large number of Allied and Axis troops serving in the Mediterranean region during the Second World War.[7] Several cases of imported sandfly fever have been reported in persons visiting endemic countries.[8–10]

2. WHAT IS THE CAUSATIVE AGENT?

Sandfly fever is caused by viruses classified in the *Phlebovirus* genus of the *Bunyaviridae* family. The *Phlebovirus* genus consists of over 50 distinct virus serotypes, some of which have not been fully characterized. According to the 9th Report of the International Committee for Taxonomy of Viruses (ICTV), phleboviruses consist of nine viral species or serotypes (Sandfly fever Naples, Salehabad, Rift Valley fever, Uukuniemi, Bujaru, Candiru,

Chilibre, Frijoles, and Punta Toro) and several tentative species (including Sandfly fever Sicilian and Corfu).[11]

Phleboviruses possess an enveloped spherical virion approximately 100 nm in diameter; a tripartite single-stranded RNA genome having a negative and ambisense coding strategy. The structural proteins of phleboviruses comprise the surface glycoproteins (Gn and Gc) encoded by the M (medium-sized) genome segment, the nucleocapsid protein encoded by the S (small) segment, and the viral RNA-dependent RNA polymerase, encoded by the L (large) genome segment. The replication site is the cytoplasm of the infected cell and virion maturation occurs in the Golgi complex.[12,13]

Within the *Phlebovirus* genus, well-known agents responsible for febrile diseases occuring in the Old World are Sandfly fever Sicilian virus (SFSV) and Sandfly fever Naples virus (SFNV)[2]. In addition to SFSV and SFNV, other phlebovirus strains are associated with sandfly fever as well. These include Sandfly fever Cyprus virus (SFCV), which was identified in Greek troops in Cyprus, and Sandfly fever Turkey virus (SFTV), which was characterized during outbreaks in central and Aegean/Mediterranean coastal provinces in Turkey.[3,14] SFCV and SFTV are considered as SFSV variants due to significant genomic sequence homology and antigenic cross-reactions.[3] Another frequently observed clinically significant phlebovirus is Toscana virus (TOSV). Due to its distinct neurotropism, TOSV has been identified as an important agent of sporadic seasonal meningitis/meningoencephalitis in endemic countries.[15] Febrile disease without central nervous system involvement due to TOSV has also been reported.[16,17] Moreover, a surprisingly vast diversity of phleboviruses has been revealed in vector sandflies in the endemic regions. Phleboviruses characterized in sandflies include Salehabad, Karimabad, and Tehran viruses in Iran, Corfu virus in Greece, Arbia virus in Italy, Massilia virus in France, Granada virus in Spain, Punique virus in Tunisia, Adria virus in Albania, as well as several putative isolates.[18–27] Although serologic data indicate human exposure to some of these strains, their pathogenicity and association with clinical disease have not yet been fully elucidated.[20–22,24,27,28] Nevertheless, recent evidence suggests Granada virus to be responsible for mild febrile diseases.[29] In the New World, Alenquer, Chagres, Candiru, and Punta Toro phleboviruses have been detected in individuals with sporadic febrile diseases in Panama, Brasil, and in northern parts of South America. The diseases in tropical America are probably forest associated and rodents living in forests presumably act as amplifying hosts.[30–32]

3. WHAT IS THE FREQUENCY OF THE DISEASE?

The risk for infection with sandfly-transmitted phleboviruses affects large areas of the Old World including southern Europe, Africa, the Middle East, and Central and Western Asia, in association with the presence of sandfly vectors.[28,33] Recent investigations have demonstrated that phlebovirus diversity in the Mediterranean basin is higher than initially suspected, and that populations living south and east of the Mediterranean Sea have a high risk for virus exposure during their lifetime.[15,28,33,34]

Historical evidence of sandfly fever in the Mediterranean region dates back to the Napoleonic Wars and SFSV-SFNV isolation was performed following the epidemics around Italy during the Second World War.[7,35] Allied forces stationed in the Mediterranean and Middle East reported tens of thousands of cases and attack rates of 3–10% (locally up to 80%). Outbreaks of sandfly fever occurred repeatedly in the former Soviet Union in the period from 1945 to 1950, predominantly in Crimea, Romania, Moldavia, and the Central Asian republics. Epidemics related to the activity of *Phlebotomus papatasi* were reported in northern Africa, southern Europe, the Middle East, and Central Asia.[36,37] Human disease in the Americas is reported from Panama, Brazil, and northern regions of South America, although phleboviruses have been isolated as far as Rio Grande Valley in Texas and southern Brasil.[30–32]

Seroepidemiological studies have shown that the prevalence of antibodies to SFSV and SFNV in indigenous populations follows the distribution of *P. papatasi* around the Mediterranean region, the Middle East, North Africa and Western Asia.[2,18,28,38,39] Evidence for SFSV activity is present in Bangladesh, India, Greece, Cyprus, Iraq, Morocco, Saudi Arabia, Somalia, Sudan, Tunisia, Turkey, Pakistan, Croatia, Iran, the southern European and Central Asian republics of Turkmenia, Tajikistan, Uzbekistan, Azerbaijan and Moldavia, former Yugoslavia, France, and Portugal.[4,18,28,39–42] SFNV follows a similar distribution and has been detected in Bangladesh, Ethiopia, Greece, Cyprus, Iraq, Morocco, Saudi Arabia, Sudan, Djibouti, Turkey, and former Yugoslavia, Turkmenia, Tajikistan, Uzbekistan, Azerbaijan and Moldavia.[28,33,39–41,43] Serosurveillance and clinical data from areas surrounding the Mediterranean suggest that SFNV infections have significantly decreased during the last decades.[33,44] Since 2006, sandfly fever epidemics were reported from Cyprus, Iraq, and Turkey.[3,4,14,45] Infections in non-immune individuals, frequently acquired as travel-related infections, are well known.[4,5,8,9,46] Figure 11.1 shows countries with documented sandfly fever phlebovirus activity in the Old World.

4. HOW IS THE VIRUS TRANSMITTED?

The only known transmission route for viruses causing sandfly fever is via blood sucking of female phlebotomine sandflies of the *Psychodidae* family.[1,33,36] Direct human-to-human or parenteral transmission has not been demonstrated to date. Many phleboviruses have been characterized in Africa, Europe, and Central Asia mainly from *Phlebotomus* and also from *Sergentomyia* species, and in the Americas from sandflies belonging to the

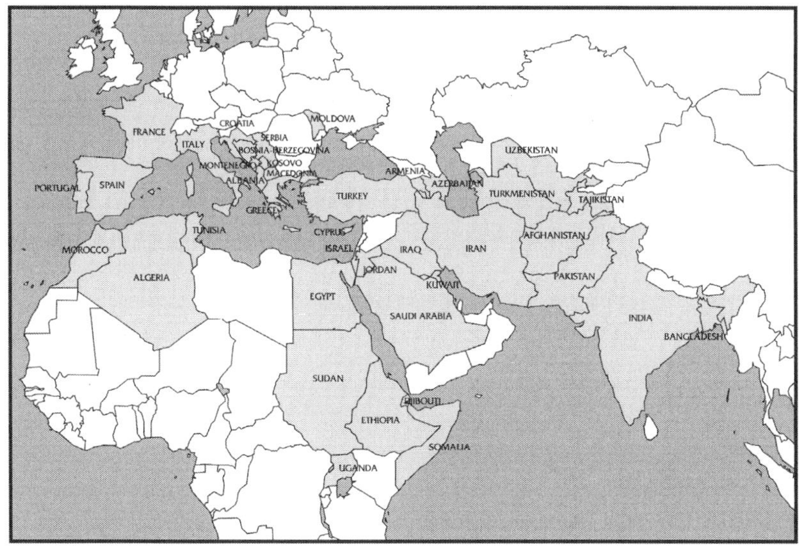

FIGURE 11.1 Countries with documented sandfly fever phlebovirus activity in the Old World.

genera *Lutzomyia sensu lato*[1,33]. *P. papatasi* has been widely acknowledged as the vector for SFSV and SFNV.[37,47,48] Furthermore, virus sequences have been detected in *P. major*, *P. ariasi*, *P. perniciosus*, *P. longicuspis*, *P. perfiliewi*, *P. neglectus*, and *Sergentomyia minuta*, implicating virus replication in a variety of sandfly species. Novel virus strains with unexplored clinical significance have also been characterized in sandflies.[18–27,49,50]

The sandfly acquires the virus as a result of biting an infected person any time from 48 hours before until 24 hours after the onset of fever. Following transmission, the virus requires an incubation period of 7–10 days, after which the sandfly remains infected for life. Many of the phleboviruses are maintained in their arthropod vectors by vertical (transovarial) transmission and that vertebrate hosts play little or no role in the basic virus maintenance cycle. This allows phleboviruses to persist during periods when adult vectors are absent or when susceptible vertebrate hosts are unavailable. Although different vertebrate species have been implicated in the SFSV life cycle, no confirmed reservoirs have been identified so far.[1,2,33,47,50,51]

Phlebotomine sandflies are present in the warm zones of Asia, Africa, Australia, southern Europe, and the Americas.[33,52] They are typically found in the moist subtropical countries of the Eastern Hemisphere between latitudes 20–45° North, particularly around the Mediterranean Sea, in the Middle East, and in parts of India. Their northward distribution reaches around the latitude of 50° North in southwest Canada and in northern France and Mongolia.[53,54] Although the southernmost distribution of sandflies extends to the latitude of 40° South, they are absent from New Zealand and the Pacific islands. The altitude range of sandfly breeding extends from 0 to 3300 meters.[55]

The epidemiological pattern of sandfly fever directly reflects the life cycle of Phlebotomine vectors. Cases usually begin to appear in April and gradually build to a peak in September. The disease may break out in epidemic form during the summer season following sandfly breeding. The factor currently known to limit the spread of sandfly fever is the distribution areas of potential vectors. The disease can emerge and/or reemerge in any geographical region, given that the sandfly vectors are active.[1,33]

5. WHICH FACTORS ARE INVOLVED IN DISEASE PATHOGENESIS? WHAT ARE THE PATHOGENIC MECHANISMS?

Since phleboviruses causing sandfly fever are not associated with human death or severe morbidity, no autopsy reports are available and pathological or histological changes in the affected individuals are not known. An adequate animal model is also lacking to study pathogenesis of the Old World sandfly fever phleboviruses, with the exception of TOSV.[56] Clinical or field isolates demonstrate poor infectivity for various laboratory animals including hamsters, mice, rats, rabbits, and guinea pigs.[7–36] No evidence of disease has been obtained with SFSV after inoculating various non-human primates, and febrile disease can only be induced in rhesus monkeys after intracerebral inoculation.[57] Adult female Syrian golden hamsters subcutaneously inoculated with Punta Toro virus develop a fulminant fatal illness characterized by hepatic and splenic necrosis and interstitial pneumonitis.[58]

The seroprevalence data from endemic countries indicate frequent virus exposure and inapparent or subclinical infections in indigenous populations. Viral replication is successfully controlled by the immune response, and specific IgM antibodies become detectable in blood during the first week after disease onset. The immunity to sandfly fever phleboviruses is type specific. Neutralizing antibodies produced during previous exposure are sufficient to suppress the occurrence of symptoms upon rechallenge with the homologous agent but do not confer cross-immunity, thus the person remains susceptible to infections with other phlebovirus serotypes.[2,34,59]

6. WHAT ARE THE CLINICAL MANIFESTATIONS?

The clinical presentation of sandfly fever due to infections with SFSV and SFNV are identical. After an incubation period of 3–6 days, the disease onset is characterized by the fever (38.8–40.3°C; 102–104.5°F), headache, retroorbital pain, photophobia, generalized aching, malaise, and chills. The face can be suffused, with injection of the conjunctivas and scleras. Photophobia may be present and can be accompanied by intense ocular pain on movement of the eyes. A faint pink erythema may be observed over the shoulders and thorax. Sandfly bites may also be recognized on the skin. Although lymphadenopathy and hepatosplenomegaly are uncommon, the spleen may be palpable in a small percentage of patients. Abdominal pain, discomfort, diarrhea, or constipation may also occur. The duration of fever is 2–4 days in 85% of cases, but may extend to 11 days. During the first day of fever, the pulse may be accelerated. Rarely, a second febrile episode can be observed. Leukopoenia is present in most cases at admission to the hospital and the lowest counts are recorded in the immediate post-febrile period. Following the febrile stage, there is fatigue and weakness, accompanied by slow pulse and frequently subnormal blood pressure. Convalescence may require a few days to several weeks and can be severely incapacitating. The prognosis is favorable without any complications or sequelae and no mortality that can be directly attributed to sandfly fever has ever been recorded.[34,59]

In experimentally-induced SFSV infections in volunteer young adults, the incubation period ranged from 50 hours to 9 days, with an average of 42–44 hours after intravenous inoculation. The onset was sudden with fever, chills, frontal headache, pain in the eyes, photophobia, backache, arthralgia, nuchal stiffness, loss of appetite, vomiting, alteration of taste, sore throat, and epistaxis. Not all patients had all symptoms. Some had giddiness and weakness, especially during recovery. Fever lasted 2–4 days. Bradycardia was noted toward the end of the febrile period, and into convalescence. Rash did not occur, although erythema of the face and conjunctival injection was common. Leukopenia was noted on the second day of fever with relative lymphopenia and neutropenia. The lymphocytes recovered more rapidly than neutrophils. The urine, liver function tests, and cerebrospinal fluid were normal. Five percent had recurrence of fever 5–7 days after initial recovery.[7,34,59,60]

In addition to the symptoms of sandfly fever, SFTV infections frequently present with aggravated gastrointestinal symptoms, elevation of hepatic enzymes, creatine kinase, and thrombocytopenia. Increased levels of alkaline phosphatase and gamma-glutamyl transpeptidase may also be noted.[3,61] Similar clinical and laboratory findings have also been documented for SFCV.[14] The post-infectious asthenia syndrome is well characterized in patients recovering from SFTV infections.[61]

In contrast to TOSV, SFSV and SFNV infections have generally not been associated with neurological manifestations and symptoms involving the central nervous system are rare or absent in the majority of the sandfly fever infections. However, cases of meningitis/meningoencephalitis due to SFSV and/or related viruses and probable central nervous system invasion of certain strains have been recognized.[1,8,27,62] Mechanisms, underlying conditions, or risk factors for central nervous system infections with phleboviruses other than TOSV require further investigation.

7. HOW DO YOU DIAGNOSE?

The diagnosis of sandfly fever is frequently based on clinical symptoms in endemic regions and laboratory confirmation is usually sought in specific incidents, travel-related/imported infections, or outbreaks. Routine laboratory examinations may reveal leukopenia, lymphopenia, neutropenia, or thrombocytopenia, elevation of hepatic enzymes, creatine kinase, alkaline phosphatase, and/or gamma-glutamyl transpeptidase, which are transient and may be pronounced, mild, or absent depending on the phase of the disease.

Serum or plasma is usually sufficient for definitive diagnostic testing for sandfly fever unless there are signs/symptoms of central nervous system involvement, when cerebrospinal fluid samples must also be submitted. Specific diagnosis can be achieved via direct viral detection or via the demonstration of virus-specific immunoglobulins.[2] Direct viral diagnosis can be performed via virus isolation on cell cultures or via viral RNA detection. Although SFSV and SFNV can be propagated in Vero, BHK-21, and LLC-MK2 cells, virus isolation is rarely attempted for diagnosis, has lower sensitivity compared to RNA detection, and is performed only in specialized laboratories.[2,63] Reverse transcription polymerase chain reaction (RT-PCR) techniques are commonly employed for viral RNA detection in clinical samples. RT-PCR applications involving various primer/probe sequences and detection methods for SFSV, TOSV, and other phleboviruses have been described.[64–67] However, since viremia is low and transient in the majority of the cases, direct methods for virus detection can only be effective during the early stages of infection (1–3 days after onset) and prior to seroconversion.[2,34,59,60]

Diagnosis via the demonstration of specific immunoglobulins is usually performed in sandfly fever. Complement fixation and hemagglutination inhibition methods have been historically used in detecting seroconversion.[36,59,60,68] Solid phase immunoassays such as immunofluorescence assay (IFA), enzyme-linked immunosorbent assay (ELISA), and immunoblot assays have been developed, which enable the detection of IgM or IgG class antibodies.[2,69,70] However, commercial tests for TOSV diagnosis are more common, and assays readily available for sandfly fever phleboviruses are currently few. Commercial or in-house solid phase immunoassays demonstrate variations of sensitivity and specificity due to the detection method and viral antigens employed in the test.[2,70] Cross-reactivity in serological assays is well known among phleboviruses, particularly among antigenically similar isolates belonging to the same genus or serocomplex such as SFNV or SFSV.[3,62,70,71] Viral neutralization test (VNT) remains as the reference method to assess the specificity of the antibody response. Thus, when precise serological characterization of a particular phlebovirus is required, reactive results in ELISA or IFA must be confirmed via VNT incorporating viruses of the same and distinct serocomplexes. Although labor intensive and only established in specialized laboratories, VNT is regarded as the gold standard assay for specificity and for calculating the antibody titer.[2,28,71,72]

8. HOW DO YOU DIFFERENTIATE THIS DISEASE FROM SIMILAR ENTITIES?

Sandfly fever presents with signs and symptoms of an undifferentiated febrile illness and a similar clinical picture can be observed in many infectious and non-infectious diseases. Since sandfly fever is non-fatal and frequently self-limited with or without supportive treatment, not all affected individuals seek medical assistance and a significant portion of symptomatic infections may be undocumented. Epidemiological data are of particular significance in the diagnosis due to the seasonality and geographical distribution of vector activity. In suspected cases, a thorough medical history should be obtained, including recent travels to the endemic areas and insect bites. Travel-related sandfly fever is usually reported during or after visits to the Mediterranean area and during summer months when the sandflies are active.[3,5,8–10,45] In the physical examination, sandfly, mosquito, or tick bites should be sought, suggesting exposure to arthropod vectors. The differential diagnosis should include other vector-borne diseases depending on the travel history, viral and parasitic infections, mild forms of hepatitis, and hematologic diseases. Although rarely observed in sandfly fever, symptoms suggesting central nervous system involvement should be recognized to differentiate and diagnose TOSV infections with similar epidemiological features and other viral encephalitides for timely intervention and proper management.

9. WHAT IS THE THERAPEUTIC APPROACH?

The treatment of phlebovirus infections is mainly symptomatic and hepatotoxic medications as well as aspirin and other non-steroidal anti-inflammatory drugs such as ibuprofen and ketoprofen are not recommended.[1] Animal models have demonstrated that viruses in the *Bunyaviridae* family are inhibited with interferon and interferon inducers. Moreover, bunyaviruses are generally considered sensitive to ribavirin, as evident in currently suggested therapies for Crimean-Congo hemorhaggic fever.[73] Ribavirin administered as an oral dose of 400 mg every 8 hours, started 1 day before experimental infection for 8 days, has been demonstrated to protect volunteers against SFSV challenge.[74] Ribavirin combined with human recombinant interferon-α was proposed as a treatment for SFSV infection, based on *in vitro* efficacy.[75] In addition to ribavirin, SFSV replication could be suppressed *in vitro* by 6-azauridine, interferon-α, glycyrrhizin, suramin sodium, dextran sulfate, and pentosan polysulphate.[75] Phleboviruses, as well as other members of the *Bunyaviridae* family, were observed to be inhibited by interferon-induced MxA protein *in vitro*.[76] Selenazole, a nucleoside carboxamide, was also effective in suppressing the replication of SFSV in cell cultures.[77] Favipiravir, a pyrazine derivative, has been found to be effective against SFNV *in vitro* and demonstrated higher therapeutic indexes compared to ribavirin in hamsters with Punta Toro virus infections, suggesting these compounds as promising antivirals.[78,79]

Although the disease course in sandfly fever is frequently benign and an evidence-based drug therapy is not possible due to the lack of randomized human trials, interferon–ribavirin combination might be considered as an emergency therapy in unique circumstances, such as a worsening disease in critically ill patients.[2]

10. WHAT ARE THE PREVENTIVE AND INFECTION CONTROL MEASURES?

Currently, no vaccines against sandfly-borne phleboviruses are available. Since the immunity is serotype specific and do not confer cross-protection with other virus serotypes, an effective vaccine must be able to mount an immune response against all relevant phlebovirus strains capable of producing clinical disease.[59] The prevention of sandfly fever relies on the control of vector proliferation in areas where people are likely to be exposed and use of individual protective measures[1]. General personal protection measures against sandfly bites include wearing of long sleeves and pants, usage of impregnated bednets, avoiding outdoor activities in the evening, and using insect repellents. Sandflies breed in vegetation within a few hundred feet of human habitations. Since these breeding sites are hard to discover, larvicidal control is impractical and control measures should be aimed at interrupting contact between humans and adult female sandflies.[80] The bloodsucking females feed only from sunset to sunrise and generally remain close to the ground. Ordinary mosquito netting and screening are of limited use, because unfed female sandflies can pass through 18-mesh squares.[33,80] Insecticide-impregnated or non-impregnated bednets with small mesh are effective in preventing sandfly bites.[81,82] Travelers to endemic countries should be encouraged to employ personal protection, especially during seasons when sandflies are active.

REFERENCES

1. Depaquit J, Grandadam M, Fouque F, et al. Arthropod-borne viruses transmitted by Phlebotomine sandflies in Europe: a review. *Euro Surveill* 2010;**11**(15):19507.
2. Dionisio D, Esperti F, Vivarelli A, Valassina M. Epidemiological, clinical and laboratory aspects of Sandfly Fever. *Curr Opin Infect Dis* 2003;**16**:383–8.
3. Carhan A, Uyar Y, Ozkaya E, et al. Characterization of a new phlebovirus related to sandfly fever sicilian virus isolated during a sandfly fever epidemic in Turkey. *J Clin Virol* 2010;**48**:264–9.
4. Ellis SB, Appenzeller G, Lee H, et al. Outbreak of sandfly fever in central Iraq, September 2007. *Mil Med* 2008;**173**:949–53.
5. Niklasson B, Eitrem R. Sandfly fever among Swedish UN troops in Cyprus. *Lancet* 1985;**1**:1212.

6. Eitrem R, Vene S, Niklasson B. Incidence of sandfly fever among Swedish United Nations soldiers on Cyprus during 1985. *Am J Trop Med Hyg* 1990;**43**:207–11.
7. Sabin AB. Experimental studies on Phlebotomus (pappataci, sandfly) fever during World War II. *Arch Gesamte Virusforsch* 1951;**4**:367–410.
8. Becker M, Zielen S, Schwartz TF, et al. Pappataci fever. *Klin Padiatr* 1997;**209**:377–9.
9. Schultze D, Korte W, Rafeiner P, Niedrig M. First report of sandfly fever virus infection imported from Malta into Switzerland, October 2011. *Euro Surveill* 2012;**17**:20209.
10. Schwarz TF, Jager G, Gilch S, Pauli C. Serosurvey and laboratory diagnosis of imported sandfly fever virus, serotype Toscana, infection in Germany. *Epidemiol Infect* 1995;**114**:501–10.
11. Plyusnin A, Beaty BJ, Elliott RM, Goldbach R, Kormelink R, Lundkvist A, et al. Virus Taxonomy: Classification and Nomenclature of Viruses. In: King AMQ, Lefkowitz E, Adams MJ, Carstens EB, editors. *Ninth Report of the International Committee on Taxonomy of Viruses*. San Diego: Elsevier; 2011. p. 693–709.
12. Elliott RM, Schmaljohn CS, Collett MS. Bunyaviridae genome structure and replication. *Curr Top Microbiol Immunol* 1991;**169**:91–141.
13. Suzich JA, Kakach LT, Collett MS. Expression strategy of a phlebovirus: biogenesis of proteins from the Rift Valley fever virus M segment. *J Virol* 1990;**64**:1549–55.
14. Papa A, Konstantinou GV, Pavlidou V, Antoniadis A. Sandfly fever virus outbreak in Cyprus. *Clin Microbiol Infect* 2006;**12**:192–4.
15. Charel RN, Gallian P, Navarro-Mari JM, Nicoletti L, et al. Emergence of Toscana virus in Europe. *Emerg Infect Dis* 2005;**11**:1657–63.
16. Hemmersbach-Miller M, Parola P, Charrel RN, Paul Durand J, Brouqui P. Sandfly fever due to Toscana virus: an emerging infection in southern France. *Eur J Intern Med* 2004;**15**:316–7.
17. Portolani M, Sabbatini AM, Beretti F, Gennari W, Tamassia MG, Pecorari M. Symptomatic infections by toscana virus in the Modena province in the triennium 1999–2001. *New Microbiol* 2002;**25**:485–8.
18. Tesh R, Saidi S, Javadian E, Nadim A. Studies on the epidemiology of sandfly fever in Iran. I. Virus isolates obtained from Phlebotomus. *Am J Trop Med Hyg* 1977;**26**:282–7.
19. Rodhain F, Madulolebond G, Hannoun C, Tesh RB. Corfou virus—a new phlebovirus isolated from phlebotomine sandflies in Greece. *Ann Inst Pasteur Virol* 1985;**136**:161–6.
20. Charel RN, Moureau G, Temmam S, et al. Massilia virus, a novel phlebovirus (*Bunyaviridae*) isolated from sandflies in the Mediterranean. *Vector Borne Zoonotic Dis* 2009;**9**:519–30.
21. Collao X, Palacios G, de Ory F, et al. Granada virus: a natural phlebovirus reassortant of the sandfly fever Naples serocomplex with low prevalence in humans. *Am J Trop Med Hyg* 2010;**83**:760–5.
22. Zhioua E, Moureau G, Chelbi I, et al. Punique virus, a novel phlebovirus, related to sandfly fever Naples virus, isolated from sandflies collected in Tunisia. *J Gen Virol* 2010;**91**:1275–83.
23. Papa A, Velo E, Bino S. A novel phlebovirus in Albanian sandflies. *Clin Microbiol Infect* 2011;**17**:585–7.
24. Moureau G, Bichaud L, Salez N, et al. Molecular and serological evidence for the presence of novel phleboviruses in sandflies from northern Algeria. *Open Virol J* 2010;**4**:15–21.
25. Peyrefitte CN, Grandadam M, Bessaud M, et al. Diversity of Phlebotomus perniciosus in Provence, Southeastern France: Detection of two putative new phlebovirus sequences. *Vector Borne Zoonotic Dis* 2013;**13**:630–6.

26. Verani P, Ciufolini MG, et al. Ecology of viruses isolated from sand flies in Italy and characterized of a new Phlebovirus (Arabia virus). *Am J Trop Med Hyg* 1988;**38**:433–9.
27. Anagnostou V, Pardalos G, Athanasiou-Metaxa M, Papa A. Novel phlebovirus in febrile child, Greece. *Emerg Infect Dis* 2011;**17**:940–1.
28. Tesh RB, Saidi S, Gajdamovic SJ, Rodhain F, Vesenjak-Hirjan J. Serological studies on the epidemiology of sandfly fever in the Old World. *Bull World Health Organ* 1976;**54**:663–74.
29. Navarro-Mari JM, Gomez-Camarasa C, Perez-Ruiz M, Sanbonmatsu-Gamez S, Pedrosa-Corral I, Jimenez-Valera M. Clinicoepidemiologic study of human infection by Granada virus, a new Phlebovirus within the Sandfly fever Naples serocomplex. *Am J Trop Med Hyg* 2013;**88**:1003–6.
30. Calisher CH, McLean RG, Smith GC, Szmyd DM, Muth DJ, Lazuick JS. Rio Grande—a new phlebotomus fever group virus from south Texas. *Am J Trop Med Hyg* 1977;**26**:997–1002.
31. Tesh RB, Chaniotis BN, Peralta PH, Johnson KM. Ecology of viruses isolated from Panamanian phlebotomine sandflies. *Am J Trop Med Hyg* 1974;**23**:258–69.
32. Trapp EE, Andrade AH, Shope RE. Itaporanga a newly recognized arbovirus from Sao Paulo state, Brazil. *Proc Soc Exp Biol Med* 1965;**118**:421–2.
33. Maroli M, Feliciangeli MD, Bichaud L, Charrel RN, Gradoni L. Phlebotomine sandflies and the spreading of leishmaniases and other diseases of public health concern. *Med Vet Entomol* 2013;**27**:123–47.
34. Ergunay K, Whitehouse C, Ozkul A. Current status of human arboviral infections in Turkey. *Vector Borne Zoonotic Dis* 2011;**11**:731–41.
35. Sabin AB, Philip CB, Paul JR. Phlebotomus (pappataci or sandfly) fever. A disease of military importance: summary of existing knowledge and preliminary report of original investigations. *JAMA* 1944;**125**:603–6.
36. Sabin AB. Recent advances in our knowledge of dengue and sandfly fever. *Am J Trop Med Hyg* 1955;**4**:198–207.
37. Hertig M, Sabin AB. Sandfly fever (pappataci, phlebotomus, three-day fever). In: Hoff EC, editor. *Preventive Medicine in World War II*, Vol. 7. Washington DC: Communicable Diseases Office of the Surgeon General, U.S. Department of the Army; 1964. p. 109–74.
38. Javadian E, Tesh R, Saidi S, Nadim A. Studies on the epidemiology of sandfly fever in Iran. Host-feeding patterns of Phlebotomus papatasi in an endemic area of the disease. *Am J Trop Med Hyg* 1977;**26**:294–8.
39. Gaidamovich SI, Obukhova VR, Sveshnikova NA, Cherednichenko IN, Kostiukov MA. Natural foci of viruses borne by Phlebotomus papatasi in the U.S.S.R. according to a serologic study of the population. *Vopr Virusol* 1978;**5**:556–60.
40. Eitrem R, Stylianou M, Niklasson B. High prevalence rates of antibody to three sandfly fever viruses (Sicilian, Naples, and Toscana) among Cypriots. *Epidemiol Infect* 1991;**107**: 685–91.
41. Goverdhan MK, Dhanda V, Modi GB, et al. Isolation of Phlebotomus (sandfly) fever virus from sandflies and humans during the same season in Aurangabad district, Maharashtra state, India. *Indian J Med Res* 1976;**64**:57–63.
42. Filipe AR. Serological survey for antibodies to arboviruses in the human population of Portugal. *Trans R Soc Trop Med Hyg* 1974;**68**:311–4.
43. Gaidamovich SY, Kurakhmedova SA, Melnikova EE. Aetiology of Phlebotomus fever in ashkhabad studied in retrospect. *Acta Virol* 1974;**18**:508–11.
44. Tesh RB, Papaevangelou G. Effect of insecticide spraying for malaria control on the incidence of sandfly fever in Athens, Greece. *Am J Trop Med Hyg* 1977;**261**:163–6.

45. Guler S, Guler E, Caglayik DY, et al. A sandfly fever virus outbreak in the East Mediterranean region of Turkey. *Int J Infect Dis* 2012;**16**:e244–6.
46. Eitrem R, Niklasson B, Weiland O. Sandfly fever among Swedish tourists. *Scand J Infect Dis* 1991;**23**:451–7.
47. Tesh RB, Modi GB. Studies on the biology of phleboviruses in sand flies (Diptera: Psychodidae). Experimental infection of the vector. *Am J Trop Med Hyg* 1984;**33**:1007–16.
48. Verani P, Lopes MC, Nicoletti L, Balducci M. Studies on Phlebotomus-transmitted viruses in Italy: I. Isolation and characterization of a Sandfly fever Naples-like virus. Arboviruses in the Mediterranean Countries. *Zentralbl Bakteriol* 1980;(Suppl 9):195–201.
49. Izri A, Temmam S, Moureau G, Hamrioui B, de Lamballerie X, et al. Sandfly fever Sicilian virus, Algeria. *Emerg Infect Dis* 2008;**14**:795–7.
50. Ergunay K, Erisoz Kasap O, Kocak Tufan Z, Turan MH, Ozkul A, Alten B. Molecular evidence indicates that Phlebotomus major sensu lato (Diptera: Psychodidae) is the vector species of the recently-identified sandfly fever Sicilian virus variant: sandfly fever turkey virus. *Vector Borne Zoonotic Dis* 2012;**12**:690–8.
51. Tesh RB, Modi GB. Maintenance of Toscana virus in *Phlebotomus perniciosus* by vertical transmission. *Am J Trop Med Hyg* 1987;**36**:189–93.
52. Killick-Kendrick R. The biology and control of phlebotomine sandflies. *Clin Dermatol* 1999;**17**:279–89.
53. Young DG, Perkins PV. Phlebotomine sandflies of North America (Diptera: Psychodidae). *J Am Mosq Control Assoc* 1984;**44**:263–304.
54. Lewis DJ. A taxonomic review of the genus Phlebotomus (Diptera: Psychodidae). *Bull Brit Mus (Nat Hist) Entomology Series* 1982;**45**:121–203.
55. Lane RP. Sandflies (Phlebotominae). In: Lane RP, Crosskey RM, editors. *Medical Insects and Arachnids*. London: Chapman & Hall; 1993. p. 78–119.
56. Cusi MG, Gori Savellini G, Terrosi C. Development of a mouse model for the study of Toscana virus pathogenesis. *Virology* 2005;**333**:66–73.
57. McClain DJ, Summers PL, Pratt WD, Davis KJ, Jennings GB. Experimental infection of nonhuman primates with sandfly fever virus. *Am J Trop Med Hyg* 1997;**56**:554–60.
58. Fisher AF, Tesh RB, Tonry J, Guzman H, Liu D, Xiao SY. Induction of severe disease in hamsters by two sandfly fever group viruses, Punta toro and Gabek Forest (Phlebovirus, Bunyaviridae), similar to that caused by Rift Valley fever virus. *Am J Trop Med Hyg* 2003;**69**:269–76.
59. Bartelloni PJ, Tesh RB. Clinical and serologic responses of volunteers infected with phlebotomus fever virus (Sicilian type). *Am J Trop Med Hyg* 1976;**25**:456–62.
60. Beisel WR, Herman YF, Sauberlich HE, Herman RH, Bartelloni PJ, Canham JE. Experimentally induced sandfly fever and vitamin metabolism in man. *Am J Clin Nutr* 1972;**25**:1165–73.
61. Kocak Tufan Z, Weidmann M, Bulut C, et al. Clinical and laboratory findings of a sandfly fever Turkey Virus outbreak in Ankara. *J Infect* 2011;**63**:375–81.
62. Ergunay K, Ismayilova V, Colpak IA, Kansu T, Us D. A case of central nervous system infection due to a novel Sandfly Fever Virus (SFV) variant: Sandfly Fever Turkey Virus (SFTV). *J Clin Virol* 2012;**54**:79–82.
63. Karabatos N, editor. *International Catalogue of Arboviruses including Certain other Viruses of Vertebrates*. 3rd ed. San Antonio, Texas, USA: American Society of Tropical Medicine and Hygiene; 1985.
64. Sanchez-Seco MP, Echevarria JM, Hernandez L, et al. Detection and identification of Toscana and other phleboviruses by RT-nested-PCR assays with degenerated primers. *J Med Virol* 2003;**71**:140–9.

65. Weidmann M, Sanchez-Seco MP, Sall AA, et al. Rapid detection of important human pathogenic Phleboviruses. *J Clin Virol* 2008;**41**:138–42.
66. Hasib L, Dilcher M, Hufert F, Meyer-König U, Weidmann M. Development of a flow-through microarray based reverse transcriptase multiplex ligation-dependent probe amplification assay for the detection of European Bunyaviruses. *Mol Biotechnol* 2011;**49**:176–86.
67. Ibrahim SM, Aitichou M, Hardick J, Blow J, O'Guinn ML, Schmaljohn C. Detection of Crimean-Congo hemorrhagic fever, Hanta, and sandfly fever viruses by real-time RT-PCR. *Methods Mol Biol* 2011;**665**:357–68.
68. Clarke DH, Cassals J. Techniques for hemagglutination and hemagglutination-inhibition with arthropod-borne viruses. *Am J Trop Med Hyg* 1958;**7**:561–73.
69. Eitrem R, Vene S, Niklasson B. ELISA for detection of IgM and IgG antibodies to sandfly fever Sicilian virus. *Res Virol* 1991;**142**:387–94.
70. Ergünay K, Litzba N, Lo MM, et al. Performance of various commercial assays for the detection of Toscana virus antibodies. *Vector Borne Zoonotic Dis* 2011;**11**:781–7.
71. Tesh RB, Peralta PH, Shope RE, Chaniotis BN, Johnson KM. Antigenic relationships among phlebotomus fever group arboviruses and their implication for the epidemiology of sandfly fever. *Am J Trop Med Hyg* 1975;**24**:135–44.
72. Calisher CH, Beaty BJ, Chandler LJ. Arboviruses. Revised and Expanded In: Lenette EH, Smith TF, editors. *Laboratory Diagnosis of Viral Infections*. 3rd ed. New York: Marcel Dekker; 1999. p. 305–32.
73. Sidwell RW, Smee DF. Viruses of the Bunya- and Togaviridae families: potential as bioterrorism agents and means of control. *Antiviral Res* 2003;**57**:101–11.
74. Huggins JW. Prospects for treatment of viral hemorrhagic fevers with ribavirin, a broad-spectrum antiviral drug. *Rev Infect Dis* 1989;**11**(Suppl 4):S750–61.
75. Crance JM, Gratier D, Guimet J, Jouan A. Inhibition of sandfly fever Sicilian virus (Phlebovirus) replication in vitro by antiviral compounds. *Res Virol* 1997;**148**:353–65.
76. Frese M, Kochs G, Feldmann H, et al. Inhibition of bunyaviruses, phleboviruses, and hantaviruses by human MxA protein. *J Virol* 1996;**70**:915–23.
77. Kirsi JJ, North JA, McKernan PA, et al. Broad-spectrum antiviral activity of 2-beta-D-ribofuranosylselenazole-4-carboxamide, a new antiviral agent. *Antimicrob Agents Chemother* 1983;**24**:353–61.
78. Gowen BB, Wong MH, Jung KH, et al. In vitro and in vivo activities of T-705 against arenavirus and bunyavirus infections. *Antimicrob Agents Chemother* 2007;**51**:3168–76.
79. Gowen BB, Wong MH, Jung KH, Smee DF, Morrey JD, Furuta Y. Efficacy of favipiravir (T-705) and T-1106 pyrazine derivatives in phlebovirus disease models. *Antiviral Res* 2010;**86**:121–7.
80. Faiman R, Cuno R, Warburg A. Control of phlebotomine sand flies with vertical fine-mesh nets. *J Med Entomol* 2009;**46**:820–31.
81. Jalouk L, Al Ahmed M, Gradoni L, Maroli M. Insecticide-treated bednets to prevent anthroponotic cutaneous leishmaniasis in Aleppo Governate, Syria: results from two trials. *Trans R Soc Trop Med Hyg* 2007;**101**:360–7.
82. Alten B, Caglar SS, Simsek FM, Kaynas S. Evaluation of protective efficacy of K-OTAB impregnated bednets for cutaneous leishmaniasis control in Southeast AnatoliaTurkey. *J Vector Ecol* 2003;**28**:97–107.

Chapter 12

Chikungunya Fever

Giovanni Rezza
Department of Infectious, Parasitic and Immunomediated Diseases, Istituto Superiore di Sanità, Rome, Italy

CASE REPRESENTATION

At the end of August 2007, the health district authority of Ravenna, a town in north-eastern Italy, informed the Istituto Superiore di Sanità (the Italian National Institute of Health) about an unusual event: a fever outbreak that they imagined to be "pappataci fever" (i.e., Toscana virus infection, which usually causes sporadic cases of fever or mild meningoencephalitis), which appeared to be localized in two small villages. The large number of cases urged a site visit to investigate the nature of the outbreak.

The clinical investigation highlighted that the patients were affected by high fever and joint pain. Headache and muscle pains were also commonly reported, whereas about half of them presented a macular skin rash, sometimes accompanied by itching. Joint pain was particularly severe, and some patients complained of persistence of pain that needed treatment for at least 1 month. The entomological investigation did not show the presence of pappataci, whereas a large number of *Aedes albopictus* mosquitoes (the so-called "tiger mosquito") were identified. Thus, on clinical and entomological grounds, chikungunya virus (CHIKV) or dengue virus (DENV) were considered as the presumptive causal agents of the illness. Serum samples were collected and analyzed, and the diagnosis of CHIK fever (CHIKF) was confirmed by the use of PCR and/or serological assays (IgM against CHIKV). The epidemiological investigation revealed that the epidemic apparently started about 10 days after a few hours' visit of a viremic person (the index case: a man coming from Kerala, India) to his relatives living in one of the originally affected villages, Castiglione di Cervia. This village, together with the contiguous village of Castiglione di Ravenna, represents the area where most of the cases subsequently occurred. The man developed high fever during the evening; *Aedes albopictus* mosquitoes, after biting the infected man, amplified the virus and the chain of transmission was initiated.

Emerging Infectious Diseases. DOI: http://dx.doi.org/10.1016/B978-0-12-416975-3.00012-1

Then, during the course of the epidemic, several minor outbreaks occurred outside the initially affected area: small clusters of cases occurred in neighboring towns, within a radius of few kilometers, whereas other clusters were detected at an appreciable distance, up to 75 km from the epicenter of the epidemic.

The implementation of vector control measures and the decline of mosquito activity at the beginning of the cold season allowed outbreak control.

(This is a published case, *Lancet* 2007.[1])

1. WHY THIS CASE WAS SIGNIFICANTLY IMPORTANT AS AN EMERGING INFECTION

CHIKF is a viral disease which is endemic in Africa and causes recurrent epidemics in Asia. In recent years, a large epidemic ravaged Indian Ocean islands and the Indian subcontinent. During the epidemic, smaller outbreaks and sporadic autochthonous cases of CHIKF have been identified in temperate areas, such as north-eastern Italy and Mediterranean France, expanding the geographical range of detection of the virus.

2. WHAT IS THE CAUSATIVE AGENT?

CHIKF is caused by the CHIKV, an arbovirus (i.e., arthropod-borne virus) belonging to the alphavirus genus of the *Togaviridae* family. CHIKV has a single-stranded RNA genome, a 60–70 nm diameter capsid, and a phospholipid envelope. The virus contains four non-structural proteins (nsP1–4) and structural proteins (C, E3, E2, 6 K, and E1). The virus is sensitive to desiccation and to temperature above 58°C[2] (Figure 12.1).

Alphaviruses are subdivided into those associated with polyarthritis and rash (predominantly Old World strains) and those associated with

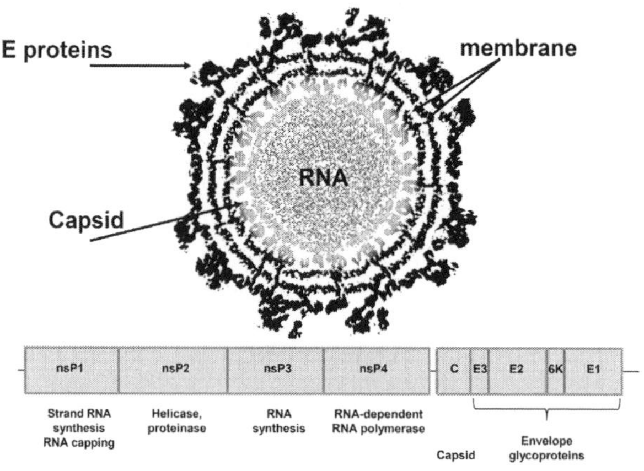

FIGURE 12.1 CHIKV characteristics and genome structure.

meningoencephalitis (predominantly New World strains). CHIKV is a member of the Old World group, along with other viruses that may cause human diseases, such as the o'nyong nyong virus (Central Africa, similar to CHIKV), Ross River and Barmah Forest viruses (Australia and the Pacific), Sindbis virus (cosmopolitan), and Mayaro virus (South America).

Three different genotypes of CHIKV have been identified: the western African, the east-central-south African (ECSA), and the Asian genotype.[2,3] Evolutionary studies have suggested an African origin for CHIKV.[4] The Asian lineage exhibits a peculiar pattern of spread, with successive epidemics detected along an eastward path. The old Asian lineage, which caused epidemics in the 1950s, was a variant of the ECSA strain that probably split into two clades: an Indian lineage, which likely went extinct, and a South-East Asian lineage.[4] The old Asian genotype has not been identified during the last epidemic in India, consistently with lack of sustainability of the human–mosquito cycle at a local scale in the absence of continued importation.[4,5]

CHIKV strains, which caused a large epidemic since 2004, spreading from Kenya toward Indian Ocean islands, the Indian subcontinent, and South-East Asia, belong to the ECSA genotype.[2,6,7] During the epidemic, two different lineages were identified, suggesting independent introductions of CHIKV strains from Kenya into the Indian Ocean islands and India.[4] A viral variant, presenting a substitution of the amino acid alanine with valine in the position 226 of the E1 protein (one of the two major envelope surface glycoproteins) of CHIKV (A226V), selected during the course of the epidemic, originated in the Indian Ocean and became predominant in specific affected areas where *Aedes albopictus* was largely predominant compared to *Aedes aegypti*, such as La Réunion and the Kerala district in India, allowing an efficient replication and dissemination of the A226V variant of CHIKV.[8]

While the 226 A strain was the only genotype observed during the first period of the outbreak (March to June 2005), the E1 A226V genotype started to be observed since September 2005. On La Réunion, the identification of A226V variant circulation preceded by at least 3 months the explosive epidemic peak of mid-December, suggesting an increase in viral transmission.[9] In the Indian region of Kerala, CHIKV strains isolated in 2007 presented the A226V mutation, which had not been found in strains isolated in Kerala and in other Indian regions in 2006.[10,11] Thus, a single amino acid substitution may have influenced vector specificity, increasing the fitness of CHIKV for specific vector species and consequently CHIKV transmission.[12] The same variant caused the outbreak propagated by *Aedes albopictus* in north-eastern Italy.[1]

3. WHAT IS THE FREQUENCY OF THE DISEASE?

CHIKV was first isolated in the Newala district of Tanzania in 1952/1953.[13] Since its identification, sporadic cases and a number of outbreaks have been reported in several African countries, on the Indian subcontinent, and in

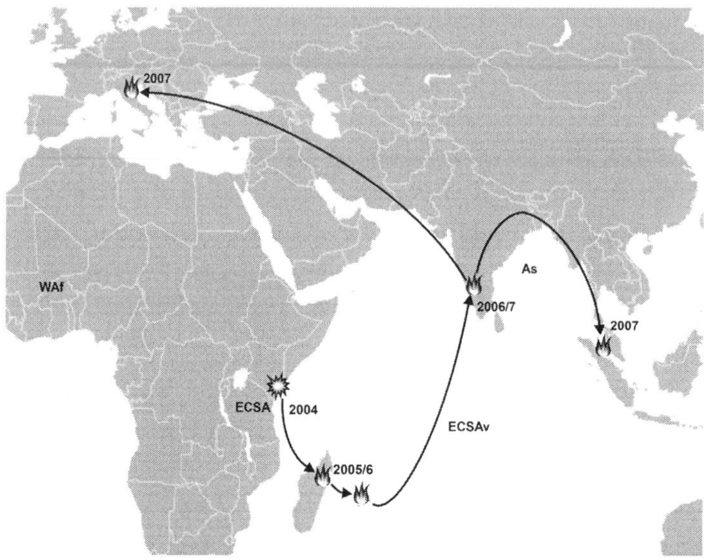

FIGURE 12.2 Geographical distribution of CHIKV genotypes and the spread of the ECSA-derived epidemic variant (ECSAv) in and from the Indian Ocean.

South-East Asia.[14,2] In the last few years, CHIKV has re-emerged, causing a series of large outbreaks, which started in Kenya in 2004, and ravaged the Comoros Islands, the island of La Réunion, and other islands in the south-west Indian Ocean in early 2005, which were followed by an epidemic in the Indian subcontinent in 2005/2006.[15,16] CHIKV caused an outbreak in the north-east of Italy in the summer of 2007.[1] Two autochthonous cases of CHIKF were also identified in Mediterranean France in the summer of 2010[17] (Figure 12.2).

4. HOW IS THE VIRUS TRANSMITTED?

CHIKV is transmitted to humans by the bite of *Aedes* spp. (i.e., *Aedes aegypti* and *Aedes albopictus*) mosquitoes. Although *Aedes aegypti* is the main vector of CHIKV, *Aedes albopictus* (the "tiger" mosquito) is becoming important in the emergence of CHIKV in temperate areas where *Aedes aegypti* is not detected.

The virus appears to be enzootic across tropical areas of Africa and Asia. In west and central Africa, CHIKV is maintained in a sylvatic cycle involving wild non-human primates and forest-dwelling *Aedes* spp. mosquitoes. However, there is little information on the vertebrate hosts involved in viral maintenance.[18] Both humans and wild primates throughout the humid forests and the semi-arid savannahs of Africa have significant levels of antibodies against CHIKV, with small-scale outbreaks following a 3–4-year cyclical

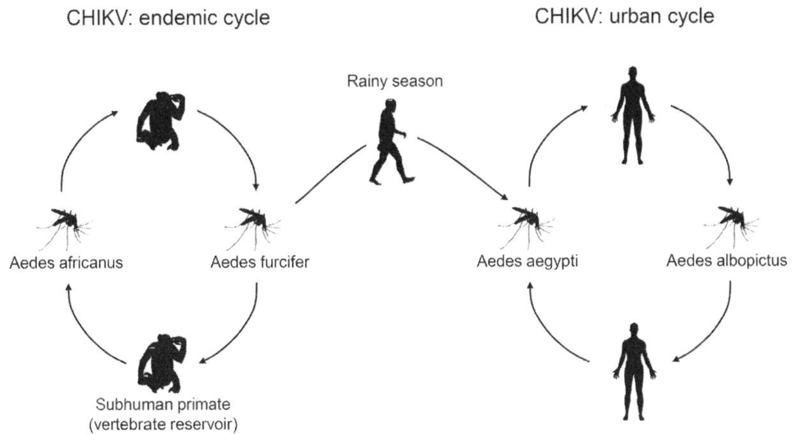

FIGURE 12.3 CHIKV endemic and urban (epidemic) cycle.

pattern consequent to the repopulation of susceptible, non-immune, wild primates.[19] In rural regions, outbreaks are more likely in the heavy rainfall season, when sylvatic mosquito density tends to increase. In Asia, the predominant vector is the urban, peri-domestic, anthropophilic *Aedes aegypti* mosquito, which is responsible for large-scale outbreaks characterized by long inter-epidemic periods, which may last several decades.[5,18] As reported in Chapter 2, since a vertebrate reservoir or a sylvan transmission cycle has not been identified in Asia, the virus is presumed to persist in a human–mosquito–human cycle.[18] However, this is still uncertain, since outbreaks of CHIKV fever in Asia have not been necessarily associated with outbreaks in Africa, which suggests an independent evolution of an African ancestor of CHIKV in Asia and the possibility of a sylvatic cycle maintaining autochthonous virus genotypes in the Asian continent[3] (Figure 12.3).

As already mentioned, although *Aedes aegypti* is the main vector of CHIKV, *Aedes albopictus* is playing an important role in the spread of the infection outside the area of activity of *Aedes aegypti* (Figure 12.4).

In Europe, *Aedes albopictus* was first detected in Albania in 1979 and established in the last decade in at least 12 countries, especially in southern Europe.[20] Although shipment of used tires infested with mosquito eggs is the main modality of spread, *Aedes albopictus* infestation has been recently associated with the shipment of a commercial plant product known as "lucky bamboo" (*Dracaena* spp.), packaged in standing water, and introduced in Europe (i.e., Netherlands) from mainland south China. The presence of *Aedes albopictus* in Italy, with the risk that this mosquito could transmit viral infections, has long been recognized, well before the occurrence of the Italian outbreak.[16,21] Two major tire retreading companies located in north-eastern Italy, importing used tires directly from the USA, were the source of mosquito

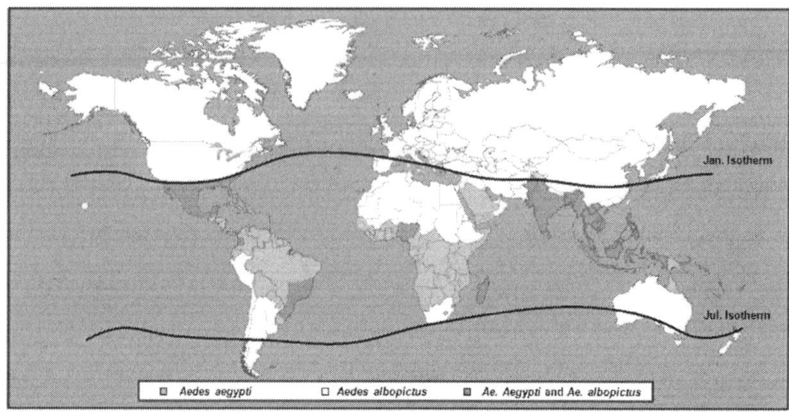

FIGURE 12.4 Geographical area of activity of *Aedes aegypti* and *Aedes albopictus*. Modified from Chevillon et al.[5]

entry to, and spread across, the country through internal trade of tires performed in collaboration with smaller companies.[22] During its further spread, *Aedes albopictus* demonstrated a high degree of fitness and capacity to breed in a huge variety of disposable containers and ground water collection sites.[23,24] In temperate areas, mosquito activity heavily declines during the cold season, thus transovarial transmission is a possible mechanism to augment the probability of virus overwintering. In La Réunion Island the virus has been isolated from field-collected *Aedes albopictus* females and in two out of 500 pools of larvae, demonstrating a low level of vertical transmission.[25]

Finally, during the recent epidemic on La Réunion, mother-to-child transmission has been documented, with an estimated vertical transmission rate of 49% in pre-term deliveries in the context of intra-partum viremia. Severe illness, mainly consisting of encephalopathy, was observed in 53% of the cases.[26]

5. WHICH FACTORS ARE INVOLVED IN DISEASE PATHOGENESIS? WHAT IS THE PATHOGENIC MECHANISM?

CHIKV is a member of the arthritogenic alphaviruses but sometimes it may cause meningoencephalitis.[27] Unlike typical encephalogenic viruses, which infect neurons, CHIKV seems to infect stromal cells of the central nervous system, in particular the lining of the choroid plexus.

Following transmission, CHIKV replicates in the skin, and then disseminates to the liver and joints, presumably through the blood. Epithelial and endothelial cells, fibroblasts, and, to a lesser extent, macrophages, are susceptible to CHIKV infection and allow viral production. CHIKV produces a marked cytopathic effect in cell cultures, and virus replication is associated with induction of apoptosis in infected cells.[28,29] Animal models show that the virus may be detected in muscle, joint, and skin fibroblasts, but also in epithelial and endothelial layers of organs like the liver, skin, and brain, as it happens in mice at higher risk of severe diseases such as those lacking the type I IFN receptor.[28] Non-human primate models also show persistent infection in splenic macrophages and endothelial cells lining the liver sinusoids. Samples from human patients affected by CHIKV myositic syndrome show viral antigen expression in skeletal muscle satellite cells but not in muscle fibers; fibroblasts may also be infected. The pathogenesis of the long-term severe debilitating arthralgia experienced by convalescent patients is still undefined, and the role of adaptive immune responses in the possible induction of autoimmunity caused by cross-reactivity between viral and host antigens. To this regard, the possibility that B and T cells responsive to CHIKV are implicated needs to be better explored.[29] However, the increased inflammatory cytokine expression (i.e., up-regulation of the pro-inflammatory cytokines interleukin-1 and interleukin-6) appears to be associated with joint disease severity.[30,31]

6. WHAT ARE THE CLINICAL MANIFESTATIONS?

The term CHIK means "to walk bent over," in the African dialect of Makonde (or Shawili), and refers to the effects of the incapacitating arthralgia.

The incubation period is short (i.e., 2–4 days on average, with a range of 1 to 12 days), and clinical onset is often abrupt. Patients affected by CHIK usually present with high fever, and severe and persistent joint pain; arthralgia/arthritis may affect about 70 to 80% of the patients and is more often localized to the extremities (i.e., ankles, wrists, phalanges), but large joints may also be affected. Headache, back and muscle pains are also common, whereas a macular skin rash is present in more than 50% of the cases, sometimes accompanied by itching. The frequency of the most common signs and symptoms of CHIKF are summarized in Table 12.1.[1,2,32,33] Lymphopenia and hypocalcemia are the most prominent laboratory findings, while thrombocytopenia is rarely observed.[34]

In most cases, CHIKF is a relatively mild disease, and severe cases, especially with neurological complications (i.e., encephalopathy, acute flaccid paralysis, or Guillain–Barré syndrome) or hemorrhagic manifestations, are relatively rare.[1,2,31] Encephalitis or meningoencephalitis is usually observed in older individuals with underlying disorders. Chronic joint pain, which may last for months or even years, is an important complication that occurs in more than one-third of the patients.

7. HOW DO YOU DIAGNOSE?

In case a person is suspected to be affected by CHIK fever, laboratory confirmation is needed, and both RT-PCR and serology (IgM or IgG, detected through HI, indirect fluorescent antibody (IFA), or enzyme-linked

TABLE 12.1 Chikungunya: Proportional Frequency of Signs and Symptoms

	Frequency (%)		
Sign/Symptom	Malaysia 1998[26]	Réunion 2005–2006[27]	Italy 2007[1]
Fever	100	100	100
Joint pain	78	100	97
Myalgia	50	60	46
Headache	50	70	51
Skin rash	50	39	52

immunoabsorbent assay (ELISA)) are to be used. RT-PCR is useful during the initial viremia phase (day 0 to 7), while IgM are detectable after an average of 2 days by ELISA or about 1 week by HI and persist for several weeks. IgG may be detected in convalescent sera.[1,2] Quantitative real-time reverse transcription-polymerase chain reaction (RT-PCR) may detect higher levels of CHIK viremia compared with other arthropod-borne infections. A four-fold increase in IgG titers in paired samples is also confirmatory of CHIKV infection. Finally, viral culture may be used as a diagnostic tool.

8. HOW DO YOU DIFFERENTIATE THIS DISEASE FROM SIMILAR ENTITIES?

CHIKF should be suspected in individuals who have recently visited areas affected by outbreaks and who present with fever and joint pain, with or without skin rash. Occasionally, joint pain may not be present or be so mild and unspecific as to be unreported, particularly in aged patients. Furthermore, other arboviral diseases, such as dengue, whose geographical distribution in Africa and Asia largely overlaps with that of CHIK virus, is responsible of similar symptoms.[35,36]

CHIKF may be misdiagnosed when one or more cases of a dengue-like syndrome occur in individuals who did not travel 2 weeks before fever onset. In this case, epidemiological (i.e., identification of index case and chain of transmission) and entomological (i.e., presence of a competent vector) investigation may corroborate clinical suspicion and guide lab diagnosis (Box 12.1).

9. WHAT IS THE THERAPEUTIC APPROACH?

There is no effective treatment for CHIKF. Specific antiviral drugs are not available and treatment is purely symptomatic. Ribavirin, which has antiviral activity against RNA viruses, is considered less active against alphaviruses,[37]

Box 12.1 Diagnostic Steps to be Adopted in Case of Identification of an Unusual Outbreak of Dengue-like Illness

The diagnosis of CHIKV fever should be taken into consideration in the presence of secondary cases of dengue-like illness and competent vectors even though travel-related cases (i.e., the index case/s) have not yet been identified.

A. One or more cases of a dengue-like illness (fever, joint pain, rash in about 50% of the cases) are identified

1. Common diseases should be excluded
2. Patient/s should be interviewed to disclose travel in CHIKV affected areas 2 weeks before fever onset
3. In the case of positive travel history, diagnostic tests for CHIKV and DENV should be performed
4. In the case of negative travel history, entomological survey should be conducted to identify presence of competent vectors (i.e., *Aedes albopictus*) in the affected area
5. If competent vectors are identified, diagnostic tests for CHIKV and DENV should be performed

B. Diagnostic tests for CHIKV and DENV

1. PCR (<1 week after fever onset)
2. IgM (≥2 days after fever onset) or a four-fold increase in IgG titer
3. IgG as detected by HI or other test (≥1 after fever onset) is to be considered diagnostic only for patients brought up in affected areas

whereas the only recommended treatments are non-steroidal anti-inflammatory drugs for joint pain.[31]

10. WHAT ARE THE PREVENTIVE AND INFECTION CONTROL MEASURES?

Measures for controlling the population of *Aedes albopictus* are of primary importance (Box 12.2). They include the use of fast-acting insecticides (synergized pyrethroids) for 3 consecutive days, applied with a truck-mounted atomizer in public spaces and a backpack mist blower in private spaces. Anti-larval measures, using formulations of insect growth regulators and *Bacillus thuringiensis* var. *israeliensis* serotype H14, are also important. Other interventions involve house-to-house interventions to control or eliminate breeding places, and encouraging community participation. For each suspected case of infection, these control measures should be done within a radius of 100 m of the individual's residence; for clusters, the control measures were extended within a 300-m radius of the most external case. Recommendations for mosquito control at community level and personal protective measures are summarized in Box 12.2.

Box 12.2 Fighting the "Tiger" Mosquito: Prevention and Outbreak Control Activities Should be Aimed at the Control of Competent Vectors

A. **Prevention Activities**

Vector control

1. Monitoring of *Aedes* indices
2. Source reduction
3. Preventive larviciding
4. Space spraying when necessary
5. Personal protection

Public awareness

Health education

Publicity

Mass media

Community participation

Routine recording and reporting

B. **Outbreak Control Activities**

Vector control

Outbreak control activities

1. Space spraying
2. Thermal fogs
3. Ultra low volume aerosols spray
4. Source reduction
5. Preventive larviciding on large-scale basis

Personal protection

1. Repellents
2. Mosquito coils
3. Window and door screens

Health education

Community participation

Mass media anti-CHIKV campaign

Cooperation between government agencies/public sectors

Follow-up activities

Routine recording and reporting

With regard to other prevention tools, safe and effective vaccines are not available yet, but research and development is ongoing with promising results.[31]

Finally, preparedness for unusual outbreaks in countries where the presence of competent vectors has been identified is key. Strengthening activities addressed toward surveillance and control of *Aedes albopictus* and early etiological identification of clusters of unusual human cases is essential to reduce the risk of outbreaks due to exotic viral agents in temperate climate countries.

REFERENCES

1. Rezza G, Nicoletti L, Angelini R, et al. Infection with Chikungunya virus in Italy: an outbreak in a temperate region. *Lancet* 2007;**370**:1840–6.
2. Pialoux G, Gauzere B-A, Jaureguiberry S, Strobel M. Chikungunya, an epidemic arbovirosis. *Lancet* 2007;**7**:319–27.
3. Ng LC, Hapuarachchi HC. Tracing the path of Chikungunya virus-Evolution and adaptation. *Infect Gen Evol* 2010;**10**:876–85.
4. Volk SM, Chen R, Tsetsarkin KA, et al. Genome-scale phylogenetic analysis of Chikungunya virus reveal independent emergences of recent epidemics and various evolutionary rates. *J Virol* 2010;**84**:6497–504.
5. Chevillon C, Briant L, Renaud F, Devaux C. The Chikungunya threat: an ecological and evolutionary perspective. *Cell* 2007;**16**:80–8.
6. Yergolkar PN, Tandale BV, Arankalle VA, et al. Chikungunya outbreaks caused by African genotype, India. *Em Infect Dis* 2006;**12**:1580–3.
7. Njenga MK, Nderitu L, Ledermann JP, et al. Tracking epidemic Chikungunya virus into Indian Ocean from East Africa. *J Gen Virol* 2008;**89**:2754–60.
8. Vazeille M, Moutailler S, Coudrier D, et al. Two Chikungunya isolates from the outbreak of La Reunion (Indian Ocean) exhibit different patterns of infection in the mosquito, *Aedes albopictus*. *PLoS One* 2007;**2**:e1168.
9. Schuffenecker I, Iteman I, Michault A, et al. Genome microevolution of Chikungunya viruses causing the Indian Ocean outbreak. *PLoS Med* 2006;**3**:e263.
10. Kumar NP, Joseph R, Kamaraj T, Jambulingam P. A226V mutation in virus during the 2007 Chikungunya outbreak in Kerala, India. *J Gen Virol* 2008;**89**:1945.
11. Bordi L, Carletti F, Castilletti C, et al. Presence of the A226V mutation in autochthonous and imported Italian Chikungunya virus strains. *Clin Infect Dis* 2008;**47**:4128–429.
12. Tsetsarkin KA, Vanlandingham DL, McGee CE, Higgs S. A single mutation in Chikungunya virus affects vector specificity and epidemic potential. *PLoS Pathog* 2007;**3**:e201.
13. Ross RW. The Newala epidemic. III. The virus: isolation, pathogenic properties and relationship to the epidemic. *J Hyg (Lond)* 1956;**54**:177–91.
14. Zuckerman AJ, Banatvala JE, Pattison JR, Griffiths PD, Schaub BD. *Principle and Practice of Clinical Virology*. 5th ed. West Sussex, England: J Wiley & Sons, Ltd.; 2005.
15. World Health Organization. Chikungunya and dengue, south-west Indian Ocean. *Wkly Epidemiol Rec* 2006;**81**:105–16.
16. Charrel RN, de Lamballerie X, Raoult D. Chikungunya outbreaks—the globalization of vectorborne diseases. *N Engl J Med* 2007;**356**:769–71.
17. Grandadam M, Caro V, Plumet S, et al. Chikungunya fever, Southeastern France. *Emerg Infect Dis* 2011;**17**:910–3.
18. Powers AM, Logue CH. Changing patterns of Chikungunya virus: re-emergence of a zoonotic arbovirus. *J Gen Virol* 2007;**88**:2363–77.
19. Jupp PG, McIntosh BM. Chikungunya virus disease. In: Monath TP, editor. *The Arbovirus: Epidemiology and Ecology*, Vol II. Boca Raton, Florida: CRC Press; 1988. . p. 137–57.
20. ECDC. Chikungunya in Italy. Mission Report 17–21 September 2007. Joint ECDC/WHO visit for a European risk assessment. <http://www.ecdc.europa.eu/documents/pdf/071020_CHK_report.pdf>.
21. Knudsen AB. Geographic spread of *Aedes albopictus* in Europe and the concern among public health authorities. Report and recommendations of a workshop, held in Rome, December 1994. *Eur J Epidemiol* 1995;**11**:345–8.

22. Romi R, Majori G. An overview of the lesson learned in almost 20 years of fight against the "tiger" mosquito. *Parassitologia* 2008;**50**:117–9.
23. Romi R. History and updating of the spread of *Aedes albopictus* in Italy. *Parassitologia* 1995;**37**:99–103.
24. Romi R, Di Luca M, Majori G. Current status of *Aedes albopictus* and *Aedes atropalpus* in Italy. *J Am Mosq Control Assoc* 1999;**15**:425–7.
25. Delatte H, et al. *Aedes albopictus* vector of chikungunya and dengue in Reunion Island: biology and control. *Parasite* 2008;**15**:3–13.
26. Gerardin P, Barau G, Michault A, et al. Multidisciplinary prospective study of mother-to-child Chikungunya virus infection on the island of La Reunion. *PLoS Med* 2008;**5**:e60.
27. Arpino C, Curatolo P, Rezza G. Chikungunya and the nervous system: what we do and do not know. *Rev Med Virol* 2009;**19**:121–9.
28. Schwartz O, Albert ML. Biology and pathogenesis of Chikungunya virus. *Nat Rev Microbiol* 2010;**8**:491–500.
29. Sourisseau M, Schilte C, Casartelli N, et al. Characterization of reemerging Chikungunya virus. *PLoS Pathog* 2007;**3**:e89.
30. Ng LF, Chow A, Sun YJ, et al. IL-beta, IL-6, and RANTES as biomarkers of Chikungunya severity. *PLoS One* 2009;**4**:4261.
31. Burt FJ, Ralph MS, Rulli NE, Mahalingam S, Heise MT. Chikungunya: a re-emerging virus. *Lancet* 2012;**379**:662–71.
32. Lam SK, Chua KB, Hooi PS, et al. Chikungunya infections—an emerging disease in Malaysia. *Southeast Asian J Trop Med Public Health* 2001;**32**:447–51.
33. Anonymous. <http://www.invs.sante.fr>.
34. Borgherini G, Poubeau P, Staikowsky F, et al. Outbreak of Chikungunya on Reunion Island. Early clinical and laboratory features in 157 adult patients. *Clin Infect Dis* 2007;**44**:1401–7.
35. Wickmann O. Severe Dengue virus infection in travellers: risk factors and laboratory indicators. *J Infect Dis* 2007;**195**:1089–96.
36. Nicoletti L, Ciccozzi M, Marchi A, et al. Chikungunya and dengue viruses in travelers. *Em Infect Dis* 2007;**14**:177–8.
37. Smee DF, Alaghamandan HA, Kini GD, Robins RK. Antiviral activity and mode of action of ribavirin 5′-sulfamate against Semliki Forest virus. *Antiviral Res* 1988;**10**:253–62.

Chapter 13

Nipah Virus Disease

Pierre E. Rollin
Viral Special Pathogens Branch, Centers for Disease Control and Prevention, Clifton Rd, Atlanta, GA, USA

CASE PRESENTATION

In February 1999, a 34-year-old man was admitted with fever (3 days since onset) drowsiness, and lethargy, but without headache at the Hospital of the University of Malaya in Kuala Lumpur, Malaysia. He had received three doses of Japanese encephalitis vaccine, the last being 3 weeks before onset of illness. A month before his illness, some of the patient's pigs became ill and died suddenly. On admission, he was oriented but drowsy, and his temperature was 38°C. Examination of the lungs showed no abnormalities, and chest radiography was normal. Blood pressure ranged from 130/70 mmHg to 170/95 mmHg with a heart rate of 100–160 beats per min. There were no focal neurological deficits and no neck stiffness. The patient refused lumbar puncture. He remained febrile and became comatose 2 days after admission, only responding to painful stimuli with facial grimacing. He developed jerking of the abdominal wall and right leg; this feature suggested segmental myoclonus. The patient was intubated and ventilated. EEG showed continuous diffuse slow waves with intermittent bi-temporal sharp waves. The next day, the doll's eye reflex was absent but his pupils remained reactive. He did not improve, and on day 7 developed hypotension (unresponsive to fluid and inopressor therapy) and died. A CSF sample was obtained at necropsy.

We did not attempt to isolate the virus from the CSF of the patient because of concerns about laboratory-acquired infection. At some stage of the illness, the patient had IgM antibodies in the CSF against Hendra viral antigens by IgM capture ELISA. IgM antibodies against the Japanese encephalitis virus were not detected in the serum or CSF of the patient.

(This is a published case report.[1])

Emerging Infectious Diseases. DOI: http://dx.doi.org/10.1016/B978-0-12-416975-3.00013-3

1. WHY THIS CASE WAS SIGNIFICANTLY IMPORTANT AS AN EMERGING INFECTION?

Nipah disease is a zoonotic viral infection first described in Malaysia in 1999, associated in the severe forms with neurologic manifestations in humans and a respiratory and neurologic disease in swine. The disease has been described in Bangladesh and India and a related virus, Hendra virus, is responsible for human and horse disease in Australia. The mortality in hospitalized patients varied between 1 and 77%. The virus persists and circulates within fruit bat (*Pteropus* sp.) populations. The wide distribution of the potential reservoirs, from Madagascar to Australia, has great implications on human and animal public health.

2. WHAT IS THE CAUSATIVE AGENT?

Nipah virus (NIPV) was first isolated from cerebrospinal fluid (CSF) specimens collected from encephalitic patients in Malaysia in 1999.[1] Morphologic, serologic, and genetic studies indicated that the virus was closely related to Hendra virus (HENV) isolated in 1994 in Australia, and both viruses (non-segmented, negative-stranded RNA viruses) form the new *Henipavirus* genus within the *Paramyxoviridae* family.[2,3] In Malaysia and Singapore in 1999, it is believed that one strain of NIPV was responsible for the epidemic. Phylogenetic analysis of sequences of the N protein open reading frame from viruses obtained from Malaysia, Cambodia, India, and Bangladesh revealed around 6% diversity within NIPV, and several co-circulating lineages of NIPV in Bangladesh.[4] Like HENV, NIPV is classified as a BSL-4 agent and its manipulation requires high-containment laboratories and trained personnel.

3. WHAT IS THE FREQUENCY OF THE DISEASE IN YOUR REGION?

During the first description and outbreak in Malaysia and Singapore, nearly 300 human cases were reported,[5] but since 1999, no other cases have been reported either in humans or in pigs from this part of South-East Asia, although the virus is known to circulate in the Pteropid bat populations. The disease has been reported twice in West Bengal State, India: Siliguri in 2001[6] and Nadia district in 2007.[7] In Bangladesh, nearly 200 human cases have been confirmed with outbreaks occurring nearly every year since 2001.[8] Because of the large geographic range of the bat reservoir and vector (Figure 13.1), the distribution of the NIPV is extensive and has already been detected in bats in Madagascar,[9] India,[10] Malaysia,[11] Cambodia,[12] Thailand,[13] Indonesia,[14] and East Timor.[15] Virus genomic sequences related to henipaviruses have also been detected in bats in Ghana.[16] Nipah antibodies have been detected in bats in Papua

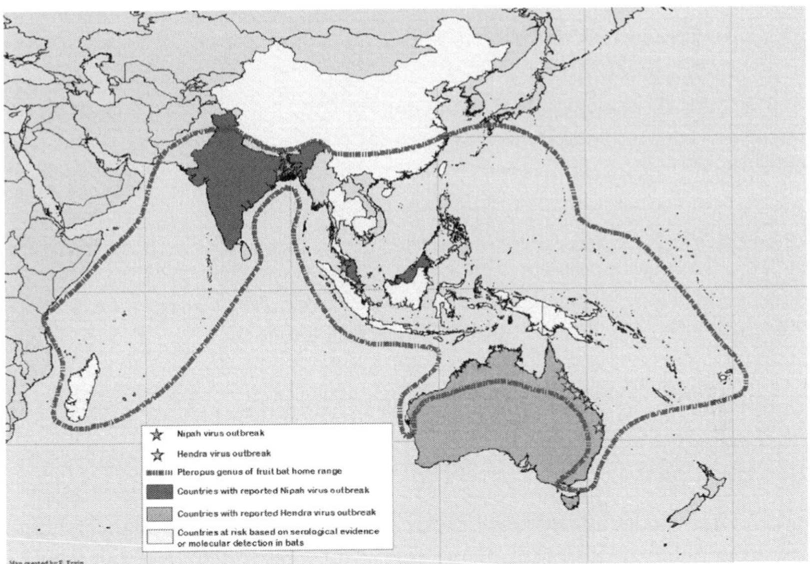

FIGURE 13.1 Geographic range of the bat reservoir and vector.

New Guinea and Australia,[15] and Ghana.[17] Antibodies to Henipa-like virus have been detected in bats in China,[18] and in pigs in Ghana.[19] Unsuspected and undiagnosed human cases in several South-East Asian countries are probable. The monthly distribution of human cases, with a peak from January to April, is certainly related to the ecology of the bat reservoirs. During the Malaysian epidemic, the human case fatality was around 39% (105/265 human cases). The reported mortality in Bangladesh is much higher (77%, 150/196 human cases), but the difference may be that encephalitis case definition selects more severe clinical forms with worst prognosis.[8]

4. HOW IS THE VIRUS TRANSMITTED?

Enzootic transmission of Nipah virus is maintained within populations of fruit bats (*Pteropus* sp.) by a mechanism still poorly understood. Only a few virus isolates are available from bats, but RNA sequences have been detected in bat urine with a seasonal peak in spring.[13] Anthropogenic changes (habitat loss, hunting) that have impacted the population dynamics of *Pteropus* species bats across much of their range are hypothesized to have facilitated emergence.[20] In Bangladesh, repeated epidemiological studies have confirmed the risk of direct transmission from bats to humans through the consumption of fresh palm sap contaminated by bat secretions including saliva.[21,22] In the original Malaysian outbreak, direct contact with Nipah virus-infected pigs (respiratory secretions or biological fluids) was definitively identified as a risk factor.[5,23]

The risk of nosocomial transmission is low and was not observed in Malaysia[24] despite the definitive presence of NIPV in saliva, throat swabs, and other patients' biological fluids.[25,26] The situation is different in Bangladesh and India where human-to-human transmission have been reported frequently.[6,27,28] Two hypotheses are discussed to explain this difference: lower standards of care and/or difference in virulence between the Malaysian and Bangladesh Nipah strains.[29] Standard precautions (contact and droplet) are always recommended when handling NIPV patients or their biological specimens.

5. WHICH FACTORS ARE INVOLVED IN DISEASE PATHOGENESIS? WHAT ARE THE PATHOGENIC MECHANISMS?

Little is known about the pathogenesis of NIPV in humans, but several animal models of infection and of the human disease (ferrets and non-human primates) are available.[30,31] In humans, infection occurs through the oral or respiratory route and is followed by an early viremia and a widespread endothelial infection. The formation of syncitia and necrosis can explain the breach of the blood–brain barrier and the subsequent central nervous system (CNS) infection followed by focal necrosis.[32] Another clinical feature of the henipaviruses infections is the possible evolution to a prolonged disease and a recrudescence of virus replication in the CNS.

The broad host range (cat, dog, pig, and human) of henipaviruses was noticed very early and could be explained by a widespread distribution among vertebrates of the ephrin-B2 and ephrin-B3 cell receptors used by NIPV.[33] The unique features of the NIPV proteins, such as the anti-interferon role of the P protein, contribute to the virulence in animals and humans. A better understanding through analysis of animal models will help the development of therapeutic and vaccine countermeasures.[34]

6. WHAT ARE THE CLINICAL MANIFESTATIONS?

It is difficult to precisely define the incubation period of Nipah disease, but in Malaysia, the mean incubation period for people moving from a pig farm after the potential infection was 10 ± 8.7 days.[35] The non-specific prodromal phase with fever, headache, myalgia, and dizziness is rapidly followed by confusion and reduced level of consciousness. Neurological symptoms usually appeared within a week post-onset and were commonly present on admission (Table 13.1). Most common neurological signs were coma, hyporeflexia or areflexia, segmental myoclonus, signs of brain-stem dysfunction with noticeable vasomotor changes consisting of hypertension, and

TABLE 13.1 Signs and Symptoms of Patients with Confirmed Nipah Infection

	Nipah				
References	Goh[36]	Sim[37]	Chong[35]	Hossain[38]	Homaira[28]
Patients in the series	$n=94$	$n=18$	$n=103$	$n=92$	$n=8$
Fever	91 (97%)	17 (94%)	100 (97%)	92 (100%)	8 (100%)
Headache	61 (65%)	12 (67%)	90 (88%)	64/88 (73%)	6 (75%)
Dizziness	34 (3%)	8 (44%)	40 (39%)		
Vomiting	25 (27%)	7 (39%)	37 (36%)	59/91 (58%)	5 (63%)
Fatigue/weakness			20 (19%)	51/76 (67%)	7 (87%)
Altered mental status/Reduced consciousness	20 (21%)	17 (94%)		82/91 (90%)	4 (50%)
Sore throat			21 (21%)		
Diarrhea		1 (5%)	21 (21%)	26/90 (29%)	
Non-productive cough	13 (14%)	3 (17%)		59/90 (62%)	5 (63%)
Respiratory difficulty/distress		17 (94%)	2 (2%)	62/90 (69%)	5 (63%)
Myalgia	11 (12%)	4 (22%)	48 (47%)		5 (63%)
Neurological signs	10 (11%)	(1–89%)	5 (5%)		
Seizure	22 (23%)	4 (22%)	27 (27%)		21/91 (23%)
Mortality	30 (32%)	11 (61%)	42 (41%)	67 (73%)	5 (63%)
Residual neurologic deficits	14 (15%)	6 (33%)	19 (19%)		

tachycardia.[36] Seizures are common. Although CNS dysfunction is a prominent clinical feature, respiratory difficulties are also noticeable, particularly in Bangladesh and India.[37–39] The majority of Bangladeshi patients who had chest radiographs performed showed abnormal findings with bilateral infiltrates or alveolar consolidation.[28,38] This was also evident in 24% of 42 patients in Malaysia[39] and eight of 11 Singapore patients.[23]

Notable laboratory findings include thrombocytopenia, elevated alanine aminotransferase and aspartate aminotransferase, and abnormal CSF values (elevated protein and white-cell count).[23,36]

In encephalitic patients, the electroencephalography (EEG) is usually abnormal with diffuse slowing in frontal and temporal areas, with a correlation between the degree of slowing and the severity of the disease.[36,40,41] Magnetic resonance imaging (MRI) usually revealed focal areas of hyperintensity on T2-weighted images in the deep white matter and subcortical areas consistent with demyelination or vasculitis.[36,41,42] Residual neurological signs were common among the survivors and relapse and late onset of Nipah encephalitis were described in Malaysia, potentially due to persistent infection in the CNS.[41]

7. HOW DO YOU MAKE THE DIAGNOSIS?

The diagnosis should be based on a combination of clinical, epidemiological, and laboratory data. Collecting the right specimen at the right time is essential. Virus can be isolated or detected by molecular techniques from throat/nasal swabs, tracheal secretions, CSF, urine, but more rarely from blood at the early stage of the disease.[25,26] Later in the course, when antibodies appear, the sensitivity of the molecular assay and virus isolation attempts will be lower than for serological assays. When the evolution is rapidly fatal, the histopathological findings and specific IHC staining are often the only means for confirming the diagnosis.[43,44]

7.1 Viral Isolation

All isolation attempts require high-containment laboratory (BSL4) and trained personnel. Virus can be isolated from acute patient samples using Vero or Vero E6 cells. Cytopathic effect, demonstrated by the formation of syncitia containing multiple nuclei, is easily seen after 3–5 days. If newborn suckling mice are available, the animals show paralysis and die 5–7 days after intracerebral inoculation.[44]

7.2 Molecular Detection: RT-PCR

Conventional and real-time reverse transcriptase polymerase chain reaction (RT-PCR) methods are available for detecting NIPV RNA from acute human samples (acute phase CSF, oropharyngeal swabs, urine, and blood), or tissues, or acute blood and secretions from naturally or experimentally infected pigs, or other experimentally infected laboratory animal models (hamster, cat, guinea pig, ferret, non-human primates).[4,31,44,45]

7.3 Immunohistochemistry

When tissues are available from autopsy or from animal models, the characteristic histopathologic and immunohistochemistry findings allow a specific diagnosis and a better understanding of the pathogenesis. In the nervous system, the most important feature is a systemic vasculitis with extensive thrombosis and adjacent parenchymal necrosis. Endothelial cell lesions and syncytial giant cell formation with presence of Nipah virus antigens seen by immunohistochemistry are particularly observed in the CNS but could be seen in other organs such as lung, spleen, lymph nodes, and kidneys.[31,32,44]

7.4 Serological Diagnosis

Several IgM and IgG ELISAs have been developed and are used in endemic areas to confirm human cases or for surveys, and Hendra or Nipah antigens could be used interchangeably.[5,44,46] The capability to detect IgM is critical for acute human cases diagnosis. In Malaysia, 88% of the first samples tested and nearly 100% a week after onset were positive for Nipah IgM.[46] The interpretation of the ELISA serological assays needs caution when NIPV and HENV are circulating in the same geographical zone; only neutralization assays can distinguish between infections with Hendra and Nipah viruses.[47]

8. HOW DO YOU DIFFERENTIATE THIS DISEASE FROM SIMILAR ENTITIES?

The clinical diagnosis cannot be based on the clinical findings alone and laboratory confirmation and epidemiological investigation are needed. During the original outbreak in Malaysia, a preliminary clinical diagnosis of Japanese encephalitis was proposed, although the epidemiological features, mostly adult non-muslim patients, was unusual for Japanese encephalitis.[2] Prodromal signs and symptoms are unspecific, as are the changes in CSF white cell count, and protein and glucose levels.[37] Distinction between Malaysia and Bangladesh clinical forms of Nipah disease may only reflect difference in selection of severe patients due to hospital case definition in the latter.[39] Antibodies compatible with Nipah infections have been found in bats in Hendra endemic areas, opening up the possibility of overlapping Nipah and Hendra infection in humans.[15]

9. WHAT IS THE THERAPEUTIC APPROACH?

Treatment is supportive with mechanical ventilatory support in patients unable to maintain correct respiratory functions. Ribavirin has been used in humans during the Nipah outbreak in Malaysia with equivocal results.[30] Passive immunization using a human monoclonal antibody targeting the Nipah G glycoprotein has been evaluated in the post-exposure therapy in the ferret model and found to be of benefit. This monoclonal was also effective in HENV post-exposure therapy in the primate model and is in the preclinical development in the US and in Australia.[30] A subunit vaccine (Hendra G protein), in use in Australia as an equine vaccine, produces cross-protective antibodies against HENV and NIPV and offers great potential for henipavirus protection in human.

10. WHAT ARE THE PREVENTIVE AND INFECTION CONTROL MEASURES?

With potential specific treatment and prophylaxis in view but not yet available, surveillance, prevention, and awareness are the only recommended measures. Controlling reservoirs or the natural cycle of the virus is impossible, but better insight into the ecology of bats is required (i.e., seasonality of the disease correlation with reproductive cycle of bats). The surveillance tools should include sensitive and specific laboratory assays for early recognition of the disease in livestock and reinforcement of infection control in commercial practices. Standard infection control practices are recommended for healthcare workers to avoid human-to-human or nosocomial infections.

REFERENCES

1. Chua KB, Goh KJ, Wong KT, et al. Fatal encephalitis due to Nipah virus among pig-farmers in Malaysia. *Lancet* 1999;**354**:1257–9.
2. Chua KB, Bellini WJ, Rota PA, et al. Nipah virus: a recently emergent deadly paramyxovirus. *Science* 2000;**288**:1432–5.
3. Murray K, Selleck P, Hooper P, et al. A morbillivirus that caused fatal disease in horses and humans. *Science* 1995;**268**:94–7.
4. Lo MK, Lowe L, Hummel KB, et al. Characterization of Nipah virus from outbreaks in Bangladesh, 2008–2010. *Emerg Infect Dis* 2012;**18**:248–52.
5. Parashar UD, Sunn LM, Ong F, et al. Case-control study of risk factors for human infection with a new zoonotic paramyxovirus, Nipah virus, during a 1998–1999 outbreak of severe encephalitis in Malaysia. *J Infect Dis* 2000;**181**:1755–9.
6. Chadha MS, Comer JA, Lowe L, et al. Nipah virus-associated encephalitis outbreak, Siliguri, India. *Emerg Infect Dis* 2006;**12**:235–40.
7. Arankalle VA, Bandyopadhyay BT, Ramdasi AY, et al. Genomic characterization of Nipah virus, West Bengal, India. *Emerg Infect Dis* 2011;**17**:876–9.
8. Rahman M, Chakraborty A. Nipah virus outbreaks in Bangladesh: a deadly infectious disease. *WHO South-East Asia J Publ Hlth* 2012;**1**:208–12.

9. Iehle C, Razafitrimo G, Razainirina J, et al. Henipavirus and Tioman virus antibodies in pteropodid bats, Madagascar. *Emerg Infect Dis* 2007;**13**:159–61.
10. Yadav PD, Raut CG, Shete AM, et al. Detection of Nipah virus RNA in fruit bat (*Pteropus giganteus*) from India. *Amer J Trop Med Hyg* 2012;**87**:576–8.
11. Johara MY, Field H, Mohd Rashdi A, et al. Nipah virus infection in bats (Order Chiroptera) in peninsular Malaysia. *Emerg Infect Dis* 2001;**7**:439–41.
12. Reynes J-M, Counor D, Ong S, et al. Nipah virus in Lyle's flying foxes, Cambodia. *Emerg Infect Dis* 2005;**11**:1042–7.
13. Wacharapluesadee S, Boongird K, Wanghongsa S, et al. A longitudinal study of the prevalence of Nipah virus in *Pteropus lylei* bats in Thailand: evidence for seasonal preference in disease transmission. *Vector-Borne Zoon Dis* 2010;**10**:183–90.
14. Sendow I, Field HE, Adjid A, et al. Screening for Nipah virus infection in west Kalimantan province, Indonesia. *Zoon Publ Hlth* 2010;**57**:499–503.
15. Breed AC, Meers J, Sendow I, et al. The distribution of henipaviruses in southeast Asia and Australasia: is Wallace's line a barrier to Nipah virus? *PLoS One* 2013;**8**: e61316.
16. Drexler JF, Corman VM, Gloza-Rausch F, et al. Henipavirus RNA in african bats. *PLoS One* 2009;**4**:e6367.
17. Hayman DT, Suu-Ire R, Breed AC, et al. Evidence of Henipavirus infection in west african fruit bats. *PLoS One* 2008;**3**:e2739.
18. Li Y, Wang J, Hickey AC, et al. Antibodies to Nipah or Nipah-like viruses in bats, china. *Emerg Infect Dis* 2008;**14**:1974–6.
19. Hayman DT, Wang LF, Barr J, et al. Antibodies to Henipavirus or Henipa-like viruses in domestic pigs in Ghana, West Africa. *PLoS One* 2011;**6**:e25256.
20. Field HE, Mackenzie JS, Daszak P. Henipaviruses: emerging paramyxoviruses associated with fruit bats. *Curr Top Microbiol Immunol* 2007;**315**:133–59.
21. Luby SP, Rahman M, Hossain MJ, et al. Foodborne transmission of Nipah virus, Bangladesh. *Emerg Infect Dis* 2006;**12**:1888–94.
22. Rahman MA, Hossain MJ, Sultana S, et al. Date palm sap linked to Nipah virus outbreak in Bangladesh, 2008. *Vector-Borne Zoon Dis* 2012;**12**:65–73.
23. Paton NI, Leo YS, Zaki SR, et al. Outbreak of Nipah-virus infection among abattoir workers in Singapore. *Lancet* 1999;**354**:1253–6.
24. Mounts AW, Kaur H, Parashar UD, et al. A cohort study of health care workers to assess nosocomial transmissibility of Nipah virus, Malaysia, 1999. *J Infect Dis* 2001;**183**:810–3.
25. Chua KB, Lam SK, Tan CT, et al. High mortality in Nipah encephalitis is associated with presence of virus in cerebrospinal fluid. *Ann Neurol* 2000;**48**:802–5.
26. Chua KB, Lam SK, Goh KJ, et al. The presence of Nipah virus in respiratory secretions and urine of patients during an outbreak of Nipah virus encephalitis in Malaysia. *J Infect* 2001;**42**:40–3.
27. Gurley ES, Montgomery JM, Hossain MJ, et al. Person-to-person transmission of Nipah virus in a Bangladeshi community. *Emerg Infect Dis* 2007;**13**:1031–7.
28. Homaira N, Rahman M, Hossain MJ, et al. Nipah outbreak with person-to-person transmission in Bangladesh, 2007. *Amer J Trop Med Hyg* 2007;**77** 5, Suppl. S:80.
29. Luby SP, Gurley ES, Hossain MJ. Transmission of human infection with Nipah virus. *Clin Infect Dis* 2009;**49**:1743–8.
30. Broder CC. Henipavirus outbreaks to antivirals: the current status of potential therapeutics. *Curr Opinion Virol* 2012;**2**:176–87.

31. Williamson M, Torres-Velez FJ. Henipavirus: a review of laboratory animal pathology. *Vet Path* 2010;**47**:871–80.
32. Wong KT, Shieh WJ, Kumar S, et al. Nipah virus infection. pathology and pathogenesis of an emerging paramyxoviral zoonosis. *Amer J Path* 2002;**161**:2153–67.
33. Negrete OA, Levroney EL, Aguilar HC, et al. Ephrinb2 is the entry receptor for Nipah virus, an emergent deadly paramyxovirus. *Nature* 2005;**436**:401–5.
34. Eaton BT, Broder CC, Middleton D, Wang LF. Hendra and Nipah viruses: different and dangerous. *Nat Rev Microbiol* 2006;**4**:23–35.
35. Chong HT, Kunjapan SR, Thayaparan T, et al. Nipah encephalitis outbreak in Malaysia, clinical features in patients from Seremban. *Can J Neurol Sci* 2002;**29**:83–7.
36. Goh KJ, Tan CT, Chew NK, et al. Clinical features of Nipah virus encephalitis among pig farmers in Malaysia. *N Engl J Med* 2000;**342**:1229–35.
37. Sim BF, Jusoh MR, Chang CC, Khalid R. Nipah encephalitis: a report of 18 patients from Kuala Lumpur hospital. *Neurol J Southeast Asia* 2002;**7**:13–8.
38. Hossain MJ, Gurley ES, Montgomery JM, et al. Clinical presentation of Nipah virus infection in Bangladesh. *Clin Infect Dis* 2008;**46**:977–84.
39. Chong HTH, Jahangir Hossain M, Tan CT. Differences in epidemiologic and clinical features of Nipah virus encephalitis between the Malaysian and Bangladesh outbreaks. *Neurol Asia* 2008;**13**:23–6.
40. Sarji SA, Abdullah BJ, Goh KJ, Tan CT, Wong KT. MR imaging features of Nipah encephalitis. *Amer J Roentgen* 2000;**175**:437–42.
41. Tan CT, Goh KJ, Wong KT, et al. Relapsed and late-onset Nipah encephalitis. *Ann Neurol* 2002;**51**:703–8.
42. Lim CCT, Lee KE, Lee WL, et al. Nipah virus encephalitis: serial MR study of an emerging disease. *Radiology* 2002;**222**:219–26.
43. Daniels P, Ksiazek T, Eaton BT. Laboratory diagnosis of Nipah and Hendra virus infections. *Microb Infect* 2001;**3**:289–95.
44. Rollin PE, Rota P, Zaki S, Ksiazek TG. Hendra and Nipah viruses. In: Versalovic J, Carroll KC, Funke G, Jorgensen JH, Landry ML, Warnock DW, editors. *Manual of Clinical Microbiology*. 10th ed Washington, DC: ASM Press; 2011. p. 1479–87.
45. Feldman KS, Foord A, Heine HG, et al. Design and evaluation of consensus PCR assays for Henipaviruses. *J Virol Methods* 2009;**161**:52–7.
46. Ramasundrum V, Tan CT, Chua KB, et al. Kinetics of IgM and IgG seroconversion in Nipah virus infection. *Neurol J Southeast Asia* 2000;**5**:23–8.
47. Crameri G, Wang LF, Morrissy C, White J, Eaton BT. A rapid immune plaque assay for the detection of Hendra and Nipah viruses and anti-virus antibodies. *J Virol Methods* 2002;**99**:41–51.
48. Tan CT, Wong KT, Chua KB. Nipah encephalitis. In: Power C, Johnson RT, editors. *Neurological Disease and Therapy*, vol. 67. Boca Raton: Taylor & Francis; 2005. p. 59–75.

Chapter 14

Middle East Respiratory Syndrome-Coronavirus (MERS-CoV) Infection

Jaffar A. Al-Tawfiq[1] **and Ziad A. Memish**[2]
[1]*Saudi Aramco Medical Services Organisation, Saudi ARAMCO, Dhahran, KSA,* [2]*Ministry of Health, Al-Faisal University, Riyadh, KSA*

CASE PRESENTATION

A 45-year-old male had a history of heavy smoking, type 2 diabetes mellitus, a history of atrophied right kidney, and ischemic heart disease. He presented with a 3-day complaint of fever of 38°C and a cough that had become productive. A chest film was unremarkable, and he was discharged home. The following day, he visited the hospital's emergency room with the same complaints. The oxygen saturation on room air and chest film was normal, and he was discharged home on oral cefuroxime. Two days later, he returned to the emergency room with worsening dyspnea and required continuous positive airway pressure (CPAP) to maintain oxygenation. Chest film revealed patchy infiltrates in his right lower lobe. Treatment with parenteral ceftriaxone, azithromycin, and oral oseltamivir were commenced after specimens were collected for diagnostic testing. He became progressively more hypoxic over the next 24 hours. Chest film revealed patchy infiltrates in his right lower lobe. Routine bacteriology, acid-fast bacillus smears, and screening influenza exams were negative. He further deteriorated and required intubation and mechanical ventilation.

Antibiotics were changed to piperacillin-tazobactam plus linezolid; treatment with corticosteroids was initiated. Immunofluorescent staining of respiratory epithelial cells for influenza A, B, respiratory syncytial virus (RSV), parainfluenza 1-3, and adenovirus were negative, and he was confirmed to be seronegative for human immunodeficiency virus (HIV), *Mycoplasma pneumoniae*, Q fever, and *Brucella*. Upper tract swabs in viral transport media were forwarded to the Saudi Ministry of Health regional

Emerging Infectious Diseases. DOI: http://dx.doi.org/10.1016/B978-0-12-416975-3.00014-5

laboratory for Middle East respiratory syndrome-coronavirus (MERS-CoV) upE reverse transcriptase polymerase chain reaction (RT-PCR). A second set of specimens including tracheal aspirate was collected. Respiratory specimens were positive for MERS-CoV.

In the intensive care unit, renal function deteriorated, and he was started on continuous renal replacement for 2 days then three hemodialysis sessions. Subsequently, oxygen requirements were moderated and he gradually defervesced, although chest radiographs continued to show infiltrates. He was then weaned off mechanical ventilation and was extubated. He was subsequently discharged home.

(This is a published case report, *Saudi Medical Journal* 2012;33:1265–9.[1])

1. WHAT IS THE CAUSATIVE AGENT?

Middle East respiratory syndrome-coronavirus (MERS-CoV) is a new human disease that was first reported from Saudi Arabia in September 2012, after identification of a novel coronavirus (CoV) from a male Saudi Arabian patient who died from severe pneumonia.[2,3] MERS-CoV had caused a significant mortality of about 50% since that time.[4]

The MERS-CoV is a novel coronavirus that was initially designated HCoV-EMC.[3] The virus was later designated after global consensus as MERS-CoV.[5] Coronaviruses are common viruses that usually cause mild to moderate upper-respiratory tract illnesses in humans. The viruses have crown-like spikes on their surfaces and hence the name coronavirus. Human coronaviruses, enveloped RNA viruses, are not new and were first identified in the mid-1960s. There are four virus clusters within the *Coronavirinae* subfamily. These are alphacoronavirus, betacoronavirus, and gammacoronavirus. The fourth cluster is a provisionally assigned new group called delta coronaviruses. All known human coronaviruses belong to the genera *Alphacoronavirus* (HCoV-229E and HCoV-NL63) and *Betacoronavirus* (HCoV-OC43, HCoV-HKU1, and SARSCoV).[3] MERS-CoV, formerly HCoV-EMC, is the first human coronavirus in lineage C of the *Betacoronavirus* genus.[3]

2. WHAT IS THE FREQUENCY OF THE DISEASE? PREVALENCE, INCIDENCE, BURDEN, AND IMPACT OF THE DISEASE

Between April 2012 and February 7, 2014 there were 182 documented cases of MERS-CoV infection worldwide.[5a] The majority of these occurred in the Kingdom of Saudi Arabia where 148 cases were reported. MERS-CoV appears to have a predilection for individuals with underlying medical comorbidities.[1,4,6–8]

3. WHAT ARE THE TRANSMISSION ROUTES?

The main modes of transmission are contact transmission, droplet transmission, and person-to-person transmission as supported by epidemiologic and phylogenetic analyses.[4] Currently, the MERS-CoV seems to have three epidemiological patterns of the disease. There are sporadic cases occurring in the communities of different Middle East countries, mainly the Kingdom of Saudi Arabia, Qatar, United Arab Emirates, and Jordan. The second pattern is nosocomial transmission within healthcare facilities to healthcare workers and other patients.[4] Intrafamilial transmission of MERS-CoV was also described.[1,4,7,9–11]

4. WHICH FACTORS ARE INVOLVED IN DISEASE PATHOGENESIS? WHAT ARE THE PATHOGENIC MECHANISMS?

The pathogenesis of the disease has been elucidated in recent studies. MERS-CoV has spike glycoprotein (S) that targets the cellular receptor, dipeptidyl peptidase 4 (DPP4).[12,13] This viral spike has a putative receptor-binding domain (RBD).[13] MERS-CoV RBD has a core and a receptor-binding subdomain, which interacts with DPP4 β-propeller MERS-CoV RBD.[13]

The MERS-CoV spike protein interacts with CD26 (also known as DPP4) and causes viral attachment to host cells and virus-cell fusion.[14] This is thought to be the first step in viral infection. The MERS-CoV infection results in profound apoptosis of infected respiratory cells within 24 hr.[15]

5. WHAT ARE THE CLINICAL MANIFESTATIONS?

MERS-CoV causes respiratory tract infection that ranges in severity from mild to fulminant respiratory infection. Mild respiratory illness was described in patients from Tunisia[16] and from the United Kingdom.[11]

The clinical presentation of MERS-CoV is similar to SARS3. The initial phase is non-specific fever and mild, non-productive cough lasting several days, followed by progressive pneumonia.[4,6] In MERS-CoV infections, most patients present with serious respiratory disease, resulting in a high mortality rate of 60%.[6] The mean age of affected patients was 56 years with a range of 14–94 years.[6] A recent case of a 2-year-old patient was described.[17] The most common symptoms are fever (87%), cough (87%), and shortness of breath (48%).[4,6] About 35% of patients had accompanying gastrointestinal symptoms, including diarrhea (22%) and vomiting (17%). Of the total cases, 50% had two medical co-morbidities, diabetes and chronic renal disease.[6]

Important laboratory abnormalities in patients with MERS-CoV include: leucopenia (14%), lymphopenia (34%), thrombocytopenia 36%, increased lactate dehydrogenase (LDH) (49%), increased alanine aminotransferase

(ALT) (11%), and increased aspartate aminotransferase (AST) (15%).[6] Chest radiographic abnormalities include: increased bronchovascular markings (17%), unilateral infiltrate (43%), bilateral infiltrates (22%), and diffuse reticulonodular pattern (4%).[4]

6. HOW DO YOU DIAGNOSE?

Laboratory testing for MERS-CoV is a challenge. Currently, there are no validated serologic assays. The main testing method relies on identification of MERS-CoV using real-time reverse transcriptase-polymerase chain reaction (RT-PCR) from respiratory tract secretions. It is not clear at this point of time whether sputum or nasopharyngeal samples are superior to throat swabs.[4]

7. HOW DO YOU DIFFERENTIATE THIS DISEASE FROM SIMILAR ENTITIES?

To date, there are no specific laboratory abnormalities or clinical data that differentiate pneumonia due to MERS-CoV from pneumonia caused by other viruses or other bacterial pathogens. The primary diagnosis of MERS-CoV infection relies on the identification of the virus in respiratory secretions using real-time RT-PCR.

8. WHAT IS THE THERAPEUTIC APPROACH?

The main therapeutic options for MERS-CoV infection are not known. There is no specific therapy for MERS-CoV infection. Recently, *in vitro* studies showed that MERS-CoV is 50–100 times more sensitive to alpha interferon (IFN-α) treatment than SARS-CoV.[18] In a recent decision support document, convalescent plasma was given an order of recommendation of 1, followed by interferon, protease inhibitors (order of recommendations of 2), and intravenous globulin (order of recommendations of 3).[19] Further randomized controlled trials of these agents are needed to establish the efficacy and side effects.

9. WHAT ARE THE PREVENTIVE AND INFECTION CONTROL MEASURES

The main infection control measures to prevent the transmission of MERS-CoV include contact isolation, standard precautions, droplet isolation, and airborne infection isolation precautions especially when healthcare workers perform aerosol generating procedures.[20] Droplet precautions include wearing a medical mask when in close contact (within 1 meter) and upon entering the room or cubicle of the patient. The Centers for Disease Control and Prevention (CDC) recommends placing patients with suspected or confirmed MERS-CoV infection in an airborne infection isolation room (AIIR).[21] If an AIIR is not available, the patient should be transferred as soon as is feasible to a facility where an AIIR is available. Pending transfer, place a facemask on the patient and isolate him/her in a single-patient room with the door closed.[21] Performing hand hygiene in accordance with the World Health Organization's (WHO) 5 moments of hand hygiene is of paramount importance and could not be stressed more. Additional measures include wearing a particulate respirator when performing aerosol-generating procedures in addition to other precautions. In a recent MERS-CoV outbreak in a healthcare setting, there was evidence of person-to-person transmission and the outbreak was aborted by the implementation of infection control measures.[4]

REFERENCES

1. AlBarrak AM, Stephens GM, Hewson R, Memish ZA. Recovery from severe novel coronavirus infection. *Saudi Med J* 2012;**33**:1265–9.
2. Centers for Disease Control and Prevention (CDC). Severe respiratory illness associated with a novel coronavirus: Saudi Arabia and Qatar, 2012. *MMWR Morb Mortal Wkly Rep* 2012;**61**:820.
3. Zaki AM, van Boheemen S, Bestebroer TM, Osterhaus AD, Fouchier RA. Isolation of a novel coronavirus from a man with pneumonia in Saudi Arabia. *N Engl J Med* 2012;**367**:1814–20.
4. Assiri A, McGeer A, Perl TM, et al. Hospital outbreak of Middle East respiratory syndrome coronavirus. *N Engl J Med* 2013;**369**:407–16.
5. de Groot RJ, Baker SC, Baric RS, et al. Middle East respiratory syndrome coronavirus (MERS-CoV): announcement of the Coronavirus Study Group. *J Virol* 2013;**87**:7790–2.

5a. Middle East respiratory syndrome coronavirus (MERS-CoV) – update. Available at: <http://www.who.int/csr/don/2014_02_07mers/en/>.

6. Assiri A, Al-Tawfiq JA, Al-Rabeeah AA, Al-Rabiah FA, Al-Hajjar S, Al-Barrak A, et al. Epidemiological, demographic, and clinical characteristics of 47 cases of Middle East respiratory syndrome coronavirus disease from Saudi Arabia: a descriptive study. *Lancet Infect Dis* **13**:752–61.
7. Bermingham A, Chand MA, Brown CS, Aarons E, Tong C, Langrish C, et al. Severe respiratory illness caused by a novel coronavirus, in a patient transferred to the United Kingdom from the Middle East, September 2012. *Euro Surveill* 2012;**17**:20290.
8. Buchholz U, Müller MA, Nitsche A, Sanewski A, Wevering N, Bauer-Balci T, et al. Contact investigation of a case of human novel coronavirus infection treated in a German hospital, October–November 2012. *Euro Surveill* 2013;**18**: pii: 20406.
9. Memish ZA, Zumla AI, Al-Hakeem RF, Al-Rabeeah AA, Stephens GM. Family cluster of Middle East respiratory syndrome coronavirus infections. *N Engl J Med* 2013;**368**:2487–94.
10. Hijawi B, Abdallat M, Sayaydeh A, et al. Novel coronavirus infections in Jordan, April 2012: epidemiological findings from a retrospective investigation. *East Mediterr Health J* 2013;**19**(Suppl. 1):S12–8.
11. Health Protection Agency (HPA) UK Novel Coronavirus Investigation Team. Evidence of person-to-person transmission within a family cluster of novel coronavirus infections, United Kingdom, February 2013. *Euro Surveill* 2013;**18**:20427.

12. Mou H, Raj VS, van Kuppeveld FJ, Rottier PJ, Haagmans BL, Bosch BJ. The receptor binding domain of the new MERS coronavirus maps to a 231-residue region in the spike protein that efficiently elicits neutralizing antibodies. *J Virol* 2013;**87**:9379–83.
13. Wang N, Shi X, Jiang L, Zhang S, Wang D, Tong P, et al. Structure of MERS-CoV spike receptor-binding domain complexed with human receptor DPP4. *Cell Res* 2013;**23**:986–93.
14. Lu G, Hu Y, Wang Q, Qi J, Gao F, Li Y, et al. Molecular basis of binding between novel human coronavirus MERS-CoV and its receptor CD26. *Nature* 2013;**500**:227–31.
15. Tao X, Hill TE, Morimoto C, Peters CJ, Ksiazek TG, Tseng CT. Bilateral entry and release of middle east respiratory syndrome-coronavirus induces profound apoptosis of human bronchial epithelial cells. *J Virol* 2013;**87**:9953–8.
16. ProMED-mail. MERS-CoV—Eastern Mediterranean (07): Tunisia ex Saudi Arabia/Qatar, fatal, WHO. May 22, 2013. <http://www.promedmail.org/direct.php?id=20130522.1730663>; [accessed 24.07.13].
17. WHO. Global alert and response (GAR): Middle East respiratory syndrome coronavirus (MERS-CoV)—update. July 7, 2013. <http://www.who.int/csr/don/2013_07_07/en/index.html>; [accessed 25.07.13].
18. de Wilde AH, Raj VS, Oudshoorn D, Bestebroer TM, van Nieuwkoop S, Limpens RW, et al. MERS-coronavirus replication induces severe in vitro cytopathology and is strongly inhibited by cyclosporin A or interferon-α treatment. *J Gen Virol* 2013;**94**:1749–60.
19. ISARIC (International Severe Acute Respiratory & Emerging Infection Consortium). Clinical Decision Making Tool for Treatment of MERS-CoV v.1.0, 18 June, 2013. Available at: <http://www.hpa.org.uk/webc/HPAwebFile/HPAweb_C/1317139281416>; [last accessed 26.07.2013].
20. World Health Organization. Infection prevention and control during health care for probable or confirmed cases of novel coronavirus (nCoV) infection interim guidance: 6 May 2013. Available at: <http://www.who.int/csr/disease/coronavirus_infections/IPCnCoVguidance_06May13.pdf>; [last accessed 25.07.2013].
21. CDC. Interim Infection Prevention and Control Recommendations for Hospitalized Patients with Middle East Respiratory Syndrome Coronavirus (MERS-CoV). Available at: <http://www.cdc.gov/coronavirus/mers/infection-prevention-control.html>; [last accessed 26.07.2013].

Chapter 15

Human Bocavirus

Tina Uršič and Miroslav Petrovec
Institute of Microbiology and Immunology, Faculty of Medicine, University of Ljubljana, Zaloška, Ljubljana, Slovenia

CASE PRESENTATION

A twenty-month-old female child was hospitalized due to acute respiratory infection with signs of respiratory distress. She was diagnosed as having acute bronchiolitis. She was prematurely born after 27 weeks of gestation and her previous medical history was unremarkable.

After her admission she was treated with short-acting beta-2 agonist. Her medical condition failed to improve, so chest radiography was performed, and hyperinflation with infiltrates in the left lower lung field was observed. Parenteral steroid and amoxicillin-clavulanic acid were commenced. In the next 24 h, neck emphysema was observed and due to the imminent respiratory failure, the girl was intubated and transferred to the intensive care unit (ICU). The child was sedated, relaxed, and mechanically ventilated. The leukocyte count was 22.0×10^9/liter, the C-reactive protein level was 14 mg/liter, and hemoglobin concentration was 9.4 g/dl. Her medical condition failed to improve and her circulatory stability was supported by intermittent intravenous saline boluses with continuous infusion of dopamine. Repeated chest radiography revealed pneumothorax of the left and the right lung, which were immediately drained. Bronchoscopy performed through endotracheal tube showed edema and inflammation of the lower respiratory tract with a large amount of mucus. Carbon dioxide partial pressure increased, and after 18 h after ICU admission reached 19.6 kPa, with excessive respiratory acidosis (pH 6.92).

A nasopharyngeal swab (at admission), tracheal aspirate (after intubation), and blood sample (at admission) were tested by real-time PCR for the presence of 15 respiratory viruses. HBoV was the only respiratory pathogen that was detected in nasopharyngeal swab (8.6×10^9 copies/ml), tracheal aspirate (2.1×10^{10} copies/ml), and plasma sample (1.8×10^6copies/ml). At admission to the ICU, blood was collected for hemoculture; bronchoalveolar fluid (BAL), tracheal aspirate, and thoracal drainage fluid were collected

Emerging Infectious Diseases. DOI: http://dx.doi.org/10.1016/B978-0-12-416975-3.00015-7

and the routine cultures were set up for detection of bacterial and fungal pathogens and remained negative during hospitalization. Human bocavirus particles were visualized by electron microscopy and immunoelectron microscopy confirmed the immune response against virus in patient plasma.

Assisted controlled ventilation with positive inspiratory pressure was effective in lowering the $PaCO_2$. Subcutaneous emphysema occurred on the child's head, cheeks, neck, and chest, because of pronounced air leak. Pneumoperitoneum was also observed on the radiograph. Her clinical condition substantially stabilized after surgical incision in the neck, and after insertion of a new thoracic drain on the left side partially relieved pneumomediastinum. Air leak decreased steadily, whereas subcutaneous emphysema persisted. Chest drains were removed on the seventh day and the girl was extubated on the ninth day in the ICU and was discharged 4 days later. The HBoV viral load decreased slowly, and on the day of her discharge, the viral load in nasopharyngeal swab was 9.9×10^4 copies/ml and 3.6×10^4 copies/ml in the plasma sample.

(This is a published case report, *Journal of Clinical Microbiology*, 2011.[1])

1. WHY THIS CASE WAS SIGNIFICANTLY IMPORTANT AS AN EMERGING INFECTION

Human bocavirus 1 (HBoV1) is a recently described respiratory virus and is detected worldwide.[2,3] It is recognized as one of the most frequently detected respiratory viruses in children with upper and lower respiratory infections. The severe clinical course of HBoV1 infection can be seen in prematurely born children or children, but rarely adults, with other underlying conditions.[1,4–7]

2. WHAT IS THE CAUSATIVE AGENT?

Human bocavirus 1 belongs to the family *Parvoviridae*, subfamily *Parvovirinae*, and genus *Bocavirus*.[2] The HBoV1 genome phylogenetic analysis showed that it is most closely related to bovine parvovirus (BPV1) and minute virus of canines (MVC), after which it was named.[2] The HBoV1 virions are icosahedral, non-enveloped and small, approximately 18–26 nm in diameter and their linear single-stranded DNA genome is 5543 bp in length[1,2,8–10] (accession no. JQ923422).

Figure 15.1 shows human bocavirus 1 virions examined with an electron microscope.

The HBoV1 genome, like all members of the *Bocavirus* genus, contains three open reading frames (ORF).[9] The left- and the middle-hand ORFs at the 5′ end encode for NS1 and NP1, two non-structural proteins, whereas the right-hand ORF, at the 3′ end, encodes for the two structural capsid viral proteins, VP1 and VP2.[2,8,9] The capsid of HBoV is assembled from 60 copies of VP2 protein and about five copies of VP1 protein and is most similar to that of parvovirus B19.[9,11]

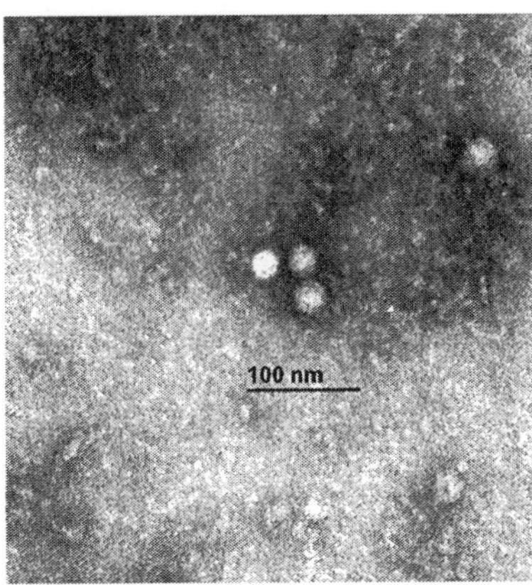

FIGURE 15.1 EM image of HBoV1 in a direct preparation from nasopharyngeal swab suspension, negatively stained with 2% phosphotungistic acid (PTA; pH 4.5) and examined with a transmission electron microscope (JEM 1200 EX II; Jeol, Japan) at a magnification of ×200,000.[1]

Three additional enteric HBoV-like viruses, HBoV2, HBoV3, and HBoV4, were identified recently in stool samples of children with gastroenteritis, but their clinical significance in symptomatic infections remains uncertain.[3] The HBoV2, HBoV3, and HBoV4 possess the same genome and capsid organization as HBoV1.[11–14]

3. WHAT IS THE FREQUENCY OF THE DISEASE?

Human bocavirus 1 was first described in 2005 in nasopharyngeal aspirates of Swedish children with acute respiratory tract infections[2] (ARTI). Since then, by using polymerase chain reaction (PCR), the HBoV1 has been found in approximately 10% of respiratory samples of children with upper or lower respiratory tract infections (URTI, LRTI) worldwide,[3,15] and in 1–9% of children with respiratory infection with or without gastrointestinal symptoms.[12,16,17]

Figure 15.2 shows the geographic distribution of HBoV1 in Europe and reported frequency of infection in children.[18–36]

The HBoV1 is one of the most frequently detected respiratory viruses in children with ARTI below 5 years of age, and is mainly detected in children between 6 and 24 months.[3,15,32,34,37–40] The reports on HBoV1 infection in adults and elderly are rare and the prevalence of HBoV1 in those age groups is rather low;[3,41] however, the infections caused by HBoV1 in immunocompromised individuals could be severe and may be even lethal.[3,42,43]

4. HOW IS THE VIRUS TRANSMITTED?

The transmission routes of HBoV1 are unknown. The HBoV1 virus is ubiquitous. It is likely that HBoV1 is transmitted similarly, as other parvoviruses, by

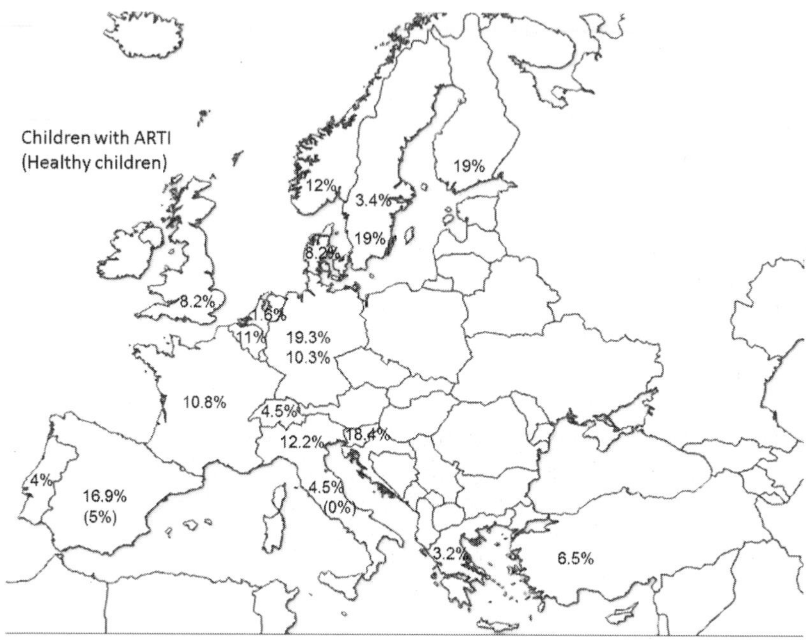

FIGURE 15.2 Distribution and frequencies of HBoV1 infection in Europe in children with ARTI (healthy children).

inhalation and by contact with infectious respiratory secretions. Respiratory infections due to HBoV1 are systemic as its DNA is frequently detected in many human secretions.[3,44] The seroepidemiological studies consistently show that most of the children are infected with HBoV1 by the age of 6[44,45] and that the IgG antibodies remain present for life.

Since the prevalence of IgG antibodies against HBoV1 is high among pregnant woman, the intrauterine infections are unlikely.[3,46]

5. WHICH FACTORS ARE INVOLVED IN DISEASE PATHOGENESIS? WHAT ARE THE PATHOGENIC MECHANISMS?

Until recently, the replication and pathogenesis of HBoV1 was not known, since no *in vitro* or animal models were available. However, in 2009 the HBoV1 was successfully cultured in primary airway epithelial cells differentiated into pseudo-stratified human airway epithelium.[47] It has been demonstrated that the transcription profile of HBoV1 displays features similar to bovine parvovirus 1 (BPV1) and minute virus of canines (MVC), which also belong to genus *Bocavirus*.[47] In 2012 the establishment of a reverse genetic system for studying HBoV1 in human airway epithelia (HAE) has enabled better understanding of HBoV1 replication and pathogenesis.[10]

The full-length genome of HBoV1 was sequenced for the first time, and cloned into a plasmid. It was demonstrated that the infectious clone was capable of replicating and producing progeny virions in the absence of a helper virus in polarized differentiated human embrionic kidney cells (HEK 293). The infectious virions are able to infect polarized differentiated human airway epithelia from both apical and basolateral surfaces.[10,48,49] HBoV1 infection disrupts the integrity of HAE by disruption of the tight barrier junction and leads to loss of cilia and airway epithelial cell hypertrophy. The permissiveness of the HBoV1 infection is dependent on various steps including virus attachment, entry, intracellular trafficking, and DNA replication, and all of these steps need to be investigated.[10] It has been shown that infection caused by MVC, which is most closely related to HBoV1, triggers the intra-S-phase arrest to slow down the host cellular DNA replication and, by activation of host DNA damage response, recruit the cellular DNA replication system for viral DNA replication.[50] It was demonstrated that HBoV1 NP1 induce apoptosis and cell cycle arrest in Hela cells, suggesting that amino acids at the N-terminal domain of NP1 protein may be critical for the cell cycle arrest and apoptosis induction in mammalian cells. The apoptotic cell death is mediated by mitochondrion pathway independent of viral genome replication or viral protein expression.[49]

In general, parvoviruses replicate their genomes via a "rolling hairpin" mechanism, a variant of a "rolling circle" replication.[11] Whether the HBoV replicates by the rolling circle or alternative rolling hairpin mechanism still needs to be answered.[11] In HBoV1, head to tail concatemeric intermediates were identified instead of head to head or tail to tail, by PCR and sequencing, suggesting that HBoV1 is replicating via a "rolling circle" mechanism.[51] Whether the persistence of HBoV1 up to several months is due to persistent replication and shedding, passive persistence after primary infection or recurrent mucosal contamination, remains unknown.[3] However, the HBoV1 persistence can be explained by the long-lasting apical and basolateral shedding of the virus, as it has been shown on *in vitro* HAE cultures. It is also possible that the HBoV1 genome can be presented as an episome with prolonged gene expression and replication.[10,51,52]

Only recently has it been demonstrated that HBoV1 productively infects two commercially available cell cultures, MucilAirway HAE (MatTek Co., Ashland, MA, USA) and EpiAirway HAE (Epithelix SaRL, Geneva, Switzerland). The HBoV1 infection resulted in destruction of the epithelial tight junction, loss of cillia and enlargement of the nucleus in the infected cells, HBoV1 persistence, indicating that those two *in vitro* cell culture systems are suitable for further HBoV1 replication and pathogenesis investigation.[53]

Although HBoV1 is respiratory virus, its DNA can be found in blood of patients with severe acute respiratory infection. After infection, the IgG and IgM against HBoV1 VP2 protein can be detected in serum, showing that HBoV1 causes systemic infections. It has been shown that HBoV1 VP2 virus-like particles (VLPs) can elicit typical virus-induced immune response involving Th1 and Th2 cell cytokines.[54,55] In immunocompromised individuals HBoV1 dissemination can occur and the HBoV1 DNA could be detected in respiratory secretions, blood, feces, and urine.[42,56]

6. WHAT ARE THE CLINICAL MANIFESTATIONS?

Since the discovery of HBoV1 in 2005, the virus was most often detected in children with acute wheezing, bronchiolitis, and pneumonia as well as in children with common cold, asthma, upper respiratory infections, and acute otitis media.[3,15,20,34,40,57–59] Prematurely born children or children with other underlying diseases exposed to HBoV1 infection often have a severe clinical course of disease, which requires treatment in an ICU with the possibility of respiratory complications and even respiratory failure.[1,6,7]

Taken together, the prevalence of respiratory manifestations in HBoV1 PCR-positive children with respiratory infections included cough 79%, fever 67%, rhinorrhea 66%, hypoxia 40%, tachypnea 35%, and wheezing 27%. Conjunctivitis, vomiting, diarrhea, and rash are present less frequently.[3,15]

In one study, the HBoV1 and HBoV2 DNA have been detected in the cerebrospinal fluid (CSF) of four out of 67 children with encephalitis.[60]

7. HOW DO YOU DIAGNOSE?

The routine laboratory diagnostics of HBoV1 infections is almost exclusively based on detection of HBoV1 DNA in respiratory samples by PCR.[3,15] Since the HBoV1 DNA in respiratory secretions can persist for a long time after primary infection, diagnostics of HBoV1 primary infection should be supplemented with the detection of HBoV1 DNA in plasma and HBoV1 serodiagnosis.[3]

7.1 Molecular Detection: PCR

Real-time PCR is a highly sensitive method and is routinely used for the detection of HBoV1 DNA in respiratory secretions and plasma. Different primers and probes have been developed for the detection of the HBoV1 genome, including the primer/probe sets for the detection of the NS1, NP1, VP1 and/or VP2 gene region, which appears to be highly sensitive and specific.[2,61–63]

Due to frequent detection of HBoV1 in samples from healthy children and the high co-detection rate of HBoV1 with other respiratory viruses in respiratory samples by PCR, the accurate HBoV1 diagnosis requires detection of HBoV1 DNA or specific IgM response in plasma samples.[19,22,26]

7.2 Serological Diagnosis: ELISA

The serological diagnosis of HBoV1 infection is based on the detection of specific IgM and IgG antibodies against HBoV1 capsid protein VP2, which is the major component of the HBoV capsid and is recognized as the predominant antigen. Detection of IgM antibodies and/or four-fold increase in IgG titer between acute and convalescent serum indicates recent infection. It has been shown that primary infections diagnosed serologically or by the presence of HBoV1 DNA in serum have been linked to respiratory symptoms.[3,22,64] It has been shown that HBoV1 mono-infection, high HBoV1 viral load determined by PCR, and viremia are associated with respiratory tract infection.[64] In one study, 94% of wheezing children with serologically verified HBoV1 infection were viremic.[22] In wheezing children with high HBoV1 DNA load in the nasopharynx, the HBoV1 IgM or an increase of IgG was detected in 96%, compared with 38% of those with low HBoV1 DNA load.[22]

Recently, IgG avidity EIA has been set up. Since the IgG avidity increases along the maturation of B cells, one can distinguish between acute and past infections or between primary and secondary infections.[3,65] However, no commercial kits for detection of HBoV specific antibodies are available at the time of this writing.

8. HOW DO YOU DIFFERENTIATE THE DISEASE FROM SIMILAR ENTITIES?

Like other respiratory viruses, HBoV1 is reported in the context of acute respiratory illnesses, including common cold, acute otitis media, exacerbation of asthma or wheezing, bronchiolitis, and pneumonia. Although some respiratory viruses are more strongly associated with specific signs and symptoms, there is usually an overlap in the clinical presentation and it is therefore not possible to distinguish between viral pathogens.[66] When the etiology of respiratory disease needs to be resolved, the differential diagnosis list should include all well-known respiratory viruses, including respiratory syncytial virus, human metapneumovirus, rhinoviruses, influenza virus A and B, adenovirus, parainfluenza virus 1–4, enteroviruses, and human coronaviruses.[66]

9. WHAT IS THE THERAPEUTIC APPROACH?

Most HBoV infections are probably self-limiting and uncomplicated. In children with severe clinical course of the HBoV infection, supportive care is the treatment of choice. The only randomized controlled study on wheezing children with serologically confirmed HBoV1 infection found prednisolone not to be effective.[3,67]

10. WHAT ARE THE PREVENTIVE AND INFECTION CONTROL MEASURES?

The preventive and infection control measures for HBoV1 are the same as for other respiratory viruses. As these viruses infect the respiratory tract, the viruses are disseminated into the air by coughing, and although the major mode of transmission of respiratory viruses is through large droplets, the transmission through contact and infectious respiratory aerosols of various sizes may also occur. However, adequate hand hygiene, use of medical masks and gloves, and isolation precautions are general infection control measures for all respiratory viral infections.[68]

REFERENCES

1. Ursic T, Steyer A, Kopriva S, Kalan G, Krivec U, Petrovec M. Human bocavirus as the cause of a life-threatening infection. *J Clin Microbiol* 2011;**49**:1179–81.
2. Allander T, Tammi MT, Eriksson M, Bjerkner A, Tiveljung-Lindell A, Andersson B. Cloning of a human parvovirus by molecular screening of respiratory tract samples. *Proc Natl Acad Sci USA* 2005;**102**:12891–6.
3. Jartti T, Hedman K, Jartti L, Ruuskanen O, Allander T, Soderlund-Venermo M. Human bocavirus-the first 5 years. *Rev Med Virol* 2012;**22**:46–64.
4. Terrosi C, Fabbiani M, Cellesi C, Cusi MG. Human bocavirus detection in an atopic child affected by pneumonia associated with wheezing. *J Clin Virol* 2007;**40**:43–5.
5. Oikawa J, Ogita J, Ishiwada N, et al. Human bocavirus DNA detected in a boy with plastic bronchitis. *Pediatr Infect Dis J* 2009;**28**:1035–6.
6. Edner N, Castillo-Rodas P, Falk L, Hedman K, Soderlund-Venermo M, Allander T. Life-threatening respiratory tract disease with human bocavirus-1 infection in a 4-year-old child. *J Clin Microbiol* 2012;**50**:531–2.
7. Korner RW, Soderlund-Venermo M, van Koningsbruggen-Rietschel S, Kaiser R, Malecki M, Schildgen O. Severe human bocavirus infection, Germany. *Emerg Infect Dis* 2012;**17**:2303–5.
8. Brieu N, Gay B, Segondy M, Foulongne V. Electron microscopy observation of human bocavirus (HBoV) in nasopharyngeal samples from HBoV-infected children. *J Clin Microbiol* 2007;**45**:3419–20.
9. Gurda BL, Parent KN, Bladek H, et al. Human bocavirus capsid structure: insights into the structural repertoire of the parvoviridae. *J Virol* 2010;**84**:5880–9.
10. Huang Q, Deng X, Yan Z, et al. Establishment of a reverse genetics system for studying human bocavirus in human airway epithelia. *PLoS Pathog* 2012;**8**:e1002899.
11. Schildgen O, Qiu J, Soderlund-Venermo M. Genomic features of the human bocaviruses. *Future Virol* 2012;**7**:31–9.
12. Arthur JL, Higgins GD, Davidson GP, Givney RC, Ratcliff RM. A novel bocavirus associated with acute gastroenteritis in Australian children. *PLoS Pathog* 2009;**5**:e1000391.
13. Kapoor A, Slikas E, Simmonds P, et al. A newly identified bocavirus species in human stool. *J Infect Dis* 2009;**199**:196–200.
14. Kapoor A, Simmonds P, Slikas E, et al. Human bocaviruses are highly diverse, dispersed, recombination prone, and prevalent in enteric infections. *J Infect Dis* 2010;**201**:1633–43.
15. Chow BD, Esper FP. The human bocaviruses: a review and discussion of their role in infection. *Clin Lab Med* 2009;**29**:695–713.
16. Lee JI, Chung JY, Han TH, Song MO, Hwang ES. Detection of human bocavirus in children hospitalized because of acute gastroenteritis. *J Infect Dis* 2007;**196**:994–7.
17. Jartti L, Langen H, Soderlund-Venermo M, Vuorinen T, Ruuskanen O, Jartti T. New respiratory viruses and the elderly. *Open Respir Med J* 2011;**5**:61–9.

18. Manning A, Russell V, Eastick K, et al. Epidemiological profile and clinical associations of human bocavirus and other human parvoviruses. *J Infect Dis* 2006;**194**:1283–90.
19. Christensen A, Nordbo SA, Krokstad S, Rognlien AG, Dollner H. Human bocavirus commonly involved in multiple viral airway infections. *J Clin Virol* 2008;**41**:34–7.
20. Allander T, Jartti T, Gupta S, et al. Human bocavirus and acute wheezing in children. *Clin Infect Dis* 2007;**44**:904–10.
21. Foulongne V, Olejnik Y, Perez V, Elaerts S, Rodiere M, Segondy M. Human bocavirus in french children. *Emerg Infect Dis* 2006;**12**:1251–3.
22. Soderlund-Venermo M, Lahtinen A, Jartti T, et al. Clinical assessment and improved diagnosis of bocavirus-induced wheezing in children, Finland. *Emerg Infect Dis* 2009;**15**:1423–30.
23. von Linstow ML, Hogh M, Hogh B. Clinical and epidemiologic characteristics of human bocavirus in Danish infants: results from a prospective birth cohort study. *Pediatr Infect Dis J* 2008;**27**:897–902.
24. Monteny M, Niesters HG, Moll HA, Berger MY. Human bocavirus in febrile children, the Netherlands. *Emerg Infect Dis* 2007;**13**:180–2.
25. De Vos N, Vankeerberghen A, Vaeyens F, Van Vaerenbergh K, Boel A, De Beenhouwer H. Simultaneous detection of human bocavirus and adenovirus by multiplex real-time PCR in a Belgian paediatric population. *Eur J Clin Microbiol Infect Dis* 2009;**28**:1305–10.
26. Bonzel L, Tenenbaum T, Schroten H, Schildgen O, Schweitzer-Krantz S, Adams O. Frequent detection of viral coinfection in children hospitalized with acute respiratory tract infection using a real-time polymerase chain reaction. *Pediatr Infect Dis J* 2008;**27**: 589–94.
27. Weissbrich B, Neske F, Schubert J, et al. Frequent detection of bocavirus DNA in German children with respiratory tract infections. *BMC Infect Dis* 2006;**6**:109.
28. Brieu N, Guyon G, Rodiere M, Segondy M, Foulongne V. Human bocavirus infection in children with respiratory tract disease. *Pediatr Infect Dis J* 2008;**27**:969–73.
29. Antunes H, Rodrigues H, Silva N, et al. Etiology of bronchiolitis in a hospitalized pediatric population: prospective multicenter study. *J Clin Virol* 2010;**48**:134–6.
30. Garcia-Garcia ML, Calvo C, Pozo F, et al. Human bocavirus detection in nasopharyngeal aspirates of children without clinical symptoms of respiratory infection. *Pediatr Infect Dis J* 2008;**27**:358–60.
31. Regamey N, Frey U, Deffernez C, Latzin P, Kaiser L. Isolation of human bocavirus from Swiss infants with respiratory infections. *Pediatr Infect Dis J* 2007;**26**:177–9.
32. Midulla F, Scagnolari C, Bonci E, et al. Respiratory syncytial virus, human bocavirus and rhinovirus bronchiolitis in infants. *Arch Dis Child* 2010;**95**:35–41.
33. Maggi F, Andreoli E, Pifferi M, Meschi S, Rocchi J, Bendinelli M. Human bocavirus in italian patients with respiratory diseases. *J Clin Virol* 2007;**38**:321–5.
34. Ursic T, Jevsnik M, Zigon N, et al. Human bocavirus and other respiratory viral infections in a 2-year cohort of hospitalized children. *J Med Virol* 2012;**84**:99–108.
35. Haidopoulou K, Goutaki M, Damianidou L, Eboriadou M, Antoniadis A, Papa A. Human bocavirus infections in hospitalized Greek children. *Arch Med Sci* 2012;**6**:100–3.
36. Midilli K, Yilmaz G, Turkoglu S, et al. Detection of human bocavirus DNA by polymerase chain reaction in children and adults with acute respiratory tract infections. *Mikrobiyol Bul* 2012;**44**:405–13.
37. Allander T. Human bocavirus. *J Clin Virol* 2008;**41**(1):29–33.
38. Moriyama Y, Hamada H, Okada M, et al. Distinctive clinical features of human bocavirus in children younger than 2 years. *Eur J Pediatr* 2010;**169**:1087–92.

39. Calvo C, Pozo F, Garcia-Garcia ML, et al. Detection of new respiratory viruses in hospitalized infants with bronchiolitis: a three-year prospective study. *Acta Paediatr* 2010;**99**: 883–7.
40. Miron D, Srugo I, Kra-Oz Z, et al. Sole pathogen in acute bronchiolitis: is there a role for other organisms apart from respiratory syncytial virus? *Pediatr Infect Dis J* 2010;**29**: e7–10.
41. Muller A, Klinkenberg D, Vehreschild J, et al. Low prevalence of human metapneumovirus and human bocavirus in adult immunocompromised high risk patients suspected to suffer from pneumocystis pneumonia. *J Infect* 2009;**58**:227–31.
42. Schenk T, Strahm B, Kontny U, Hufnagel M, Neumann-Haefelin D, Falcone V. Disseminated bocavirus infection after stem cell transplant. *Emerg Infect Dis* 2007;**13**:1425–7.
43. Sadeghi M, Kantola K, Finnegan DP, et al. Possible involvement of human bocavirus-1 in the death of a middle-aged immunosuppressed patient. *J Clin Microbiol* 2013;**51**(10):3461–3.
44. Kantola K, Hedman L, Allander T, et al. Serodiagnosis of human bocavirus infection. *Clin Infect Dis* 2008;**46**:540–6.
45. Endo R, Ishiguro N, Kikuta H, et al. Seroepidemiology of human bocavirus in Hokkaido prefecture, Japan. *J Clin Microbiol* 2007;**45**:3218–23.
46. Riipinen A, Vaisanen E, Lahtinen A, et al. Absence of human bocavirus from deceased fetuses and their mothers. *J Clin Virol* 2010;**47**:186–8.
47. Dijkman R, Koekkoek SM, Molenkamp R, Schildgen O, van der Hoek L. Human bocavirus can be cultured in differentiated human airway epithelial cells. *J Virol* 2009;**83**:7739–48.
48. Deng X, Yan Z, Luo Y, et al. In vitro modeling of human bocavirus 1 infection of polarized primary human airway epithelia. *J Virol* 2013;**87**:4097–102.
49. Sun B, Cai Y, Li Y, et al. The nonstructural protein NP1 of human bocavirus 1 induces cell cycle arrest and apoptosis in Hela cells. *Virology* 2013;**440**:75–83.
50. Luo Y, Chen AY, Qiu J. Bocavirus infection induces a DNA damage response that facilitates viral DNA replication and mediates cell death. *J Virol* 2011;**85**:133–45.
51. Lusebrink J, Schildgen V, Tillmann RL, et al. Detection of head-to-tail DNA sequences of human bocavirus in clinical samples. *PLoS One* 2011;**6**:e19457.
52. Kapoor A, Hornig M, Asokan A, Williams B, Henriquez JA, Lipkin WI. Bocavirus episome in infected human tissue contains non-identical termini. *PLoS One* 2011;**6**:e21362.
53. Deng X, Li Y, Qiu J. Human bocavirus 1 infects commercially-available primary human airway epithelium cultures productively. *J Virol Methods* 2013;**195**:112–9.
54. Chung JY, Han TH, Kim JS, Kim SW, Park CG, Hwang ES. Th1 and Th2 cytokine levels in nasopharyngeal aspirates from children with human bocavirus bronchiolitis. *J Clin Virol* 2008;**43**:223–5.
55. Lindner J, Zehentmeier S, Franssila R, et al. CD4+ T helper cell responses against human bocavirus viral protein 2 viruslike particles in healthy adults. *J Infect Dis* 2008;**198**:1677–84.
56. de Vries JJ, Bredius RG, van Rheenen PF, et al. Human bocavirus in an immunocompromised child presenting with severe diarrhea. *J Clin Microbiol* 2009;**47**:1241–3.
57. Gendrel D, Guedj R, Pons-Catalano C, et al. Human bocavirus in children with acute asthma. *Clin Infect Dis* 2007;**45**:404–5.
58. Beder LB, Hotomi M, Ogami M, et al. Clinical and microbiological impact of human bocavirus on children with acute otitis media. *Eur J Pediatr* 2009;**168**:1365–72.
59. Don M, Soderlund-Venermo M, Valent F, et al. Serologically verified human bocavirus pneumonia in children. *Pediatr Pulmonol* 2010;**45**:120–6.

60. Mitui MT, Tabib SM, Matsumoto T, et al. Detection of human bocavirus in the cerebrospinal fluid of children with encephalitis. *Clin Infect Dis* 2012;**54**:964–7.
61. Lu X, Chittaganpitch M, Olsen SJ, et al. Real-time PCR assays for detection of bocavirus in human specimens. *J Clin Microbiol* 2006;**44**:3231–5.
62. Choi JH, Chung YS, Kim KS, et al. Development of real-time PCR assays for detection and quantification of human bocavirus. *J Clin Virol* 2008;**42**:249–53.
63. Tozer SJ, Lambert SB, Whiley DM, et al. Detection of human bocavirus in respiratory, fecal, and blood samples by real-time PCR. *J Med Virol* 2009;**81**:488–93.
64. Christensen A, Nordbo SA, Krokstad S, Rognlien AG, Dollner H. Human bocavirus in children: mono-detection, high viral load and viraemia are associated with respiratory tract infection. *J Clin Virol* 2010;**49**:158–62.
65. Hedman L, Soderlund-Venermo M, Jartti T, Ruuskanen O, Hedman K. Dating of human bocavirus infection with protein-denaturing IgG-avidity assays-secondary immune activations are ubiquitous in immunocompetent adults. *J Clin Virol* 2010;**48**:44–8.
66. Pavia AT. Viral infections of the lower respiratory tract: old viruses, new viruses, and the role of diagnosis. *Clin Infect Dis* 2011;**52**(Suppl. 4):S284–9.
67. Jartti T, Soderlund-Venermo M, Allander T, Vuorinen T, Hedman K, Ruuskanen O. No efficacy of prednisolone in acute wheezing associated with human bocavirus infection. *Pediatr Infect Dis J* 2011;**30**:521–3.
68. Seto WH, Conly JM, Pessoa-Silva CL, Malik M, Eremin S. Infection prevention and control measures for acute respiratory infections in healthcare settings: an update. *East Mediterr Health J* 2013;**19**(Suppl. 1):S39–47.

Chapter 16

Norovirus Gastroenteritis

Momoko Mawatari and Yasuyuki Kato

Division of Preparedness and Emerging Infections, Disease Control and Prevention Center, National Center for Global Health and Medicine, Toyama, Shinjuku-ku, Tokyo, Japan

CASE PRESENTATION

A 50-year-old female with a past history of uterus cancer 15 years ago was admitted to the hospital in Tokyo, Japan. She had arrived in Tokyo from the island of Kyusyu with her husband 5 days ago to meet her daughter. Her complaints had started 1 day ago, which were fatigue, thirstiness, and drowsiness. When her daughter came home at night, she found her mother lying on the floor. She also found that her mother had diarrhea and had vomited and then called an ambulance.

On admission, her level of consciousness was unclear; Glasgow Coma Scale (GCS) was 11 (E3V2M5). Body temperature was 41°C; blood pressure 86/54 mmHg; heart rate 110 bpm; and respiratory rate 16/min. She had severe watery diarrhea, dry mouth and skin, and general muscle hypertonia. The white blood cell count was 15,600/μL, hemoglobin level 12.9 g/dL, creatinine 1.25 mg/dL, and glucose 136 mg/dL. She underwent a lumbar puncture, which revealed normal initial pressure, nucleated cell count, and glucose level in the CSF. A rapid immunochromatographic test for norovirus was positive in her stool sample. She was diagnosed with norovirus gastroenteritis, but was still suspected of bacterial sepsis. Therefore, a bolus infusion of extracellular fluid was given and ceftriaxone 2 g QD was started empirically. Whole body CT imaging revealed edema of the intestinal wall, but no evidence of other abnormalities.

Her consciousness recovered gradually and became clear on the fourth day of hospitalization. Ceftriaxone was stopped on the second day. On the fifth day, she was able to have meals and discharged. Her symptoms were probably caused by norovirus infection.

Emerging Infectious Diseases. DOI: http://dx.doi.org/10.1016/B978-0-12-416975-3.00016-9

1. WHY THIS CASE WAS SIGNIFICANTLY IMPORTANT AS AN EMERGING INFECTION

Norovirus gastroenteritis is the most common acute non-bacterial gastroenteritis worldwide.[1] The pathogen norovirus is a highly contagious agent. Sometimes, it causes severe illness, even encephalopathy, and also chronic gastroenteritis in immunocompromised patients.

2. WHAT IS THE CAUSATIVE AGENT?

The disease is caused by norovirus, which belongs to the family *Caliciviridae*. Caliciviruses contain a single-stranded RNA genome and have a relatively simple structure, containing one major (VP1) and one minor (VP2) capsid protein. The viruses are not enveloped and are resistant to ethanol. Expression of the VP1 in experimental systems produces empty capsids or virus-like particles (VLPs) that work as immune reagents and help investigate the pathogenesis. Norovirus has five genogroups based on sequence homology.[2] Genogroups GI, GII, and GIV are human pathogens and have multiple genotypes within each group. There have been at least eight distinct genotypes within genogroup GI and 27 genotypes within GII.[3] Frequent recombination of genomes between strains also contributes to genetic diversity of the virus.[4,5]

3. WHAT IS THE FREQUENCY OF THE DISEASE?

Cases of norovirus gastroenteritis increase in temperate regions in cold weather season. In developed countries, serum antibody to the norovirus can be detected at the age of 3 to 4 years, with antibody prevalence gradually rising to greater than 50% by the fifth decade of life.[6–9] Studies using recombinant antigen of the virus suggest that a significant increase in seroprevalence occurs in infancy, on entry into primary schools, and in young adulthood.[10] The epidemiology in developing countries remains unknown.

4. HOW IS THE VIRUS TRANSMITTED?

Norovirus is mainly transmitted by the ingestion of contaminated food or water. The virus is extremely contagious with an estimated infectious dose as low as 18 viral particles.[11] Shellfish, such as clams and oysters, are extremely common vehicles in such outbreaks. This is particularly true of oysters since they are filter feeders that concentrate on particulate matter from the environment.[12] The virus is also transmitted person to person or via contaminated environmental surfaces, and fomites, such as shared toilet facilities or elevator buttons.[13,14] Each gram of feces (about one-quarter teaspoon) during peak shedding contains the virion, which can infect 5 billion

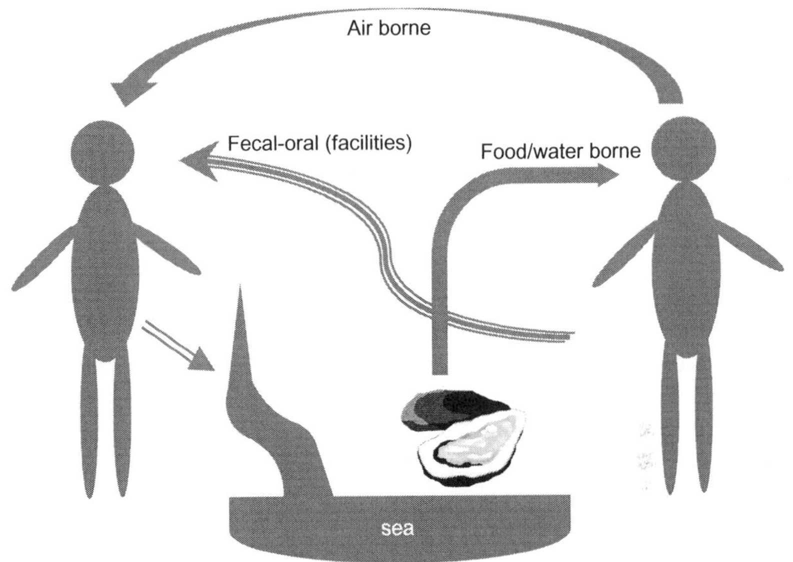

FIGURE 16.1 Transmission routes of norovirus.

people. Since it is stable in the environment the virus causes outbreaks at hospitals, long-term care facilities, schools, and aboard cruise ships. In these situations, airborne transmission is occasionally implicated (Figure 16.1).

5. WHICH FACTORS ARE INVOLVED IN DISEASE PATHOGENESIS? WHAT ARE THE PATHOGENIC MECHANISMS?

Acute infection with norovirus results in a reversible histopathologic lesion in the jejunum,[15,16] with the stomach[17] and rectum spared. The histopathological changes include blunting of villi, shortening of microvilli, dilation of endoplasmic reticulum, and an increase in intracellular multivesiculate bodies. These changes appear within 24 hours after infection and persist for a variable period of time after the illness. The lesions usually disappear within 2 weeks after the onset of illness although some jejunal lesions have been noted as late as 6 weeks after infection.

The precise mechanisms of diarrhea and vomiting remain unknown. Diarrhea induced by the norovirus is associated with a transient malabsorption of D-xylose and fat[18] and with decreased activity of brush-border enzymes, including alkaline phosphatase and trehalase.[15]

Susceptibility to infection with certain norovirus strains is related to the presence of several blood group carbohydrate antigens, which are also expressed on a variety of cells including gastrointestinal epithelial cells.[19–23]

These findings might explain why some individuals show long-term resistance to norovirus infection.[24]

Norovirus has been increasingly recognized as an important pathogen of chronic gastroenteritis in immunocompromised patients, such as allogeneic hematopoietic stem cell transplant recipients[25–26] and HIV-infected patients.[27,28] In a case series of 13 patients who underwent hematopoietic stem cell transplantation, the median duration of viral excretion was 150 days (range 60–380).[26]

6. WHAT ARE THE CLINICAL MANIFESTATIONS?

Symptoms of the illness caused by each genogroup are indistinguishable. The incubation period for norovirus gastroenteritis is generally 24 to 48 hours, with a range from 18 to 72 hours. The onset of symptoms can be either gradual or abrupt, and most persons initially complain of abdominal cramps or nausea. Both vomiting and diarrhea generally develop, although either can be present alone. Myalgias, malaise, and headaches are also seen. Fever (with temperatures of 38.3 to 38.9°C [101 to 102°F]) develops in about half of cases. Diarrheal stool is usually moderate in amounts, with four to eight stool passes over a period of 24 hours. The stool does not typically contain blood, or mucous, and may be loose to watery; fecal leukocytes are not seen. Disease manifestations generally last 48 to 72 hours with a full and rapid recovery.

More severe disease and death have been reported in elderly patients.[29,30] Afebrile seizure sometimes occurs in children. Norovirus is more commonly associated with benign convulsion than rotavirus (8.67% [15/173] vs. 1.29% [3/232], $P < 0.001$).[31] Norovirus-related encephalopathy rarely causes severe neurologic sequelae, which includes severe delays in mental and motor development, and tonic seizures even 2 years after the onset of illness.[32]

There are limited reports on norovirus infection in immunocompromised individuals. The viruses may be shed for prolonged periods of time in persons with severe compromise of cellular immunity.

Laboratory tests generally reveal normal results. The peripheral white blood cell count is within normal ranges or slightly elevated, and relative lymphopenia might be observed at the height of the illness. Elevation of BUN and serum creatinine level means severe dehydration although this condition rarely occurs.

7. HOW DO YOU DIAGNOSE?

The signs and symptoms of illness are not specific, but predominance of vomiting suggests that viral gastroenteritis is more likely than bacterial diarrhea in adults. Routine laboratory tests are also not helpful in making a diagnosis of norovirus infection.

The absence of fecal leukocytes, as determined by microscopic examination of stools stained with methylene blue,[33] is a useful tool with which to exclude infection with enteroinvasive pathogens such as *Shigella* or *Campylobacter*.

Immune electron microscopy, in which immune sera are used to aggregate and highlight virions in stool suspensions, was the method used originally to identify the virus. However, it is relatively insensitive for diagnosis of norovirus infection, because of the small numbers of intact virion particles in diarrheal stools.[34]

Enzyme immunoassays (EIAs) to detect norovirus antigens in stool have been developed and are commercially available. The sensitivity and specificity of the EIAs range from 36 to 75% and 83 to 99%, respectively.[35] These assays are sensitive enough to detect homologous viruses but might not detect antigenic variants.[36] The sensitivity of EIAs might also vary according to genogroup.

Reverse transcriptase-polymerase chain reaction (RT-PCR) is highly sensitive and specific in terms of detection of noroviruses.[37–41]

Several factors can affect the sensitivity and specificity of RT-PCR assays and should be carefully paid attention to, including the viral RNA extraction method, the primers used in amplification, and the methods used for confirmation of test results.[42]

Serum norovirus-specific antibodies can be used in outbreak investigation.[43] Serum antibody titer rises can be detected within 10 to 14 days after the onset of illness.[44,45]

8. HOW DO YOU DIFFERENTIATE THIS DISEASE FROM SIMILAR ENTITIES?

The symptoms are non-specific and it is impossible to distinguish the illness due to other causes (infectious or non-infectious). The distinct seasonality can be observed in winter season in temperate regions. Compared with norovirus, rotavirus causes a higher maximum temperature and more diarrhea.[31] If diarrhea persists over a week in an individual with a history of travel, hiking, or oral–anal sexual activity, the patient should be evaluated for protozoa such as entamoeba, giardia, or cryptosporidium.[46] Recent antibiotic use or hospitalization should prompt consideration of *Clostridium difficile* infection. The patient age ranges from infant to adult while rotavirus gastroenteritis predominates in pediatric population. The specific diagnosis largely depends upon microbiological or serological diagnosis.

Vomiting and diarrhea sometimes occur in systemic illnesses and should be carefully evaluated; for example, sepsis, pyelonephritis, cholangitis, cholecystitis, pancreatitis, appendicitis, and gynecologic diseases.

9. WHAT IS THE THERAPEUTIC APPROACH?

Norovirus gastroenteritis is usually self-limited and is treated with supportive measures (fluid repletion and unrestricted nutrition). For adults with mild to moderate hypovolemia, oral rehydration solutions may be superior to sports drinks in maintaining electrolyte balance along with hydration. Patients with severe dehydration require intravenous fluids. Analgesics and antiemetics can be given as symptomatic treatment of headache, myalgia, and nausea. Although antiperistaltic agents are frequently prescribed to control diarrhea, their effect on the disease course and on excretion of virus has not been strictly evaluated.

Specific antiviral therapy is not available. There was a randomized double-blind placebo-controlled trial, which evaluated nitazoxanide, a thiazolide anti-infective agent, in treating viral gastroenteritis in adults and adolescents.[47] Significant reductions in time to resolution of symptoms were observed for all patients analyzed ($P < 0.0001$) and for subsets of patients with rotavirus ($P = 0.0052$) and norovirus ($P = 0.0295$). This trial, however, included small numbers of participants (13 norovirus patients were enrolled, six of 13 took nitazoxanide, and seven took placebo).

10. WHAT ARE THE PREVENTIVE AND INFECTION CONTROL MEASURES

There are currently no licensed vaccines against norovirus. A randomized double-blind placebo-controlled trial, which examined immunogenicity of a vaccine based on virus-like particles (VLPs) as well as protection against challenge with live virus, was conducted. Seroresponse to the vaccine, as measured by IgA, occurred in 70% of participants at day 42. In the per-protocol analysis, 69% of participants receiving placebo developed viral gastroenteritis due to Norwalk virus compared with 37% of participants receiving vaccine ($P = 0.006$).[48]

The most effective and feasible preventive measure is adequate hand hygiene and avoidance of close contact with people with symptoms of gastroenteritis. Individuals should wash hands carefully with soap and water especially after using the toilet and changing diapers, and always before eating, preparing, or handling food. The efficacy of alcohol-based hand sanitizers against norovirus remains controversial with mixed evidence depending on the product formulation and evaluation methodology. In finger pad studies, soap and water used for 20 seconds have been demonstrated to reduce norovirus by $0.7-1.2 \log_{10}$ by RT-PCR assay, whereas alcohol-based hand sanitizers did not demonstrate any appreciable reduction of viral RNA.[49]

Patients with norovirus gastroenteritis should not prepare food for others or provide healthcare during the disease course and for at least several days after the resolution of symptoms. This precaution also applies to sick workers in settings such as schools and daycare centers.

In outbreak settings in healthcare facilities, cruise ships, and college dormitories, exclusion and isolation of infected persons are the most practical means of interrupting transmission of virus and limiting contamination of the environment. Patients should be placed on isolation for a minimum of 48 hours after the resolution of symptoms. Healthcare workers should wear appropriate personal protective equipment (gloves, gown, mask, and goggles) based on careful risk assessment of the patient's signs and symptoms. Units in a healthcare facility may be closed to new admissions (ward closure) to prevent the introduction of new, susceptible patients.[50] School closure is sometimes necessary to prevent the further transmission in school settings.

Disinfection of the environment is one of the key approaches to interrupt norovirus spread. Sodium hypochlorite has been widely recommended to disinfect norovirus on the surfaces and its efficacy has been well documented.[51–54] The use of upholstered furniture and rugs or carpets in patient care areas should be avoided because these items are difficult to clean and disinfect properly.[55]

When outbreaks spread across a community and are difficult to contain, social measures, which restrict social gatherings, might be necessary as in pandemic influenza.

REFERENCES

1. Dolin R, Treanor JJ. Norovirus and other caliciviruses. In: Mandell LG, Bennett JE, Dolin R, editors. *Principles and practice of infectious diseases*. 7th ed. Philadelphia: Churchill Livingstone Elsevier; 2010. p. 2399–405. Chapter 175.
2. Hutson AM, Atmar RL, Estes MK. Norovirus disease: changing epidemiology and host susceptibility factors. *Trends Microbiol* 2004;**12**:279–87.
3. Zheng DP, Ando T, Fankhauser RL, et al. Norovirus classification and proposed strain nomenclature. *Virology* 2006;**346**:312–23.
4. Etherington GJ, Dicks J, Roberts IN. High throughput sequence analysis reveals hitherto unreported recombination in the genus norovirus. *Virology* 2006;**345**:88–95.
5. Ambert-Balay K, Bon F, Le Guyader F, et al. Characterization of new recombinant noroviruses. *J Clin Microbiol* 2005;**43**:5179–86.
6. Greenberg HB, Valdesuso J, Kapikian AZ, et al. Prevalence of antibody to the Norwalk virus in various countries. *Infect Immun* 1979;**26**:270–3.
7. Dolin R, Roessner KD, Treanor J, et al. Radioimmunoassay for detection of snow mountain agent of viral gastroenteritis. *J Med Virol* 1985;**19**:11–8.
8. Hinkula J, Ball JM, Lofgren S, et al. Antibody prevalence and immunoglobulin IgG subclass pattern to norwalk virus in Sweden. *J Med Virol* 1995;**47**:52–7.
9. Numata K, Nakata S, Jiang X, et al. Epidemiologic study of Norwalk virus infections in Japan and Southeast Asia by enzyme-linked immunosorbent assays with norwalk virus capsid protein produced by the baculovirus expression system. *J Clin Microbiol* 1994;**32**:121–6.
10. Gray JJ, Jiang X, Morgan-Capner P, et al. Prevalence of antibodies to norwalk virus in England: detection by enzyme-linked immunosorbent assay using baculovirus-expressed norwalk virus capsid antigen. *J Clin Microbiol* 1993;**31**:1022–5.

11. Teunis PF, Moe CL, Liu P, et al. Norwalk virus: how infectious is it? *J Med Virol* 2008;**80**:1468–76.
12. Centers for Disease Control and Prevention (CDC). Notes from the field: norovirus infections associated with frozen raw oysters—Washington, 2011. *MMWR Morb Mortal Wkly Rep* 2012;**61**:110.
13. Ho MS, et al. Viral gastroenteritis aboard a cruise ship. *Lancet* 1989;**2**:961–5.
14. Wu HM, et al. A norovirus outbreak at a longterm-care facility: the role of environmental surface contamination. *Infect Control Hosp Epidemiol* 2005;**26**:802–10.
15. Agus SG, Dolin R, Wyatt RG, et al. Acute infectious nonbacterial gastroenteritis: intestinal histopathology. *Ann Intern Med* 1973;**79**:18–25.
16. Schreiber DS, Blacklow NR, Trier JS. The mucosal lesion of the proximal small intestine in acute infectious nonbacterial gastroenteritis. *N Engl J Med* 1973;**288**:1318–23.
17. Widerlite L, Trier J, Blacklow N, et al. Structure of the gastric mucosa in acute infectious nonbacterial gastroenteritis. *Gastroenterology* 1975;**70**:321–5.
18. Blacklow NR, Dolin R, Feson DS, et al. Acute infectious nonbacterial gastroenteritis: Etiology and pathogenesis. *Ann Intern Med* 1972;**76**:993–1000.
19. Marionneau S, Ruvoen N, Le Moullac-Vaidye B, et al. Norwalk virus binds to histo-blood group antigens present on gastroduodenal epithelial cells of secretor individuals. *Gastroenterology* 2002;**122**:1967–77.
20. Hutson AM, Atmar RL, Marcus DM, et al. Norwalk virus-like particle hemagglutination by binding to histo-blood group antigens. *J Virol* 2003;**77**:405–15.
21. Huang P, Farkas T, Marionneau S, et al. Noroviruses bind to human ABO, Lewis, and secretor histo-blood group antigens: identification of 4 distinct strain-specific patterns. *J Infect Dis* 2003;**188**:19–31.
22. Tan M, Huang P, Meller J, et al. Mutations within the P2 domain of norovirus capsid affect binding to human histo-blood group antigens: evidence for a binding pocket. *J Virol* 2003;**77**:12562–71.
23. Hutson AM, Atmar RL, Graham DY, et al. Norwalk virus infection and disease is associated with ABO histo-blood group type. *J Infect Dis* 2002;**185**:1335–7.
24. Parrino TA, Schreiber DS, Trier JS, et al. Clinical immunity in acute gastroenteritis caused by norwalk agent. *N Engl J Med* 1977;**291**:86–9.
25. Roddie C, Paul JP, Benjamin R, et al. Allogeneic hematopoietic stem cell transplantation and norovirus gastroenteritis: a previously unrecognized cause of morbidity. *Clin Infect Dis* 2009;**49**:1061–8.
26. Saif MA, Bonney DK, Bigger B, Forsythe L, Williams N, Page J, et al. Chronic norovirus infection in pediatric hematopoietic stem cell transplant recipients: a cause of prolonged intestinal failure requiring intensive nutritional support. *Pediatr Transplant* 2011;**15**:505–9.
27. Wingfield T, Gallimore CI, Xerry J, Gray JJ, Klapper P, Guiver M, et al. Chronic norovirus infection in an HIV-positive patient with persistent diarrhoea: a novel cause. *J Clin Virol* 2010;**49**:219–22.
28. Cegielski JP, Msengi AE, Miller SE. Enteric viruses associated with HIV infection in Tanzanian children with chronic diarrhea. *Pediatr AIDS HIV Infect* 1994;**5**:296–9.
29. Lopman BA, Reacher MH, Vipond IB, et al. Clinical manifestation of norovirus gastroenteritis in health care settings. *Clin Infect Dis* 2004;**39**:318–24.
30. Mattner F, Sohr D, Heim A, et al. Risk groups for clinical complications of norovirus infections: an outbreak investigation. *Clin Microbiol Infect* 2006;**12**:69–74.
31. Chan CM, Chan CW, Ma CK, Chan HB. Norovirus as cause of benign convulsion associated with gastro-enteritis. *J Paediatr Child Health* 2011;**47**:373–7.

32. Obinata K, Okumura A, Nakazawa T, Kamata A, Niizuma T, Kinoshita K, et al. Norovirus encephalopathy in a previously healthy child. *Pediatr Infect Dis J* 2010;**29**:1057–9.
33. Harris JC, DuPont HL, Hornick RB. Fecal leukocytes in diarrheal illness. *Ann Intern Med* 1972;**76**:697–703.
34. Marshall JA, Bruggink LD. Laboratory diagnosis of norovirus. *Clin Lab* 2006;**52**:571–81.
35. Gray JJ, Kohli E, Ruggeri FM, et al. European multicenter evaluation of commercial enzyme immunoassays for detecting norovirus antigen in fecal samples. *Clin Vaccine Immunol* 2007;**14**:1349–55.
36. Jiang X, Wang J, Estes MK. Characterization of SRSVs using RT-PCR and a new antigen ELISA. *Arch Virol* 1995;**140**:363–74.
37. Jiang X, Wang J, Graham DY, Estes MK. Detection of norwalk virus in stool by polymerase chain reaction. *J Clin Microbiol* 1992;**30**:2529–34.
38. Moe CL, Gentsch J, Ando T, et al. Application of PCR to detect Norwalk virus in fecal specimens from outbreaks of gastroenteritis. *J Clin Microbiol* 1994;**32**:642–8.
39. Schwab KJ, Estes MK, Neill FH, Atmar RL. Use of heat release and an internal RNA standard control in reverse transcription-PCR detection of norwalk virus from stool samples. *J Clin Microbiol* 1997;**35**:511–4.
40. Ando T, Monroe SS, Noel JS, Glass RI. A one-tube method of reverse transcription-PCR to efficiently amplify a 3-kilobase region from the RNA polymerase gene to the poly(A) tail of small round-structured viruses (Norwalk-like viruses). *J Clin Microbiol* 1997;**35**:570–7.
41. Kundu S, Lockwood J, Depledge DP, et al. Next-generation whole genome sequencing identifies the direction of norovirus transmission in linked patients. *Clin Infect Dis* 2013;**57**:407–14.
42. Vinjé J, Vennema H, Maunula L, et al. International collaborative study to compare reverse transcriptase PCR assays for detection and genotyping of noroviruses. *J Clin Microbiol* 2003;**41**:1423–33.
43. Gary GW, Anderson LJ, Keswick BH, et al. Norwalk virus antigen and antibody response in an adult volunteer study. *J Clin Microbiol* 1987;**25**:2001–3.
44. Brinker JP, Blacklow NR, Estes MK, et al. Detection of norwalk virus and other genogroup 1 human caliciviruses by a monoclonal antibody, recombinant-antigen-based immunoglobulin M capture enzyme immunoassay. *J Clin Microbiol* 1998;**36**:1064–9.
45. Brinker JP, Blacklow NR, Jiang X, et al. Immunoglobulin M antibody test to detect genogroup II norwalk-like virus infection. *J Clin Microbiol* 1999;**37**:2983–6.
46. Thielman NM, Guerrant RL. Clinical practice. Acute infectious diarrhea. *N Engl J Med* 2004;**350**:38–47.
47. Rossignol JF, El-Gohary YM. Nitazoxanide in the treatment of viral gastroenteritis: a randomized double-blind placebo-controlled clinical trial. *Aliment Pharmacol Ther* 2006;**24**:1423–30.
48. Atmar RL, Bernstein DI, Harro CD, et al. Norovirus vaccine against experimental human norwalk virus illness. *New Engl J Med* 2011;**365**:2178–87.
49. Liu P, Yuen Y, Hsiao HM, Jaykus LA, Moe C. Effectiveness of liquid soap and hand sanitizer against Norwalk virus on contaminated hands. *Appl Environ Microbiol* 2010;**76**:394–9.
50. Harris JP, Lopman BA, O'Brien SJ. Infection control measures for norovirus: a systematic review of outbreaks in semi-enclosed settings. *J Hosp Infect* 2010;**74**:1–9.
51. Doultree JC, Druce JD, Birch CJ, Bowden DS, Marshall JA. Inactivation of feline calicivirus, a Norwalk virus surrogate. *J Hosp Infect* 1999;**41**:51–7.

52. Duizer E, Bijkerk P, Rockx B, De Groot A, Twisk F, Koopmans M. Inactivation of caliciviruses. *Appl Environ Microbiol* 2004;**70**:4538–43.
53. Girard M, Ngazoa S, Mattison K, Jean J. Attachment of noroviruses to stainless steel and their inactivation using household disinfectants. *J Food Prot* 2010;**73**:400–4.
54. Park GW, Boston DM, Kase JA, Sampson MN, Sobsey MD. Evaluation of liquid- and fog-based application of sterilox hypochlorous acid solution for surface inactivation of human norovirus. *Appl Environ Microbiol* 2007;**73**:4463–8.
55. MacCannell T, Umscheid CA, Agarwal RK, Lee I, Kuntz G, Stevenson KB. Healthcare Infection control practices advisory committee-HICPAC. Guideline for the prevention and control of norovirus gastroenteritis outbreaks in healthcare settings. *Infect Control Hosp Epidemiol* 2011;**32**:939–69.

Chapter 17

Enterohemorrhagic *Escherichia coli* (EHEC): Hemorrhagic Colitis and Hemolytic Uremic Syndrome (HUS)

Jakob P. Cramer

Section Tropical Medicine and Infectious Diseases, University Medical Center Hamburg-Eppendorf and Bernhard Nocht Institute for Tropical Medicine, Bernhard-Nocht-Strasse 74, Hamburg, Germany

CASE PRESENTATION

On Friday May 27, 2011, during the large serotype O104:H4 outbreak in northern Germany, a 44-year-old woman presented to the emergency department in Hamburg, Germany, with bloody diarrhea and abdominal cramps, but no fever, for 5 days. She reported a weight gain of 2–3 kg in the past days and the occurrence of lid edema. Blood analysis revealed the following alterations: thrombocytes 28,000/μl, bilirubin 2.6 mg/dl (normal <1.2), creatinine 2.7 mg/dl (normal 0.5–1.0), C-reactive protein 12 mg/l (normal <5), and lactate dehydrogenase 898 U/l (normal 135–214). One day earlier (May 26) she had contacted her general practitioner and all parameters had been within normal ranges except for a slightly elevated lactate dehydrogenase of 279 U/l. The patient was very sportive and said that healthy living and food conditions were very important for her and her family. She and her daughter, who was also hospitalized with hemolytic uremic syndrome, had eaten out at a bagel shop a few days earlier. Both recovered but the mother required prolonged hemodialysis for several weeks.

1. WHY THIS CASE WAS SIGNIFICANTLY IMPORTANT AS AN EMERGING INFECTION

Enterohemorrhagic *Escherichia coli* (EHEC) infections—mostly serotype O157:H7—are continuously recognized in industrialized countries. The above

Emerging Infectious Diseases. DOI: http://dx.doi.org/10.1016/B978-0-12-416975-3.00017-0

FIGURE 17.1 **Ruminants live in close contact with humans.** There is a high risk for food or drinking water contamination in sub-Saharan Africa.

described case, in contrast, was one of 855 cases suffering from the complication hemolytic uremic syndrome (HUS), out of 3816 persons infected with the strain O104:H4 during one of the largest EHEC epidemics which originated in northern Germany between May and June 2011.[1] New strains emerge and adapt their virulence profile, for example, by lateral gene transfer between different potentially pathogenic *E. coli* bacteria colonizing a host's intestinal tract. The generation of new strains may in particular occur in settings where humans live in close contact with ruminants (asymptomatic carrier of EHEC) and where food contaminations occur frequently (Figure 17.1). International food production and distribution routes may contribute to the spread of new EHEC strains.[2] The above-mentioned EHEC strain O104:H4, as an example, likely evolved from an ancestor strain that had first been isolated in Central Africa.[3,4]

EHEC are a pathogenic subgroup of Shiga toxin-producing *E. coli* (STEC) causing intestinal disease as non-bloody diarrhea (gastroenteritis) as well as bloody diarrhea (hemorrhagic colitis), which potentially result in extraintestinal complications, in particular the HUS. The first outbreaks were described in Oregon and Michigan, USA, in 1982.[5] Since then, many different serotypes have been isolated. Infections have mostly been linked with contaminated food or water.[6–8]

2. WHAT IS THE CAUSATIVE AGENT?

EHEC are Gram-negative, facultative anaerobic Enterobacteriaceae asymptomatically colonizing the intestinal tract of several ruminants with cattle being the main reservoir and source for direct or indirect human infections.[9] More than 200 different serotypes are known. The most frequently reported serotype in North America, Japan, and Europe is the non-sorbitol-fermenting strain O157:H7. Other comparatively common serotypes include O26, O45, O91, O103, O111, O128, and O145.[6,8–11]

3. WHAT IS THE FREQUENCY OF THE DISEASE?

EHEC have been linked with hemorrhagic colitis around the world. Single cases/case series and smaller outbreaks are continuously reported in particular from countries with well-functioning medical infrastructure and public health surveillance systems. For 2012, the Centers of Disease Control and Prevention (CDC) calculated an overall incidence rate of 2.28 cases per 100,000 population in the USA, while the incidence in Germany was officially published to be 1.9 cases per 100,000 population.[11,12] Case monitoring, however, requires elaborate bacteriologic infrastructure to differentiate pathogenic from commensal *E. coli* and to characterize serotypes as well as specific virulence factors. Hence, limited data are available from resource-poor countries.[13]

4. HOW ARE EHEC TRANSMITTED?

The disease is a zoonosis and infections usually occur by food- or waterborne as well as, infrequently, by person-to-person transmissions during outbreaks.[8,14] Fewer than 10 microorganisms have been calculated to be sufficient to cause infection and disease.[15] In the past, foodborne outbreaks have mainly been linked to food sources of animal origin like meat and meat products/ground beef as well as unpasteurized milk.[16–19] Incorrect meat processing and cooking, for example, caused a large outbreak related to hamburgers served in fast-food restaurants and hospitals in the USA in 1994.[16] Food contaminations—in particular related to meat production—have been observed in many different settings. Recent publications demonstrated the presence of EHEC on 14% of retail meat samples in Egypt, on 5% of cattle hides in a Mexican meat production plant, on 14% of living cattle in Peru, and in 16% of slaughter area air samples at US commercial beef processing plants.[20–23] This may also explain why EHEC infections and their associated complication HUS occur comparatively frequently in countries with traditionally high meat consumption—for example, Argentina.[24] In recent years, the proportion of infections associated with fresh produce including cheese, fruit (apple) juices, bean, lettuce, and other vegetables seems to increase—possibly via direct contaminations through animal dung or indirectly via EHEC-containing water.[25–27] For the large German outbreak in 2011, sprouts from fenugreek seeds were identified as the source of infection and traced back to a single producer farm in the north of the country.[1] Sprouts, which are usually consumed uncooked and served "hidden" on sandwiches or salads, have been repeatedly linked with EHEC outbreaks.[28,29] Such "stealth food" items may cause difficulties (recall bias) in identifying the infectious source applying standard case–control studies in outbreak investigations. Other transmissions occurred via petting zoos as well as indirectly via contaminated drinking water and swimming pools.[30–32]

5. WHICH FACTORS ARE INVOLVED IN DISEASE PATHOGENESIS? WHAT ARE THE PATHOGENIC MECHANISMS?

The key virulence factor is the Shiga toxin (Stx), also known as vero(cyto)toxin (VT). Stx was first purified from *Shigella dysenteriae*. This phage-encoded holotoxin consists of five B subunits (B pentamer) and a single A subunit. The B pentamer mediates the binding of the toxin to the glycolipid globotriaosylceramide (Gb3) present on the surface of target cells. The A subunit cleaves ribosomal RNA, causing protein synthesis to cease.[10,33] While Gb3 functioning as an Stx receptor is expressed by Paneth cells in the human intestinal tract, cattle lack this receptor, which explains their natural resistance to clinical disease.[34,35] The Stx family consists of two subgroups—Stx1 and Stx2—sharing approximately 55% amino acid homology.[10,33,36] As EHEC lack a secretory mechanism for Stx, it is released by lambdoid-phage-mediated lysis triggered by DNA damage and a subsequent SOS response.[10,33,36,37] It is thought that this lytic cycle is induced by host stress and/or immune defense mechanisms as well as antibiotics.[38] The toxin and putatively additional bacterial virulence factors are ultimately responsible for endothelial damage and microvascular dysfunction as well as subsequent hemolysis.[10,33,39]

However, not all Stx-producing *E. coli* are associated with human disease, as additional virulence factors shared with other pathogenic *E. coli*-like enteropathogenic *E. coli* (EPEC) or enteroaggregative *E. coli* (EAEC) facilitating the colonization of the large intestine are necessary. Some EHEC strains show the histopathologic feature, termed attaching and effacing (A/E) characteristic for EPEC, which is mediated by adhesins (e.g., intimin) encoded by genes on a pathogenicity island called the locus of enterocyte effacement (LEE). Other EHEC possess the autoaggregative properties of EAEC mediated by aggregative adherence fimbriae (AAFs) allowing the formation of thick layers on the intestinal mucosa.[6,10,33] Possibly, intestinal aggregation and cytopathogenic properties are a prerequisite for hemorrhagic colitis, which is in some cases followed by Stx-triggered hemolysis and kidney failure after the initiation of the lytic cycle. Moreover, adherence to epithelial cells may facilitate Stx absorption.[40] To distinguish *E. coli* that produce Stx in general from the subset that is associated with additional pathogenic factors necessary to cause hemorrhagic colitis and HUS, the terms Shiga toxin-producing *E. coli* (STEC) or vero(cyto) toxin-producing *E. coli* (VTEC) for the former and EHEC for the latter are commonly used in the literature.

It is reasonable to assume that EHEC strains associated with outbreaks of hemorrhagic colitis and HUS in the past have emerged from EPEC or EAEC strains that obtained their ability to produce Stx by horizontal gene transfer by a lambda-like bacteriophage.[3,4,41,42] Some recent EHEC strains have also been characterized by plasmid-mediated extended spectrum betalactamase (ESBL) antibiotic resistance profile.[40]

6. WHAT ARE THE CLINICAL MANIFESTATIONS?

Initially, the disease is characterized by watery diarrhea typically accompanied by abdominal cramps beginning 2–3 days after ingestion of the pathogen. Another 2–4 days later, almost all adult patients but only half of infected children subsequently develop bloody diarrhea often associated with nausea and vomiting. Fever is not a typical symptom.[1,43] After about 1 week of diarrhea, the majority of patients recover without further complications, while a minority experiences severe complications with around 10–15% (mostly data from children) developing HUS.[16,18,43] In the 2011 outbreak in Germany, 22% of EHEC-infected patients developed HUS, predominantly adults. This was an unusually high proportion possibly explained by the above-described combination of adherence properties and the ability to produce Stx by the outbreak strain.[1,40] Figure 17.2 illustrates the clinical course of the disease. Table 17.1 summarizes the most important symptoms in adults and children observed during the German outbreak in 2011.

The HUS is characterized by acute renal failure, edema, and hemolysis leading to the typical triad thrombocytopenia, elevated creatinine, and lactate dehydrogenase. Often, elevated leukocytes and C-reactive protein indicate systemic inflammation (Table 17.1). Clinical deterioration/onset of HUS occur suddenly, often within 24–36 hours at a time when some patients actually fail to recover from bloody diarrhea. Figure 17.3 illustrates the sudden onset of the HUS during the German outbreak. Risk factors as shown in Table 17.2 in EHEC-infected patients for developing HUS are bloody

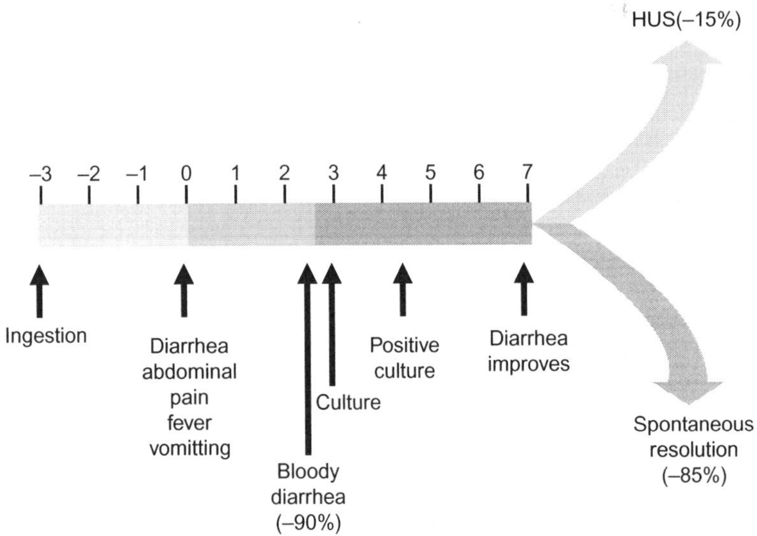

FIGURE 17.2 Clinical course of EHEC-infected patients.[43] *Reprinted with permission.*

TABLE 17.1 Symptoms in Adults and Children Observed during the 2011 Outbreak in Northern Germany caused by an O104:H4 EHEC Strain (Modified from[1])

Symptom	Adults			Children		
	Without HUS ($n = 119$)	With HUS ($n = 23$)	p-Value	Without HUS ($n = 6$)	With HUS ($n = 17$)	p-Value
Bloody diarrhea (%)	106/116 (91)	20/22 (91)	0.94	3/5 (60)	11/17 (65)	0.85
Abdominal pain (%)	102/115 (89)	19/22 (86)	0.76	5/5 (100)	8/9 (89)	0.44
Nausea (%)	23/84 (27)	9/19 (47)	0.09	3/4 (75)	3/7 (43)	0.30
Vomiting (%)	14/94 (15)	6/20 (30)	0.11	4/5 (80)	9/13 (69)	0.65
Temperature, °C	36.6 ± 0.5	36.8 ± 0.5	0.31	36.9 ± 0.5	37.0 ± 0.6	0.80
Hemoglobin, g/dl	14.2 ± 1.3	11.8 ± 2.8	<0.001	13.7 ± 1.6	10.1 ± 1.9	<0.001
Leukocytes, × 109/l	11.0 ± 3.5	12.4 ± 5.8	0.11	12.2 ± 5.3	12.1 ± 5.0	0.98
Platelets, × 109/l	245.2 ± 51.3	111.7 ± 91.9	<0.001	295.7 ± 35.5	63.6 ± 56.8	<0.001
Creatinine, mg/dl	0.8 ± 0.2	2.0 ± 1.7	<0.001	0.7 ± 0.3	4.7 ± 4.3	0.04
Lactate dehydrogenase, U/l	193 ± 92	671 ± 464	<0.001	235 ± 39	1707 ± 860	<0.001
C-reactive protein, mg/l	14.4 ± 26.0	29.9 ± 32.6	0.02	39.0 ± 61.2	18.5 ± 18.9	0.22

Reprinted with permission.

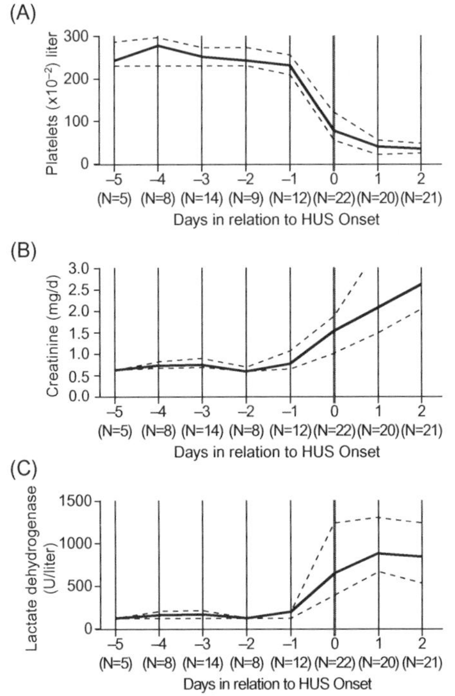

FIGURE 17.3 Sudden onset of HUS. The figure shows selected laboratory values from up to 22 patients followed prospectively. Parameters are shown relative to the onset of the HUS.[1] *Reprinted with permission.*

diarrhea, vomiting, older age, and elevated leukocytes.[18,44] EHEC infections are the major cause of acute renal failure and, ultimately, a leading cause of kidney transplantation in children in particular in populations frequently consuming beef or unpasteurized dairy products.[24,43]

Intestinal complications observed during the 2011 German outbreak were peritonitis, (sub)ileus, as well as perforation eventually requiring surgery. Neurological complications included cognitive impairment, aphasia, epileptic seizures, oculomotor disturbances, myoclonus, and headache with no morphologic signs of microbleeds, thrombotic vessel occlusion, or ischemic infarction.[45] About 1–4% of patients with symptomatic EHEC infection die from complications, predominantly the elderly or those with concomitant chronic diseases.[1,8]

7. HOW IS THE DISEASE DIAGNOSED?

Hemorrhagic colitis, in particular when occurring in clusters or in patients lacking travel history for other potential infectious agents like *Entamoeba*

TABLE 17.2 Risk Factors for Hemolytic Uremic Syndrome and Risk Score to Predict Development of HUS in Patients with Confirmed O104:H4 EHEC Infection During the Outbreak in Northern Germany in 2011 (Multivariable Logistic Regression Model, Add Points to Get Total Risk Score)[44]

Symptom	OR	95% CI Lower	Upper	P	Risk Score Points
Vomiting	3.48	1.88	6.53	<0.001	1.2
Age ≥ 75 years*	3.27	1.12	9.70	0.030	1.2
Visible blood in stools	3.91	1.20	16.01	0.036	1.4
Leukozytens (per 1000 cells/mm^3 higher)	1.12	1.10	1.31	<0.001	0.2

**Compared to patients <75 years.*
OR = odds ratio.
95% Ci = 95% confidence interval.

histolytica, should prompt the clinician to demand specific bacteriologic diagnosis of EHEC. Rapid diagnosis is important for individual clinical management as well as for transmission prevention during outbreaks.

Chromogenic agar mediums/sorbitol-MacConkey agars for Enterobacteriaceae can be used for culture. However, sorbitol-fermenting EHEC strains exist. To identify *E. coli*, mass spectrometry can also be used. Stx production can be detected by enzyme-linked immunosorbent assays (ELISA) or polymerase chain reaction (PCR). In case of ESBL resistance patterns, respective characteristics can be used for selective culture. Additional virulence factors are ascertained by (multiplex-) PCR (e.g., genes of the LEE, *eae*, virulence factors typical for EAEC), which can also be used to characterize the specific Stx (*stx1*, *stx2*, subgroups). Antimicrobial resistance can be tested by microdilution assays applying minimal inhibitory concentrations according to respective guidelines. Serotyping, for example by pulsed-field gel electrophoresis, is important to characterize an outbreak strain but does not play a role in individual diagnostic procedures once the outbreak strain is known.[40,46,47]

8. HOW DO YOU DIFFERENTIATE THIS DISEASE FROM SIMILAR ENTITIES?

The most important differential diagnoses of EHEC-associated bloody diarrhea include infections as well as autoimmune diseases. Infectious hemorrhagic colitis/enteritis can be caused by other bacteria-like *Campylobacter*, *Shigella*, *Salmonella*, enteroinvasive *E. coli* (EIEC) or *Clostridium difficile*.[48] Other infectious agents include ameba (*Entameba histolytica*) and—in the immunocompromised—cytomegalovirus (CMV). Non-infectious diseases that need to be taken into account are chronic inflammatory bowel disease like Crohn's disease and colitis ulcerosa. However, EHEC-associated bloody diarrhea should in particular be suspected when several patients accrue in a short period of time indicating an outbreak. The treating physician needs to specifically demand for bacteriologic diagnosis of EHEC as this pathogen is usually not tested for/separated from commensal *E. coli* in routine microbiologic procedures.

9. WHAT IS THE THERAPEUTIC APPROACH?

The therapeutic approach of gastroenteritis and hemorrhagic colitis is based on symptomatic therapy, and prevention of as well as monitoring for potential complications. Once complications like HUS are present, the medical focus lies on organ support and symptomatic approaches. As EHEC infections occur sporadically and outbreaks begin suddenly, it is difficult to prospectively assess the effectiveness of specific therapeutic options in appropriate randomized controlled trials. Therefore, almost none of the therapeutic options are based on sound evidence. This also includes the use of immunomodulatory agents to specifically treat HUS or severe neurological symptoms.[49]

Uncomplicated gastroenteritis as well as clinically stable patients with hemorrhagic colitis should be followed up continuously on an outpatient basis until complete recovery. For the first (at least) 7 days after onset of symptoms, clinical examination (edema) as well as blood analysis (including creatinine, blood count, and lactate dehydrogenase) should be performed daily to detect complications like HUS as early as possible. Increased fluid intake to prevent dehydration and to support the kidney function should be recommended. Attempts have been made to reduce intestinal EHEC overgrowth by bowel cleansing (with polyethylene glycol as applied for colonoscopy preparation) and probiotics.

Antibiotics may trigger the bacterium's lytic cycle/Stx release and are, therefore, feared to induce the onset of HUS in previously uncomplicated EHEC infections. Therefore, the use of antibiotics is generally discouraged in patients still at risk for developing HUS.[43,50,51] *In vitro*, some antibiotics did not up-regulate Stx-transcription or bacteriophage induction but these observations may be strain specific.[52,53] Antibiotics, however, do not seem to worsen the clinical course of patients in whom HUS is already established.[49] In these patients, antibiotics (like azithromycin) may shorten EHEC shedding.[54,55] In patients at risk for HUS, antibiotics should be considered in cases with peritonitis/systemic signs of bacterial infections or under controlled clinical trial conditions only.

It is clear that HUS patients with renal failure require hemodialysis. Traditionally, plasmapheresis has been initiated after HUS onset but questionable beneficial effects on the one hand as well as patient impact of this repeated procedure on the other do not support its general use. Similarly, glucocorticoids lack proof of clinical effects. Eculizumab, a monoclonal antibody against the complement protein C5, has been successfully used in patients with paroxysmal nocturnal hemoglobinuria and atypical HUS.[56,57] Triggered by a promising case series in children with post-infectious HUS published shortly before the 2011 EHEC outbreak in Germany,[58] eculizumab was broadly used in patients with HUS or severe neurological complications during the German outbreak for the first time. Available data do not provide sufficient evidence to generally recommend eculizumab at this time as a beneficial effect compared to spontaneous improvement/current supportive therapy remains to be established.[49,59] Other approaches have been assessed in patients with HUS including IgG depletion through immunoadsorption[60] but any of these interventions as well as future approaches should only be applied under clinical trial conditions if feasible in an outbreak situation.

Severely ill patients will have to be monitored under intensive care conditions. Neurologic complications like coma and seizures will have to be treated accordingly. Severe colitis complicated by colon perforation will require surgery.

10. WHAT ARE THE PREVENTIVE AND INFECTION CONTROL MEASURES?

Preventive measures begin with controlled food production processes and do not end with personal food hygiene when preparing food in one's own kitchen. If it is specifically looked for, EHEC will eventually be found on vegetables and other potentially contaminated food. It is, therefore, essential to thoroughly wash cucumbers and tomatoes, etc. and rinse lettuce and sprouts before consumption. Also, raw meat (before cooking) should not be cut followed by vegetable preparation on the same surface (in particular salad items consumed uncooked) as this facilitates cross-contamination. Often, however, the individual has no control over the food production processes: several outbreaks or infections have been associated with professional food preparation and eating in (children's daycare, school, hospital, enterprise) canteens or restaurants.[1,16,28,61]

Once an outbreak has been recognized, the public health agencies will undertake efforts to identify the source. In most circumstances, however, the final identification of the source will no longer influence the epidemic curve. Outbreaks are self-limiting and in the past the source has generally been identified after the peak infection wave.

During an outbreak, patients requiring hospitalization should be isolated. In large outbreaks, cohort isolation is possible and their management should be separated from routine patient flow. Patients in compensated health condition not requiring continuous medical care should be instructed to stay at home, not to prepare food for others, and to follow strict toilet hygiene (thorough hand washing, separate towels, etc.). It is in particular important for laboratory personnel to follow hygiene standards to avoid laboratory infections.

The handling of infected patients with prolonged shedding of EHEC in their feces is difficult. On the one hand, staff should not continue to work in child care or gastronomy during the period of positive stool samples. On the other hand, discontinuing work can seriously affect their career as well as psychological status as some individuals continue shedding EHEC for weeks and months.[54,55,61] Whether or not some antibiotics like azithromycin are a safe option to shorten EHEC shedding and whether or not findings from the O104 outbreak in Germany can be generalized to other EHEC strains, in particular to O157 strains, remain controversial.[51]

The use of antibiotics on animal farms and in the meat industry is unlikely to facilitate the generation of new EHEC strains but may influence their antibiotic resistance profile for example by selecting plasmid-encoded ESBL resistance.[62,63]

REFERENCES

1. Frank C, Werber D, Cramer JP, Askar M, Faber M, an der Heiden M, et al. HUS investigation team. Epidemic profile of shiga-toxin-producing Escherichia coli O104:H4 outbreak in Germany. *N Engl J Med* 2011;**365**:1771–80.
2. Bavaro MF. E. coli O157:H7 and other toxigenic strains: the curse of global food distribution. *Curr Gastroenterol Rep* 2012;**14**:317–23.
3. Rasko DA, Webster DR, Sahl JW, Bashir A, Boisen N, Scheutz F, et al. Origins of the E. coli strain causing an outbreak of hemolytic-uremic syndrome in Germany. *N Engl J Med* 2011;**365**:709–17.
4. Bezuidt O, Pierneef R, Mncube K, Lima-Mendez G, Reva ON. Mainstreams of horizontal gene exchange in enterobacteria: consideration of the outbreak of enterohemorrhagic E. coli O104:H4 in Germany in 2011. *PLoS One* 2011;**6**:e25702.
5. Centers for Disease Control (CDC). Isolation of E. coli O157:H7 from sporadic cases of hemorrhagic colitis—united states. *MMWR Morb Mortal Wkly Rep* 1982;**31**(580):585.
6. Kaper JB, Nataro JP, Mobley HL. Pathogenic Escherichia coli. *Nat Rev Microbiol* 2004;**2**:123–40.
7. Mellmann A, Bielaszewska M, Köck R, Friedrich AW, Fruth A, Middendorf B, et al. Analysis of collection of hemolytic uremic syndrome-associated enterohemorrhagic escherichia coli. *Emerg Infect Dis* 2008;**14**:1287–90.
8. Pennington H. Escherichia coli O157. *Lancet* 2010;**376**:1428–35.
9. Croxen MA, Finlay BB. Molecular mechanisms of Escherichia coli pathogenicity. *Nat Rev Microbiol* 2010;**8**:26–38.
10. Melton-Celsa A, Mohawk K, Teel L, O'Brien A. Pathogenesis of Shiga-toxin producing escherichia coli. *Curr Top Microbiol Immunol* 2012;**357**:67–103.
11. Robert-Koch-Institut (RKI). Infektionsmedizinisches Jahrbuch meldepflichtiger Infektionskrankheiten für 2012. Berlin; 2013.
12. Centers for Disease Control and Prevention (CDC). Incidence and trends of infection with pathogens transmitted commonly through food—foodborne diseases active surveillance network, 10 U.S. sites, 1996–2012. *MMWR Morb Mortal Wkly Rep* 2013;**62**:283–7.

13. Okeke IN. Diarrheagenic Escherichia coli in sub-Saharan Africa: status, uncertainties and necessities. *J Infect Dev Ctries* 2009;**3**:817–42.
14. Snedeker KG, Shaw DJ, Locking ME, Prescott RJ. Primary and secondary cases in Escherichia coli O157 outbreaks: a statistical analysis. *BMC Infect Dis* 2009;**9**:144.
15. Teunis P, Takumi K, Shinagawa K. Dose response for infection by Escherichia coli O157: H7 from outbreak data. *Risk Anal* 2004;**24**:401–7.
16. Bell BP, Goldoft M, Griffin PM, Davis MA, Gordon DC, Tarr PI, et al. A multistate outbreak of Escherichia coli O157:H7-associated bloody diarrhea and hemolytic uremic syndrome from hamburgers. The Washington experience. *JAMA* 1994;**272**:1349–53.
17. Keene WE, Hedberg K, Herriott DE, Hancock DD, McKay RW, Barrett TJ, et al. A prolonged outbreak of Escherichia coli O157:H7 infections caused by commercially distributed raw milk. *J Infect Dis* 1997;**176**:815–8.
18. Dundas S, Todd WT, Stewart AI, Murdoch PS, Chaudhuri AK, Hutchinson SJ. The central Scotland Escherichia coli O157:H7 outbreak: risk factors for the hemolytic uremic syndrome and death among hospitalized patients. *Clin Infect Dis* 2001;**33**:923–31.
19. Guh A, Phan Q, Nelson R, Purviance K, Milardo E, Kinney S, et al. Outbreak of Escherichia coli O157 associated with raw milk, Connecticut, 2008. *Clin Infect Dis* 2010;**51**:1411–7.
20. Sallam KI, Mohammed MA, Ahdy AM, Tamura T. Prevalence, genetic characterization and virulence genes of sorbitol-fermenting Escherichia coli O157:H- and E. coli O157:H7 isolated from retail beef. *Int J Food Microbiol* 2013;**165**:295–301.
21. Narvaez-Bravo C, Miller MF, Jackson T, Jackson S, Rodas-Gonzalez A, Pond K, et al. Salmonella and Escherichia coli O157:H7 prevalence in cattle and on carcasses in a vertically integrated feedlot and harvest plant in Mexico. *J Food Prot* 2013;**76**: 786–95.
22. Rivera FP, Sotelo E, Morales I, Menacho F, Medina AM, Evaristo R, et al. Short communication: detection of shiga toxin-producing Escherichia coli (STEC) in healthy cattle and pigs in Lima, Peru. *J Dairy Sci* 2012;**95**:1166–9.
23. Schmidt JW, Arthur TM, Bosilevac JM, Kalchayanand N, Wheeler TL. Detection of escherichia coli O157:H7 and salmonella enterica in air and droplets at three U.S. commercial beef processing plants. *J Food Prot* 2012;**75**:2213–8.
24. Rivero MA, Passucci JA, Rodriguez EM, Parma AE. Role and clinical course of verotoxigenic Escherichia coli infections in childhood acute diarrhoea in Argentina. *J Med Microbiol* 2010;**59**:345–52.
25. Cody SH, Glynn MK, Farrar JA, Cairns KL, Griffin PM, Kobayashi J, et al. An outbreak of Escherichia coli O157:H7 infection from unpasteurized commercial apple juice. *Ann Intern Med* 1999;**130**(3):202–9.
26. Honish L, Predy G, Hislop N, Chui L, Kowalewska-Grochowska K, Trottier L, et al. An outbreak of E. coli O157:H7 hemorrhagic colitis associated with unpasteurized gouda cheese. *Can J Public Health* 2005;**96**:182–4.
27. Söderström A, Lindberg A, Andersson Y. EHEC O157 outbreak in Sweden from locally produced lettuce, August–September 2005. *Euro Surveill* 2005;**10**:E050922.1.
28. Michino H, Araki K, Minami S, Takaya S, Sakai N, Miyazaki M, et al. Massive outbreak of Escherichia coli O157:H7 infection in schoolchildren in Sakai City, Japan, associated with consumption of white radish sprouts. *Am J Epidemiol* 1999;**150**:787–96.
29. Breuer T, Benkel DH, Shapiro RL, Hall WN, Winnett MM, Linn MJ, et al. Investigation Team. A multistate outbreak of Escherichia coli O157:H7 infections linked to alfalfa sprouts grown from contaminated seeds. *Emerg Infect Dis* 2001;**7**:977–82.

30. Centers for Disease Control and Prevention (CDC). Outbreaks of Escherichia coli O157:H7 associated with petting zoos—North Carolina, Florida, and Arizona, 2004 and 2005. *MMWR Morb Mortal Wkly Rep* 2005;**54**:1277–80.
31. Lienemann T, Pitkänen T, Antikainen J, Mölsä E, Miettinen I, Haukka K, et al. Shiga toxin-producing Escherichia coli O100:H$^-$: stx2e in drinking water contaminated by waste water in Finland. *Curr Microbiol* 2011;**62**:1239–44.
32. Verma A, Bolton FJ, Fiefield D, Lamb P, Woloschin E, Smith N, et al. An outbreak of E. coli O157 associated with a swimming pool: an unusual vehicle of transmission. *Epidemiol Infect* 2007;**135**:989–92.
33. Bergan J, Dyve Lingelem AB, Simm R, Skotland T, Sandvig K. Shiga toxins. *Toxicon* 2012;**60**:1085–107.
34. Pruimboom-Brees IM, Morgan TW, Ackermann MR, Nystrom ED, Samuel JE, Cornick NA, et al. Cattle lack vascular receptors for Escherichia coli O157:H7 Shiga toxins. *Proc Natl Acad Sci USA* 2000;**97**:10325–9.
35. Schüller S, Heuschkel R, Torrente F, Kaper JB, Phillips AD. Shiga toxin binding in normal and inflamed human intestinal mucosa. *Microbes Infect* 2007;**9**:35–9.
36. O'Brien AD, Tesh VL, Donohue-Rolfe A, Jackson MP, Olsnes S, Sandvig K, et al. Shiga toxin: biochemistry, genetics, mode of action, and role in pathogenesis. *Curr Top Microbiol Immunol* 1992;**180**:65–94.
37. Toshima H, Yoshimura A, Arikawa K, Hidaka A, Ogasawara J, Hase A, et al. Enhancement of Shiga toxin production in enterohaemorrhagic Escherichia coli serotype O157:H7 by DNase colicins. *Appl Environ Microbiol* 2007;**73**:7582–8.
38. Hughes DT, Sperandio V. Inter-kingdom signalling: communication between bacteria and their hosts. *Nat Rev Microbiol* 2008;**6**:111–20.
39. Karch H, Friedrich AW, Gerber A, Zimmerhackl LB, Schmidt MA, Bielaszewska M. New aspects in the pathogenesis of enteropathic hemolytic uremic syndrome. *Semin Thromb Hemost* 2006;**32**:105–12.
40. Bielaszewska M, Mellmann A, Zhang W, Köck R, Fruth A, Bauwens A, et al. Characterisation of the Escherichia coli strain associated with an outbreak of haemolytic uraemic syndrome in Germany, 2011: a microbiological study. *Lancet Infect Dis* 2011;**11**:671–6.
41. Reid SD, Herbelin CJ, Bumbaugh AC, Selander RK, Whittam TS. Parallel evolution of virulence in pathogenic escherichia coli. *Nature* 2000;**406**:64–7.
42. Rohde H, Qin J, Cui Y, Li D, Loman NJ, Hentschke M, et al. E. coli O104:H4 genome analysis crowd-sourcing consortium. Open-source genomic analysis of Shiga-toxin-producing E. coli O104:H4. *N Engl J Med* 2011;**365**:718–24.
43. Tarr PI, Gordon CA, Chandler WL. Shiga-toxin-producing escherichia coli and haemolytic uraemic syndrome. *Lancet* 2005;**365**:1073–86.
44. Zoufaly A, Cramer JP, Vettorazzi E, Sayk F, Bremer JP, Koop I, et al. Risk factors for development of hemolytic uremic syndrome in a cohort of adult patients with STEC 0104: H4 infection. *PLoS One* 2013;**8**:e59209.
45. Magnus T, Röther J, Simova O, Meier-Cillien M, Repenthin J, Möller F, et al. The neurological syndrome in adults during the 2011 northern German E. coli serotype O104:H4 outbreak. *Brain* 2012;**135**:1850–9.
46. Ribot EM, Fair MA, Gautom R, Cameron DN, Hunter SB, Swaminathan B, et al. Standardization of pulsed-field gel electrophoresis protocols for the subtyping of eschaerichia coli O157:H7, Salmonella, and shigella for pulsenet. *Foodborne Pathog Dis* 2006;**3**:59–67.

47. Vallières E, Saint-Jean M, Rallu F. Comparison of three different methods for detection of Shiga toxin-producing Escherichia coli in a tertiary pediatric care center. *J Clin Microbiol* 2013;**51**:481–6.
48. Ina K, Kusugami K, Ohta M. Bacterial hemorrhagic enterocolitis. *J Gastroenterol* 2003;**38**:111–20.
49. Menne J, Nitschke M, Stingele R, Abu-Tair M, Beneke J, Bramstedt J, et al. EHEC-HUS consortium. Validation of treatment strategies for enterohaemorrhagic Escherichia coli O104:H4 induced haemolytic uraemic syndrome: case-control study. *BMJ* 2012;**345**: e4565.
50. Wong CS, Mooney JC, Brandt JR, Staples AO, Jelacic S, Boster DR, et al. Risk factors for the hemolytic uremic syndrome in children infected with Escherichia coli O157:H7: a multivariable analysis. *Clin Infect Dis* 2012;**55**:33–41.
51. Seifert ME, Tarr PI. Therapy: azithromycin and decolonization after HUS. *Nat Rev Nephrol* 2012;**8**:317–8.
52. Bielaszewska M, Idelevich EA, Zhang W, Bauwens A, Schaumburg F, Mellmann A, et al. Effects of antibiotics on Shiga toxin 2 production and bacteriophage induction by epidemic Escherichia coli O104:H4 strain. *Antimicrob Agents Chemother* 2012;**56**:3277–82.
53. Corogeanu D, Willmes R, Wolke M, Plum G, Utermöhlen O, Krönke M. Therapeutic concentrations of antibiotics inhibit Shiga toxin release from enterohemorrhagic E. coli O104: H4 from the 2011 German outbreak. *BMC Microbiol* 2012;**12**:160.
54. Nitschke M, Sayk F, Härtel C, Roseland RT, Hauswaldt S, Steinhoff J, et al. Association between azithromycin therapy and duration of bacterial shedding among patients with Shiga toxin-producing enteroaggregative Escherichia coli O104:H4. *JAMA* 2012;**307**:1046–52.
55. Vonberg RP, Höhle M, Aepfelbacher M, Bange FC, Belmar Campos C, Claussen K, et al. Duration of fecal shedding of Shiga toxin-producing Escherichia coli O104:H4 in patients infected during the 2011 outbreak in Germany: a multicenter study. *Clin Infect Dis* 2013;**56**:1132–40.
56. Nürnberger J, Philipp T, Witzke O, Opazo Saez A, Vester U, Baba HA, et al. Eculizumab for atypical hemolytic-uremic syndrome. *N Engl J Med* 2009;**360**:542–4.
57. Kelly RJ, Hill A, Arnold LM, Brooksbank GL, Richards SJ, Cullen M, et al. Long-term treatment with eculizumab in paroxysmal nocturnal hemoglobinuria: sustained efficacy and improved survival. *Blood* 2011;**117**:6786–92.
58. Lapeyraque AL, Malina M, Fremeaux-Bacchi V, Boppel T, Kirschfink M, Oualha M, et al. Eculizumab in severe Shiga-toxin-associated HUS. *N Engl J Med* 2011;**364**:2561–3.
59. Kielstein JT, Beutel G, Fleig S, Steinhoff J, Meyer TN, Hafer C, et al. Collaborators of the DGfN STEC-HUS registry. Best supportive care and therapeutic plasma exchange with or without eculizumab in Shiga-toxin-producing E. coli O104:H4 induced haemolytic-uraemic syndrome: an analysis of the German STEC-HUS registry. *Nephrol Dial Transplant* 2012;**27**:3807–15.
60. Greinacher A, Friesecke S, Abel P, Dressel A, Stracke S, Fiene M, et al. Treatment of severe neurological deficits with IgG depletion through immunoadsorption in patients with Escherichia coli O104:H4-associated haemolytic uraemic syndrome: a prospective trial. *Lancet* 2011;**378**:1166–73.
61. Wahl E, Vold L, Lindstedt BA, Bruheim T, Afset JE. Investigation of an Escherichia coli O145 outbreak in a child day-care centre—extensive sampling and characterization of eae- and stx1-positive E. coli yields epidemiological and socioeconomic insight. *BMC Infect Dis* 2011;**11**:238.

62. Mora A, Blanco JE, Blanco M, Alonso MP, Dhabi G, Echeita A, et al. Antimicrobial resistance of Shiga toxin (verotoxin)-producing Escherichia coli O157:H7 and non-O157 strains isolated from humans, cattle, sheep and food in Spain. *Res Microbiol* 2005;**156**:793–806.
63. Ju W, Shen J, Li Y, Toro MA, Zhao S, Ayers S, et al. Non-O157 Shiga toxin-producing Escherichia coli in retail ground beef and pork in the Washington D.C. area. *Food Microbiol* 2012;**32**:371–7.

Chapter 18

Emerging *Clostridium difficile* Infections

Itaru Nakamura and Yasutaka Mizuno
Department of Infection Control and Infectious Diseases, Tokyo Medical University Hospital, Tokyo, Japan

CASE PRESENTATION

A 32-year-old woman, who was a pharmacist, status post-cesarean section, was admitted to the Tokyo Medical University Hospital (TMUH) for severe colitis. She had no significant past medical history, was not taking any addictive medicines, had no recent antibiotic treatment (including quinolones) prior to the cesarean section, and had not traveled overseas. Before her admission to TMUH, she was hospitalized in an obstetrics and gynecology clinic for delivery. A prophylactic antimicrobial agent for surgical site infection, cefazolin, was used from the day of her cesarean operation. Two days after this, fever and diarrhea started, and a high white blood cell count and high C-reactive protein level were noted. Although cefozopran (2.0 g/day) and isepamicin (400 mg/day) were administered for 5 days, clinical symptoms and abnormal laboratory data became more pronounced. However, renal function remained normal during this period.

Nine days after the operation, an abnormally high white blood cell count of 42,200 per cubic millimeter and a high C-reactive protein level of 31.5 mg/dl were noted. The serum albumin level was 2.1 mg/dl. In addition, she developed a high fever, severe abdominal pain, and severe diarrhea (15 loose stools per day). She was admitted to our hospital for treatment of severe colitis. Upon admission, her circulatory status was stable and her body temperature was 37.4°C. Although she had no manifestations of peritoneal irritation or surgical site infection, an abdominal computed tomography revealed a massive amount of ascites fluid; peritonitis and ileus were presumed (Figure 18.1). A rapid test of stool for the presence of *C. difficile* toxin A and toxin B was performed, with a positive result. *C. difficile* was cultured from a stool specimen. Oral VCM at 0.5 g/day for *C. difficile*

Emerging Infectious Diseases. DOI: http://dx.doi.org/10.1016/B978-0-12-416975-3.00018-2

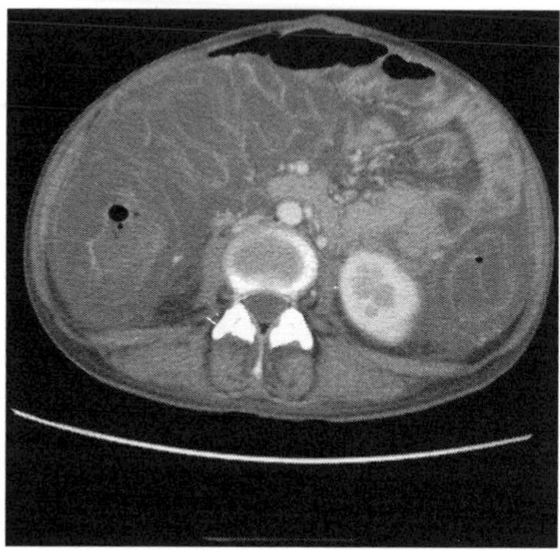

FIGURE 18.1 Enhanced abdomen CT reveals massive amount of ascites fluid and significant bowel wall thickening.

infection and ampicillin sulbactam (ABPC/SBT) at 12 g/day IV for peritonitis were administered.

Three days after admission to our hospital, although the white blood cell count decreased to 34,100 per cubic millimeter, she still had severe diarrhea 15 times per day and no improvement in the severity of abdominal pain. Oral VCM was increased to 2.0 g/day and intravenous immunoglobulin at 5 g/day (maximum dose approved by the Japanese Ministry of Health, Labour and Welfare) was initiated. However, signs of colitis did not resolve after these treatments. The culture of peritoneal fluid from the day of admission revealed *Enterococcus faecalis*.

Seven days after admission, the white blood cell count increased and abdominal pain worsened. Emergency exploratory surgery was performed on the same day (day 7). Perforation at the splenic flexure of the colon and muddy ascites were observed. A temporary colostomy was created at the site of the perforation through the skin of the abdominal wall. Enteral administration of VCM (2.0 g/day) via the colostomy, a VCM enema (2.0 g/day), and oral MTZ (1500 mg/day) were used, combined with oral VCM (2.0 /day). The culture of intraoperative ascites revealed *C. difficile*, *E. faecalis*, and *Candida* spp. ABPC/SBT (12 g/day IV) was continued and micafungin (100 mg/day IV) was added for the peritonitis.

Three days after the emergency colostomy operation, the severity of diarrhea and paralytic ileus had not improved. Linezolid (LZD) at 1200 mg/day IV was administered concurrently. After that, the clinical signs of colitis improved and laboratory data gradually normalized. The patient was treated with antibiotics against *C. difficile* infection after the operation for a total of 18 days.

The housekeeping gene (*tpi*), toxin A gene (*tcdA*), toxin B gene (*tcdB*), and binary toxin gene (*cdtA* and *cdtB*) assays were positive. DNA sequencing of *tcdC* revealed a base 117 deletion and contained an 18-bp *tcdC* deletion. The MIC values of VCM, MTZ, moxifloxacin, and LZD were 0.38 μg/ml, 0.25 μg/ml, >32 μg/ml, and <1.0 μg/ml, respectively. The antibiotic susceptibility of our isolate was in common with that of the epidemic strain characterized by the high resistance to fluoroquinolones.[1] PCR ribotyping showed the same patterns as ribotype 027. We therefore regarded our isolate as the epidemic strain ribotype 027.

(This is a published case report, *Journal of Infection and Chemotherapy*, 2014 (in press).)

1. WHY THIS CASE WAS SIGNIFICANTLY IMPORTANT AS AN EMERGING INFECTION

Clostridium difficile colitis remains the most common cause of nosocomial and antibiotic-associated diarrhea. Epidemic strain ribotype 027, not a conventional strain, had spread to Canada, the United States, England, parts of continental Europe, and Japan.[2] In addition, hospital outbreaks of unusually severe and epidemic *C. difficile* infection were noted more than ever.

Epidemic strain was reported first in 2003 from Canada.[3] A previous report revealed that disease severity is consistent with stool toxin level.[4] It is thought that the toxin is related to cell retraction and apoptosis.[5] The epidemic strain has the predisposition to produce larger quantities of toxins than other *C. difficile* strains.[6] As a result, it is presumed to be associated with the development of perforation. Here, we will present a case of fulminant colitis due to epidemic strain.

2. WHAT IS THE CAUSATIVE AGENT?

Clostridium difficile is an anaerobic Gram-positive, toxin-producing bacillus and the causative organism of antibiotic-associated enterocolitis (Figure 18.2). This organisim releases two kinds of toxin, toxin A and toxin B, which induce colitis and diarrhea. The level of toxin A and B is thought to correlate with severity of enterocolitis. Diarrhea and colitis after using antibiotic treatment are usually regarded as antibiotic-associated diarrhea, and *C. difficile* colitis remains the most common cause of nosocomial diarrhea and is implicated in 10–30% of such cases.

In 1978, *C. difficile* was identified as the pathogen of antibiotic colitis. However, *C. difficile* infection (CDI) has been observed to be more severe and refractory to conventional treatment from around 2003 in North America and Europe.[3] This was attributed to a new strain known as the epidemic strain NAP1/BI/027. This strain produces binary toxin and increases the production of toxin in comparison with conventional strains. Fluoroquinolones

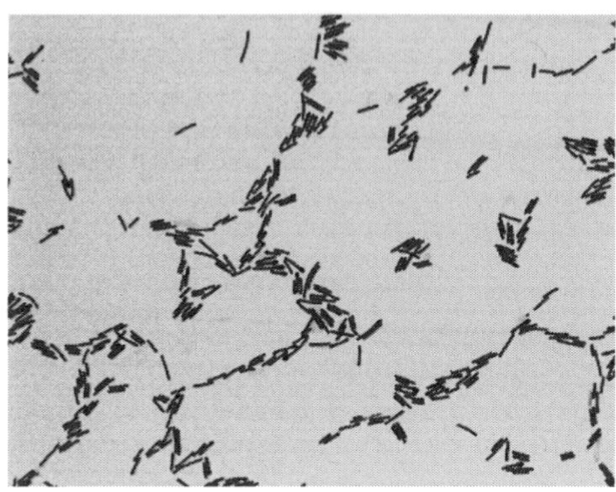

FIGURE 18.2 Gram strain of *C. difficile*.

are now major inducing agents along with cephalosporins, which presumably reflect newly acquired *in vitro* resistance and escalating rates of use. In addition to NAP1/BI/027, infection due to ribotype 078 has a feature of severe colitis, as reported in the Netherlands.[7]

3. WHAT IS THE FREQUENCY OF THE DISEASE?

About 2% healthy adult and 20% adult hospitalized patients are *C. difficile* carriers without diarrhea.[8,9] Although the prevalence is uncertain because microbiology and epidemiology are changing rapidly, incidence was reported in 25 to 43 persons among hospitalized patients in Quebec, Canada, in 2003.[10] In the United States, gastroenteritis-associated mortality has more than doubled during the past decade, primarily affecting the elderly. *C. difficile* is the main contributor to gastroenteritis-associated deaths.[11] Based upon surveys of Canadian hospitals conducted in 1997 and 2005, incidence rates range from 3.8 to 9.5 cases per 10,000 patient days, or 3.4 to 8.4 cases per 1000 admissions, in acute care hospitals.[2,12] From 1999 to 2003 in Massachusetts, a total of 55,380 inpatient days and $55.2 million were spent on the management of CDI. An estimate of the annual excess hospital costs in the USA was $3.2 billion per year for the years 2000–2002.[13]

In addition to this, the incidence of community-acquired CDI increased significantly too. Compared with those with hospital-acquired infection, patients with community-acquired infection were younger (median age 50 years compared with 72 years), more likely to be female (76% vs. 60%), had lower comorbidity scores, and were less likely to have severe infection (20% vs. 31%) or have been exposed to antibiotics (78% vs. 94%).[14]

4. HOW ARE THE BACTERIA TRANSMITTED?

The transmission route of this organism is mainly contact infection. This is important for control of nosocomial infection and *C. difficile* is present in the hospital environment. Rooms with a CDI patient were more likely to be contaminated than rooms without a CDI patient.[15] Especially for the duration of diarrhea, contact precaution is thought to be needed. Successful infection control measures designed to prevent horizontal transmission include the use of gloves in handling body substances and replacement of electronic thermometers with disposable devices.[16]

Alcohol-based hand rubs (ABHRs) are commonly an effective means of decreasing the transmission of bacterial pathogens. However, alcohol is not effective against *C. difficile* spores. Hand washing with soap and water is significantly more effective at removing *C. difficile* spores from the hands than are ABHRs.[17] Also, the Centers for Disease Control and Prevention recommends soap and water hand hygiene when caring for patients with CDI.

5. WHICH FACTORS ARE INVOLVED IN DISEASE PATHOGENESIS? WHAT ARE THE PATHOGENIC MECHANISMS?

C. difficile releases two potent exotoxins: toxin A and toxin B, which induce enterocolitis. The level of toxin A and B in stools is thought to correlate with the severity of enterocolitis. Toxin A causes inflammation leading to intestinal fluid secretion, mucosal injury, and inflammation.[18] And the human colon is approximately 10 times more sensitive to the damaging effects of toxin B than toxin A.[5]

There is an association between a systemic anamnestic response to toxin A, as evidenced by increased serum levels of IgG antibody against toxin A, and asymptomatic carriage of *C. difficile*.[8] A serum antibody response to toxin A, during an initial episode of CDI, is associated with protection against recurrence.[19] In addition, it is reported that host susceptibility to CDI is related both to a defective humoral immune response to CD toxin A and to host IL-8 AA genotype.[20]

Antimicrobial agents that induce CDI are shown in Table 18.1.

6. WHAT ARE THE CLINICAL MANIFESTATIONS?

(See line 9 in the Case Presentation.)

The main clinical manifestations of CDI are watery diarrhea, lower abdominal pain, low grade fever, and leukocytosis, although it can cause a spectrum of manifestations from asymptomatic to severe fulminant disease, relapsing, and occasionally fatal colitis. The symptom typically begins after 5–10 days of antibiotic treatment. Uncommonly, it is present as late as 10 weeks after cessation of treatment.[21]

CDI is a prominent cause of leukocytosis and this diagnosis should be considered for patients with WBC counts of ≧15,000 cells/mm^3, even in the absence of diarrheal symptoms.[22] In most of these cases diarrhea often develops in the next 1 to 2 days. Nausea, malaise, occult colonic bleeding, and

TABLE 18.1 Antimicrobial Agents to Induce CDI

More Frequently Associated Agents	Less Frequently Associated Agents
Fluoroquinolones	Chloramphenicol
Clindamycin	Metronidazole
Broad spectrum penicillin	Ripampin
Broad spectrum cephalosporin	Aminoglycosides
Macrolides	Vancomycin
Tetracyclines	
Trimethoprim-sulfamethoxazole	

dehydration have also been reported. In addition, enteric protein loss and nutritional status were measured in patients with symptoms due to *C. difficile.*[23] The protein-losing enteropathy responds to appropriate therapy.

The criteria proposed for defining severe or complicated CDI are leukocytosis with a white blood cell count of more than 15,000 cells/mL or a serum creatinine level greater than or equal to 1.5 times the premorbid level. In addition, fulminant colitis is a diagnosis based on complication with hypotension, shock, ileus, or megacolon.[2] Toxic megacolon complicating pseudomembranous enterocolitis-associated systemic toxicity and the absence of mechanical obstruction is a serious problem that carries a high morbidity and mortality rate, regardless of treatment.[24] The epidemic strain of *C. difficile* NAP1/BI/027 was associated with a high incidence of disease as well as severe and fatal disease, for example fulminant colitis or toxic megacolon. (See Case Presentation.)

7. HOW DO YOU DIAGNOSE?

CDI should be suspected in patients with diarrhea who have previously received antibiotics. Enzyme immunoassay (EIA) testing for *C. difficile* toxins A and B is rapidly and commonly used, although it is less sensitive than the cell cytotoxin assay, the gold standard for diagnosis. Commercial EIA tests have been introduced that either detect toxin A only or detect both toxins A and B. In laboratory testing that included cytotoxin and culture, the sensitivity of these tests is 63–94%, with a specificity of 75–100%.[2] Therefore, a repeat toxin EIA test can boast sensitivity. A *C. difficile* culture of a stool specimen is also most sensitive and useful in cases where confirmation of organism toxicity is optimal. GDH assay is detection by tests for the *C. difficile* common antigen. A GDH test using EIA shows a sensitivity of 85–95% and a specificity of 89–98%. Since a positive GDH test indicates existence of *C. difficile* in a stool specimen, further testing is required to determine whether the *C. difficile* strain is toxigenic. The confirmatory test has primarily been a cell cytotoxin assay or toxin EIA test.

Direct visualization of exudative plaque or pseudomembranes using endoscopy establishes the diagnosis of pseudomembranous colitis. Pseudomembrane manifestation is characteristically raised, yellow or off-white plaques up to 2 cm in diameter, scattered over the colorectal mucosa. Endoscopy is usually reserved for special situations because of its cost and risk to the patient.[25]

8. HOW DO YOU DIFFERENTIATE THIS DISEASE FROM SIMILAR ENTITIES?

Only approximately 30% of hospitalized patients who have antibiotic-associated diarrhea will have CDI.[26] Among non-infectious causes, it may be attributable to osmotic mechanisms by the alteration of microflora rather than *C. difficile* infection. Diarrhea due to osmotic mechanisms will improve by cessation of antibiotic treatment. Infection with *K. oxytoca* might be considered in patients with antibiotic-associated colitis who are negative for *C. difficile*.[27]

9. WHAT IS THE THERAPEUTIC APPROACH?

Discontinue therapy with the inciting antimicrobial agents as soon as possible, as this may influence the risk of CDI recurrence. Oral metronidazole[1] (MTZ) and oral vancomycin (VCM) have been the main antimicrobial agents in the treatment of CDI. Intravenous VCM has no effect on *C. difficile* colitis since the antibiotic is not excreted appreciably into the colon. Increasing use of oral VCM is thought to induce the emergence of vancomycin-resistant enterococci and VCM is much more expensive than MTZ. Therefore, MTZ has generally been recommended for the first line drug. VCM is mainly used after MTZ is found to be ineffective or not well tolerated. Time to resolve symptoms might be shorter with VCM than with MTZ treatment. VCM at 125 mg four times per day was superior to MTZ therapy in severe cases. Leukocytosis likely reflects the severity of colonic inflammation. In general, the definition of a severe case can be summarized by following two agents: leukocytosis with a white blood cell count of 15,000 cells/μL or higher, or a serum creatinine level greater than 1.5 times the premorbid level.

Oral MTZ is the drug of choice for the initial episode of mild-to-moderate CDI. The dosage is 500 mg orally three times per day for 10–14 days. Oral VCM is the drug of choice for an initial episode of severe CDI. The dosage is 125 mg orally four times per day for 10–14 days. (See line 25 in the Case Presentation.)

Ileus may impair the delivery of orally administered VCM to the colon. In fulminant and complicated cases, VCM, 500 mg four times per day by mouth or by nasogastric tube, plus MTZ, 500 mg every 8 hours intravenously, should be administered. If complete ileus, consider adding rectal instillation of VCM. Intravenous immunoglobulins (150–400 mg/kg) have been used for some patients not responding to other therapies.[28] Some patients with fulminant colitis might need aggressive surgical intervention—colectomy.[29] The optimal timing of surgery remains uncertain.

Historically, 6–25% of patients treated for CDI have experienced at least one additional episode. Antibiotic treatment for the first recurrence is recommended as for the initial episode. However, treatment for a second recurrence is selected with a tapering and/or pulsed regimen of oral VCM: after the usual dosage of 125 mg four times per day for 10–14 days, VCM is administered at 125 mg two times per day for a week, 125 mg once per day for a week, and then 125 mg every 2 or 3 days for 2–8 weeks.[2]

10. WHAT ARE THE PREVENTIVE AND INFECTION CONTROL MEASURES?

Patients with suspected or proven CDI should be placed on contact precaution. Healthcare workers and visitors must use gloves and gowns on entry to a room of a patient with CDI. Compliance with the practice of hand hygiene is important. Patients with CDI should be admitted in a private room with contact precautions. If single rooms are not available, a dedicated commode should be provided for each patient.[2] Contact precautions are thought to be needed for the duration of diarrhea.

Identification and removal of environmental sources of *C. difficile*, including replacement of electronic rectal thermometers with disposables, can reduce the incidence of CDI. Current evidence supports the use of chlorine-containing cleaning agents (with at least 1000 ppm available chlorine) for environmental cleaning and disinfection, particularly to address environmental contamination in areas associated with endemic or epidemic CDI.

REFERENCES

1. McDonald LC, Killgore GE, Thompson A, Owens Jr RC, Kazakova SV, Sambol SP, et al. An epidemic, toxin gene-variant strain of Clostridium difficile. *N Engl J Med* 2005;**353**:2433–41.
2. Cohen SH, Gerding DN, Johnson S, Kelly CP, Loo VG, McDonald LC, et al. Clinical practice guidelines for Clostridium difficile infection in adults: 2010 update by the society for healthcare epidemiology of America (SHEA) and the infectious diseases society of America (IDSA). *Infect Control Hosp Epidemiol* 2010;**31**:431–55.
3. Bartlett JG. Narrative review: the new epidemic of clostridium difficile-associated enteric disease. *Ann Intern Med* 2006;**145**:758–64.

4. Akerlund T, Svenungsson B, Lagergren A, Burman LG. Correlation of disease severity with fecal toxin levels in patients with Clostridium difficile-associated diarrhea and distribution of PCR ribotypes and toxin yields in vitro of corresponding isolates. *J Clin Microbiol* 2006;**44**:353–8.
5. Riegler M, Sedivy R, Pothoulakis C, Hamilton G, Zacherl J, Bischof G, et al. Clostridium difficile toxin B is more potent than toxin A in damaging human colonic epithelium in vitro. *J Clin Invest* 1995;**95**:2004–11.
6. McEllistrem MC, Carman RJ, Gerding DN, Genheimer CW, Zheng L. A hospital outbreak of Clostridium difficile disease associated with isolates carrying binary toxin genes. *Clin Infect Dis* 2005;**40**:265–72.
7. Goorhuis A, Bakker D, Corver J, Debast SB, Harmanus C, Notermans DW, et al. Emergence of Clostridium difficile infection due to a new hypervirulent strain, polymerase chain reaction ribotype 078. *Clin Infect Dis* 2008;**47**:1162–70.
8. Kyne L, Warny M, Qamar A, Kelly CP. Asymptomatic carriage of Clostridium difficile and serum levels of IgG antibody against toxin A. *N Engl J Med* 2000;**342**:390–7.
9. Aronsson B, Mollby R, Nord CE. Antimicrobial agents and Clostridium difficile in acute enteric disease: epidemiological data from Sweden, 1980–1982. *J Infect Dis* 1985;**151**:476–81.
10. Pepin J, Valiquette L, Alary ME, Villemure P, Pelletier A, Forget K, et al. Clostridium difficile-associated diarrhea in a region of Quebec from 1991 to 2003: a changing pattern of disease severity. *CMAJ* 2004;**171**:466–72.
11. Hall AJ, Curns AT, McDonald LC, Parashar UD, Lopman BA. The roles of Clostridium difficile and norovirus among gastroenteritis-associated deaths in the United States, 1999–2007. *Clin Infect Dis* 2012;**55**:216–23.
12. Miller MA, Hyland M, Ofner-Agostini M, Gourdeau M, Ishak M. Canadian Hospital Epidemiology Committee. Canadian Nosocomial Infection Surveillance P. Morbidity, mortality, and healthcare burden of nosocomial Clostridium difficile-associated diarrhea in Canadian hospitals. *Infect Control Hosp Epidemiol* 2002;**23**:137–40.
13. O'Brien JA, Lahue BJ, Caro JJ, Davidson DM. The emerging infectious challenge of clostridium difficile-associated disease in Massachusetts hospitals: clinical and economic consequences. *Infect Control Hosp Epidemiol* 2007;**28**:1219–27.
14. Khanna S, Pardi DS, Aronson SL, Kammer PP, Orenstein R, St Sauver JL, et al. The epidemiology of community-acquired Clostridium difficile infection: a population-based study. *Am J Gastroenterol* 2012;**107**:89–95.
15. Dubberke ER, Reske KA, Noble-Wang J, Thompson A, Killgore G, Mayfield J, et al. Prevalence of Clostridium difficile environmental contamination and strain variability in multiple health care facilities. *Am J Infect Control* 2007;**35**:315–8.
16. Gerding DN, Johnson S, Peterson LR, Mulligan ME, Silva Jr. J. Clostridium difficile-associated diarrhea and colitis. *Infect Control Hosp Epidemiol* 1995;**16**:459–77.
17. Jabbar U, Leischner J, Kasper D, Gerber R, Sambol SP, Parada JP, et al. Effectiveness of alcohol-based hand rubs for removal of Clostridium difficile spores from hands. *Infect Control Hosp Epidemiol* 2010;**31**:565–70.
18. Lima AA, Innes Jr DJ, Chadee K, Lyerly DM, Wilkins TD, Guerrant RL. Clostridium difficile toxin A. Interactions with mucus and early sequential histopathologic effects in rabbit small intestine. *Lab Invest* 1989;**61**:419–25.
19. Kyne L, Warny M, Qamar A, Kelly CP. Association between antibody response to toxin A and protection against recurrent Clostridium difficile diarrhoea. *Lancet* 2001;**357**:189–93.

20. Jiang ZD, Garey KW, Price M, Graham G, Okhuysen P, Dao-Tran T, et al. Association of interleukin-8 polymorphism and immunoglobulin G anti-toxin A in patients with Clostridium difficile-associated diarrhea. *Clin Gastroenterol Hepatol* 2007;**5**:964–8.
21. Tedesco FJ. Pseudomembranous colitis: pathogenesis and therapy. *Med Clin North Am* 1982;**66**:655–64.
22. Wanahita A, Goldsmith EA, Musher DM. Conditions associated with leukocytosis in a tertiary care hospital, with particular attention to the role of infection caused by clostridium difficile. *Clin Infect Dis* 2002;**34**:1585–92.
23. Dansinger ML, Johnson S, Jansen PC, Opstad NL, Bettin KM, Gerding DN. Protein-losing enteropathy is associated with Clostridium difficile diarrhea but not with asymptomatic colonization: a prospective, case-control study. *Clin Infect Dis* 1996;**22**:932–7.
24. Trudel JL, Deschenes M, Mayrand S, Barkun AN. Toxic megacolon complicating pseudomembranous enterocolitis. *Dis Colon Rectum* 1995;**38**:1033–8.
25. Fekety R. Guidelines for the diagnosis and management of Clostridium difficile-associated diarrhea and colitis. American College of Gastroenterology, Practice Parameters Committee. *Am J Gastroenterol* 1997;**92**:739–50.
26. Bartlett JG. Clinical practice. Antibiotic-associated diarrhea. *N Engl J Med* 2002;**346**:334–9.
27. Hogenauer C, Langner C, Beubler E, Lippe IT, Schicho R, Gorkiewicz G, et al. Klebsiella oxytoca as a causative organism of antibiotic-associated hemorrhagic colitis. *N Engl J Med* 2006;**355**:2418–26.
28. McPherson S, Rees CJ, Ellis R, Soo S, Panter SJ. Intravenous immunoglobulin for the treatment of severe, refractory, and recurrent Clostridium difficile diarrhea. *Dis Colon Rectum* 2006;**49**:640–5.
29. Sailhamer EA, Carson K, Chang Y, Zacharias N, Spaniolas K, Tabbara M, et al. Fulminant Clostridium difficile colitis: patterns of care and predictors of mortality. *Arch Surg* 2009;**144**:433–9. discussion 9–40.

Chapter 19

Multidrug-Resistant Tuberculosis

Jan Heyckendorf, Christoph Lange and Julia Martensen
Medical Clinic Research Center Borstel, German Center for Infection Research (DZIF) Tuberculosis Unit, Center for Infection and Inflammation (ZIEL), University of Lübeck, Parkallee, Borstel, Germany

CASE PRESENTATION

A 16-year-old girl was admitted to our hospital for diagnosis and treatment of right upper lobe infiltrations. The parents, who were admitted at the same time, presented with a 5-year history of multidrug-resistant tuberculosis (MDR-TB), which (under insufficient treatment) had evolved into extensively drug-resistant tuberculosis (XDR-TB) in both cases.

During the last 5 years, their daughter had shown a positive tuberculin skin test (TST) and a positive interferon-γ release assay (IGRA) result (performed with cells from whole blood), but repeatedly normal chest X-rays and no evidence of acid-fast bacilli (AFB) in the sputum smear microscopy. She had received preventive chemotherapy with second line antituberculosis drugs twice for durations of 6 months each. Now, as the parents' TB aggravated, her chest X-ray showed right upper lobe infiltrations for the first time.

On admission, she was asymptomatic and in good general health. A chest computed tomography (CT) scan showed discrete nodular infiltrations in the right upper lobe and no other pathologies. Sputum smear microscopy and nucleic acid amplifications for the detection of *M. tuberculosis* DNA yielded three times negative results. A bronchoscopy with collection of bronchial fluid, bronchoalveolar lavage (BAL), and transbronchial biopsies was performed. All samples were negative for AFB on microscopy and for *M. tuberculosis* by nucleic acid amplification. However, the result of an IGRA performed with mononuclear cells from the BAL was positive. There were no signs for other causes of the infiltrations such as other infections or interstitial lung disease found on microbiological cultures and histopathological examinations.

As it was crucial to ensure a diagnosis of M/XDR-TB before starting a multidrug second line antituberculosis regimen, a second bronchoscopy was

Emerging Infectious Diseases. DOI: http://dx.doi.org/10.1016/B978-0-12-416975-3.00019-4

performed. Again, BAL and transbronchial biopsies were negative for AFBs. Histopathological examinations again showed no granulomas. *Mycobacterium tuberculosis*-specific nucleic acid amplification from the BAL again showed a negative test result; however, this time *M. tuberculosis*-specific DNA could be amplified from transbronchial biopsies of the right upper pulmonary lobe. Drug resistance against rifampicin was identified by molecular methods, highly indicative of a MDR or XDR-strain of *M. tuberculosis*. Subsequently, all collected samples' cultures remained negative.

In summary, a 16-year-old female who was a household contact of two adults with AFB smear-positive pulmonary XDR-TB presented without symptoms of disease but with a small non-caveating right upper lobe pulmonary infiltrate, a positive IGRA result from the blood and the BAL, and documentation of *M. tuberculosis* nucleic acid in transbronchial biopsies from the right upper pulmonary lobe. In addition, rapid molecular testing was highly indicative for M/XDR-TB.

These findings allowed the probable diagnosis of M/XDR-TB. As transmission from the parents was very probable, we chose a treatment according to the drug susceptibility test (DST) results of the parents' *M. tuberculosis* strains. The patient started on a treatment with para-aminosalicylic acid, terizidone/cycloserine (those two being the only susceptible drugs in the parents), clofazimine, Meronem/Combactam, and pyrazinamide according to WHO recommendations. Under this treatment, a chest CT scan after 2 months showed a regression of the pulmonary infiltrations. Treatment was scheduled for a total of 20 months according to WHO recommendations.

1. WHY THIS CASE WAS SIGNIFICANTLY IMPORTANT AS AN EMERGING INFECTION

Tuberculosis is a leading cause of morbidity and mortality worldwide, ranging on the 10th position of all causes of death.[1]

Multidrug-resistant TB (MDR-TB), defined as tuberculosis caused by *M. tuberculosis* which is resistant against the two first-line drugs rifampicin and isoniazid, has dramatically increased in recent years.[2] XDR-TB is defined as an infection with an MDR-TB strain, which is additionally resistant against an injectable second line drug (amikacin, kanamycin, capreomycin) and a fluoroquinolone.

In contrary to the positive outcome in drug-susceptible TB ($>85\%$ treatment success worldwide),[2] MDR-TB results in higher mortality, fewer cure rates, longer but less effective treatment with toxic side effects, high costs, and difficult logistics in many countries. Treatment success is reported as 48–64% for MDR-TB and 20–40% for XDR-TB.[2,3]

Figure 19.1 shows the estimated absolute numbers of estimated TB cases and deaths in millions (1990–2011).[2]

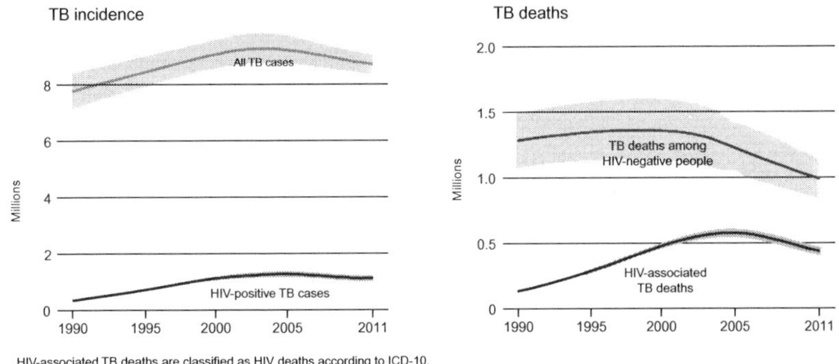

FIGURE 19.1 Estimated absolute numbers of estimated TB cases and deaths in millions (1990–2011).[2]

2. WHAT IS THE CAUSATIVE AGENT?

The causative agent for the above-described case is *Mycobacterium tuberculosis* (MTB). It was discovered by Robert Koch in 1882 and connected to TB being an infectious disease. The aerobic and asporogenous pathogen is characterized by its waxy outer coating, which is responsible for its acid fastness. Being part of the genus *Mycobacterium*, it is the only Mycobacteriaceae that is described by an unusually high genomic DNA G + C content and the production of mycolic acids. MTB is closely related to the genera *Nocardia* and *Corynebacterium*. The bacterium has a slow replication rate of about 20 hours and has the ability to persist in infected persons.

TB is caused by seven closely related pathogens forming the tuberculosis complex (*M. tuberculosis*, *M. africanum*, *M. canettii*, *M. bovis*, bacille Calmette–Guérin (BCG), *M. pinnipedii*, and *M. microti*). In the context of M/XDR-TB, the main genetic mechanisms of these pathogens have been discovered.[4] Among these "drug-resistance genes," mutations in rpoB are the main causative changes responsible for rifampicin resistance. In the case of isoniazid, inhA and katG mutations lead to poor response to the antituberculosis drug. These and other resistance gene mutations can be proven in diagnostic tests for drug susceptibility (see Chapter 7).

3. WHAT IS THE FREQUENCY OF THE DISEASE?

While tuberculosis is spread worldwide, distribution can be divided into high and low burden countries. The estimated number of tuberculosis patients worldwide is 8.6 million. The estimated global incidence rate is 122 cases per 100,000. The main tuberculosis burden occurs in Africa and Asia with a high number of HIV-TB coinfections, especially in Africa.

Drug-resistant tuberculosis arises predominantly in formerly treated patients in countries with insufficient tuberculosis treatment due to collapsing health systems, mainly Eastern Europe (former Soviet Union states) and Central Asia. While on average 3.6% of new cases and 20.2% of previously treated cases are estimated to have MDR-TB, in those 27 high burden countries, MDR-RB rates rise up to 34% in new cases and 62% in formerly treated cases. The rate of XDR-TB within MDR-TB patients is considered to be 9.6% on average.[2] However, the proportion of newly diagnosed cases with MDR-TB is increasing in some countries dramatically, indicative of active M/XDR-TB transmission.[5]

Figure 19.2 shows the percentage of new TB cases with MDR-TB,[2] Figure 19.3 the percentage of previously treated TB cases with MDR-TB,[2] and Figure 19.4 the countries that had notified at least one case of XDR-TB by the end of 2012.[2]

4. HOW ARE THE BACTERIA TRANSMITTED?

TB is usually transmitted by airborne droplet infection.[6] Coughing patients with TB produce infectious droplets carrying *M. tuberculosis* bacteria. These droplets are inhaled and therefore transported to the lungs, where they can spread resulting in pulmonary tuberculosis (see Chapter 5). Hence, smear-positive TB patients should be isolated to avoid further spreading of the pathogen (see Chapter 10). Common risk factors for TB are HIV infection, alcoholism, chronic renal failure, diabetes mellitus, and immunosuppression

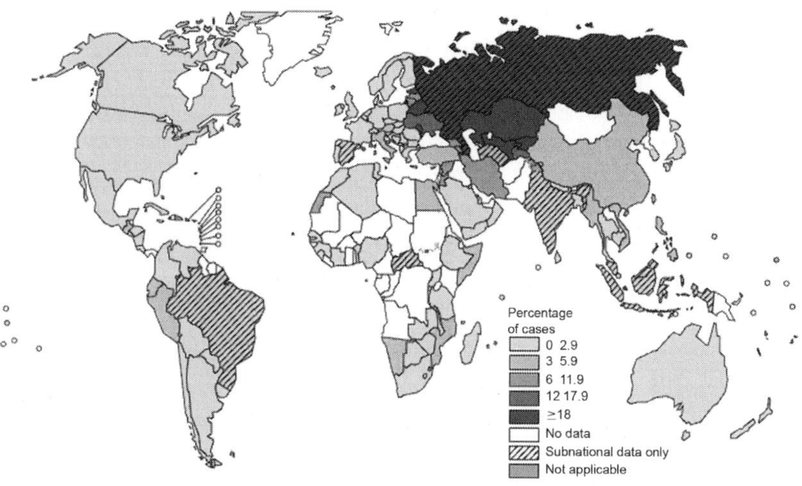

FIGURE 19.2 Percentage of new TB cases with MDR-TB.[2]

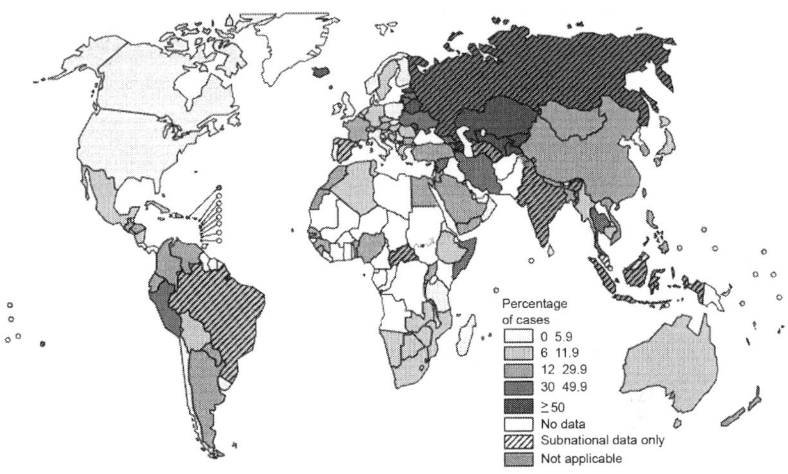

FIGURE 19.3 Percentage of previously treated TB cases with MDR-TB.[2]

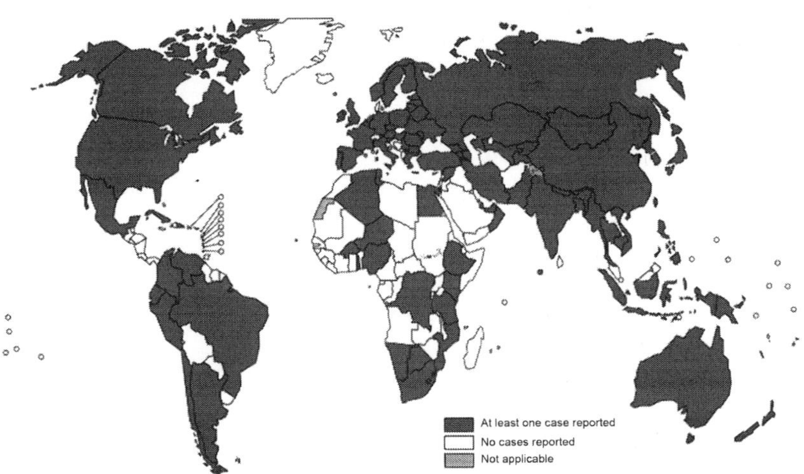

FIGURE 19.4 Countries that had notified at least one case of XDR-TB by the end of 2012.[2]

due to chronic illness or immunosuppressive pharmacotherapy such as corticosteroids or TNF-α antagonists.[7–9] Among these risk factors, immunosuppression due to HIV infection is the most relevant cause for the infection and outbreak of TB, leading to 320,000 deaths from TB patients with HIV coinfection.[2]

5. WHICH FACTORS ARE INVOLVED IN DISEASE PATHOGENESIS? WHAT ARE THE PATHOGENETIC MECHANISMS?

After inhalation of the pathogen and its reaching of alveoli, *M. tuberculosis* is first ingested by macrophages and dendritic cells.[10] The pathogen can persist in various intracellular compartments and survive and replicate in these niches. The classical histopathological habitat created by the bacteria is the granuloma.[11] These lesions are formed by macrophages, which can be multinucleated and are then termed Langhans giant cells. Moreover, TB granulomas consist of B cells, T cells, dendritic cells, and neutrophils. The granuloma usually forms a central necrosis, which differentiates TB granulomas from most other possible granulomas. The control of *M. tuberculosis* by specific T cells is mainly reached through the release of interferon-γ, but other factors such as vitamin D-dependent cathelicidin and TNF-α are important to kill the bacteria. However, the immune system is not able to develop complete sterilizing activity to clear the body from the pathogen. *M. tuberculosis* is capable of affecting any compartment of the human body by hematogenous spreading, resulting in tissue damage with clinical importance (see Chapter 6).

6. WHAT ARE THE CLINICAL MANIFESTATIONS?

(See line 13 in the Case Presentation.)

Infection with *M. tuberculosis* mainly affects the lungs, causing pulmonary tuberculosis. Symptoms of pulmonary tuberculosis are non-specific and may last for months or longer before the diagnosis is set. Patients complain about weight loss, night sweats, fever, persistent cough, dyspnea, weakness, and reduced effort tolerance. In a minority of patients, hemoptysis occurs.

The chest X-ray of a patient with pulmonary tuberculosis can show non-specific nodular, reticular, nodulo-reticular, dense, or otherwise non-specific infiltrations. When cavitations are seen, they are predominantly located in the upper lobes. On CT scan, bronchiolitic infiltrations may classically appear as a "tree-in-bud" pattern. No radiological appearance is pathognomonic for tuberculosis.

There are no serological markers, which are highly specific for tuberculosis. Laboratory abnormalities often include anemia and may include moderately to highly elevated erythrocyte sedimentation time and moderate leukocytosis. Immunological tests such as TST and IGRAs confirm sensitization with *M. tuberculosis* antigens. However, the timing of sensitization is not indicated by the test result and neither TST nor IGRAs can distinguish between latent infection and active tuberculosis, unless IGRAs are performed with cells derived from the site of infection.

In approximately 20% of cases, tuberculosis affects extra-pulmonary organs. Notable manifestations include pleural tuberculosis, lymph node

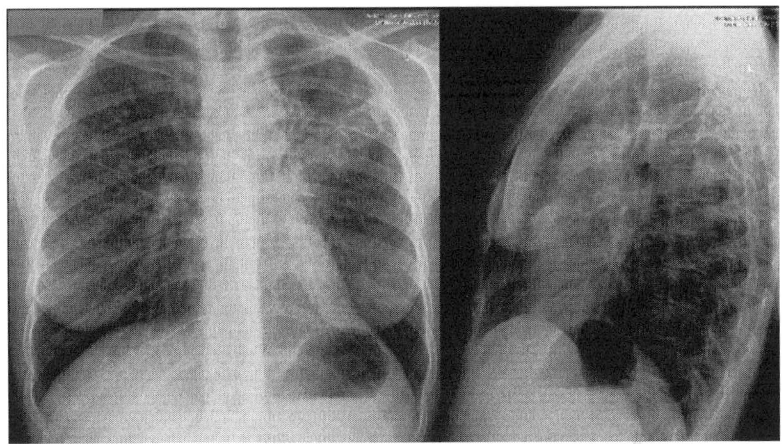

FIGURE 19.5 Chest X-ray of pulmonary tuberculosis: bilateral infiltrations and cavity in left upper lobe.

tuberculosis, central nervous system tuberculosis, abdominal tuberculosis, urogenital tuberculosis, and osteoarticular tuberculosis. A disseminated form of tuberculosis is called miliary tuberculosis since its histopathological appearance is similar to millet seeds.[12]

Figure 19.5 shows a chest X-ray of pulmonary tuberculosis with bilateral infiltrations and cavity in left upper lobe and Figures 19.6 and 19.7 show CT scans of tuberculosis.

7. HOW DO YOU DIAGNOSE?

(See lines 13–23 in the Case Presentation.)

Diagnosis of tuberculosis relies on imaging studies, molecular–biological, immunological, and microbiological methods. The gold standard is the detection of growth of *M. tuberculosis* on selective solid or liquid culture media, although culture results may take several weeks to detect *M. tuberculosis* growth.

When symptoms and X-ray/thoracic CT scan are suspicious for pulmonary tuberculosis, initially two–three sputum samples should be collected for microscopy of acid-fast bacilli and *M. tuberculosis* culture. Results of microscopy can be obtained on the same day, but around 50% of the patients with culture confirmed tuberculosis (especially children and HIV-positive patients) have undetectable acid-fast bacilli on sputum smear examinations.

Detection of *M. tuberculosis*-specific DNA by nucleic acid amplification has a very high specificity (98%) in both AFB-positive and AFB-negative smears, thus, a positive result is highly indicative of active tuberculosis.

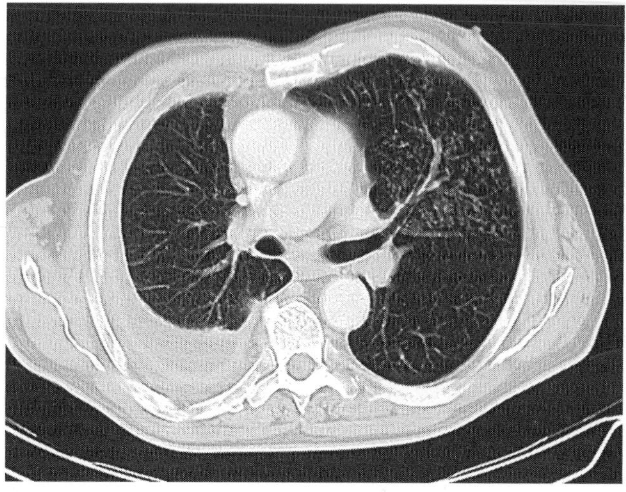

FIGURE 19.6 CT scan of tuberculosis: right side pleural effusion and tree-in-bud pattern of the left upper lobe.

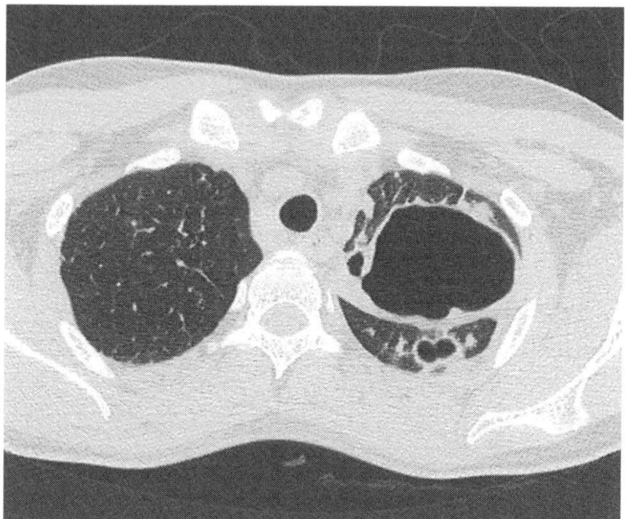

FIGURE 19.7 CT scan of tuberculosis: cavities in the left upper lobe.

The test sensitivity in AFB-positive smears is also excellent (approximately 98%). When the bacterial load is low and acid-fast bacilli are undetectable on sputum smear microscopy, the sensitivity of *M. tuberculosis*-specific DNA amplification is still around 70%.[13]

When the diagnosis cannot be confirmed by sputum smear microscopy or nucleic acid amplification techniques, bronchoscopy with bronchoalveolar lavage and transbronchial biopsies of the affected region should be performed.[12]

Histological examination should reveal caseous granuloma and seldom shows acid-fast bacilli in the tissue under these circumstances.

While tuberculin skin test and blood IGRA cannot differentiate active tuberculosis from latent TB infection (LTBI), IGRA from bronchoalveolar lavage fluid (BAL-IGRA) compared to blood IGRA may help distinguish active pulmonary from latent TB, as antigen-specific T-cells gather at the site of infection. BAL-IGRA showed sensitivity of 92% and specificity of 87% in smear-negative patients.[14]

A drug susceptibility test (DST) should be performed for first line drugs from culture from every patient's bacterial isolate; drug susceptibility testing is especially crucial when drug resistance is suspected.

Phenotypic DSTs from culture are the gold standard, but results may take up to more than 3 months. Molecular-based DSTs like the GeneXpert MTB/Rif or lineprobe assays provide a more rapid, yet less reliable result, allowing the probable diagnosis of MDR-TB or even XDR-TB within 1–2 days.[15]

It needs to be acknowledged that both phenotypic and genotypic DSTs do not show good reliability for ethambutol, pyrazinamide, and second line drugs other than fluoroquinolones and injectables.

Figure 19.8 is a flow diagram for the diagnosis of tuberculosis.[12]

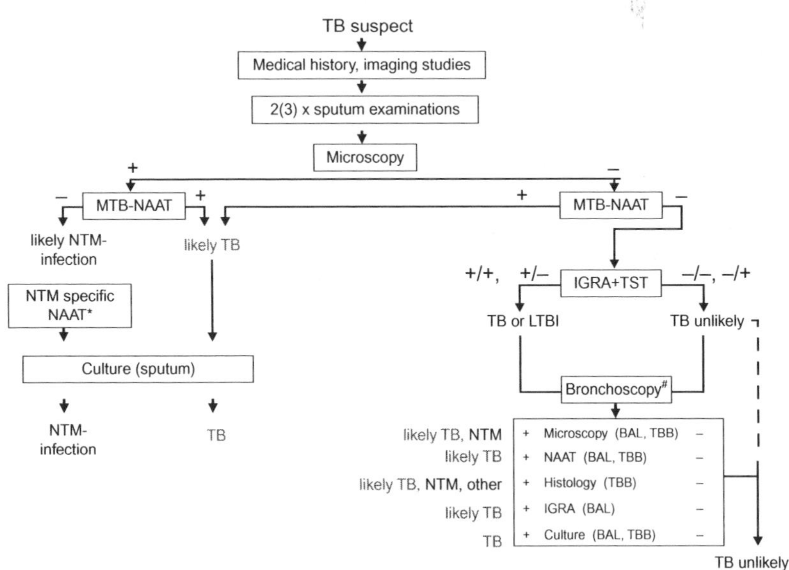

FIGURE 19.8 Flow diagram for the diagnosis of tuberculosis.[12]

TABLE 19.1 Differential Diagnosis of Pulmonary Tuberculosis

Category	Disease	Diagnostic Methods
infectious	pneumonia	clinical course, microscopy, culture
	non-tuberculous mycobacteria	culture
	nokardiosis	culture
	actinomycosis	microscopy, culture
	rhodococcosis	microscopy, culture
non-infectious	lung cancer	histology
	sarcoidosis	CD4/CD8-ratio in bronchioalveolar lavage
	vasculitis	serology (cANCA), renal impairment, non-caseous granuloma

8. HOW DO YOU DIFFERENTIATE THIS DISEASE FROM SIMILAR ENTITIES?

Once a positive *M. tuberculosis* culture is available, the diagnosis of tuberculosis is confirmed. Until then, the probability of diagnoses other than tuberculosis depends on the prevalence of the disease. In high burden countries (mainly in Africa, Central and South-East, Asia and Russia) tuberculosis is the most likely diagnosis when a cough for >2 weeks, fever, weight loss, and night sweats occur, especially in HIV-positive patients.

In low incidence countries, especially in Western Europe and North America, differential diagnoses must be considered as there are several other diseases more likely to cause wasting, cough, cavities, and hemoptysis than tuberculosis. Likely differential diagnoses are infectious diseases caused by non-tuberculous mycobacteria (NTM) or necrotizing pneumonia, and non-infectious diseases such as lung cancer, sarcoidosis, or vasculitis such as granulomatous angiitis (formerly Wegener's disease).

Microscopy and culture, nucleic acid amplification methods, bronchoscopy with biopsies for histology, and BAL-IGRA will lead the way to the right diagnosis.[12]

9. WHAT IS THE THERAPEUTIC APPROACH?

(See paragraph 7 in the Case Presentation.)

While drug-susceptible pulmonary tuberculosis is generally treated successfully with 6 months' therapy of a combination of four drugs (rifampicin, isoniazid, ethambutol, and pyrazinamide for 2 months, followed by rifampicin and isoniazid for another 4 months), treatment of MDR-TB and XDR-TB is far more difficult, costly, and generally recommended for 20 months with a combination of ≥4 drugs shown to be active by DST testing.

The WHO recommends dividing anti-tuberculosis drugs into five groups: first line drugs, second line injectable drugs, fluoroquinolones, oral bacteriostatic second line drugs, and drugs with unclear efficacy (see Table 19.2).

As MDR-TB strains are resistant against two of the most effective drugs, rifampicin and isoniazid, second line drugs have to be added to the therapeutic regimen. While the initial drug regimen may be standardized (e.g., following results of molecular testing), it is crucial to adjust the drug regimen once the results of culture-based DST become available.

There are no randomized trials on MDR-TB treatment with existing drug regimens available. Current recommendations are based on systematic reviews based on observational studies,[16,17] and a large retrospective meta-analysis on >9000 individual patient data on which WHO recommendations are also based.[18]

In their 2011 guidelines, WHO recommends a combination of at least four second line drugs for MDR treatment. A combination with a second line injectable drug and a later generation fluoroquinolone is favorable. When

TABLE 19.2 Anti-Tuberculosis Agents According to WHO

	Group Name	Agent
1	**first line drugs**	rifampicin isoniazid ethambutol pyrazinamide
2	**injectible drugs**	amikacin capreomycin kanamycin
3	**fluoroquinolones**	levofloxacin moxifloxacin gatifloxacin ofloxacin
4	**oral bacteriostatic second line drugs**	ethionamide prothionamide cycloserine/terizidone paraaminosalicylic acid
5	**drugs with unclear efficacy**	amoxicillin/clavulanate clarithromycin clofazimine imipenem/meropenem linezolid thioacetone

TABLE 19.3 M/XDR-Treatment[18a]

Composition of MDR-TB Drug Regimen
1. Drug susceptibility testing on liquid or solid media
2. Choose, if possible: a. injectible second line drug (e.g. amikacin) b. later generation fluoroquinolone (e.g. levofloxacin) c. ethionamide or protionamide d. cycloserine or terizidone
3. Choose at least 4 drugs (it is unclear whether all patients with M/XDR-TB should be treated with pyrazinamide)
4. Choose group 5 drugs only if needed to sum up to at least four active drugs
5. Treat for a total of 20 months with an intensive phase (incl. Injectible drug) of 8 months
6. Prolongation of the duration of therapy should be considered when the clinical course indicates so
7. For XDR-TB, proceed accordingly to DST and MDR-recommendations

possible, according to the DST result, ethionamide or prothionamide should be the first drug of choice from WHO group IV. WHO also recommends that pyrazinamide should be added to all >four drug MDR-TB drug regimens, irrespective of the DST test result, although this recommendation is not universally accepted. Group V drugs should only be used if needed to sum up the regimen to ≥ 4 active drugs. If group V drugs have to be used, a higher total number of drugs should be considered.[19] The duration of the therapy should last 20 months with an intensive phase (i.e., including the parental agent) of 8 months, although it can be modified according to the patient's clinical course.[18,19]

The long duration of treatment for M/XDR-TB is consuming considerable resources and negatively affects compliance. Shorter and standardized drug regimens would be highly welcome, especially in low resource settings. Promising results have been reported about a 9-month fixed drug/fixed dosage regimen from a trial in Bangladesh that is currently undergoing validation at different international sites.[20] In selected cases with localized disease, surgery may be considered additionally to medical therapy.[21]

For XDR-TB, the same recommendations as for MDR-TB apply, as there is no specific evidence for XDR treatment available. In XDR-TB, the individual patient's DST is even more important.

Recently, new antituberculosis drugs are being developed in clinical studies. For the first time in almost 50 years a new antituberculosis drug, Bedaquiline, was provisionally granted approval by the FDA for combination therapy of MDR-TB in the USA. Addition of Bedaquiline to a standard combination of drugs resulted in faster sputum culture conversion in a higher number of patients and less development of resistance in co-administered drugs during a 24-week period.[22]

The application for licensing of Delamanid, another new antituberculosis drug, has been submitted to the EMA in Europe; however, the EMA has adopted a negative opinion recommending refusal of the marketing authorization for this drug.[23,24]

Of course, recommendations on new drugs are yet to come and administration should be done wisely within a combination therapy to prevent development of *M. tuberculosis* drug resistance against these agents.

10. WHAT ARE THE PREVENTIVE AND INFECTION CONTROL MEASURES?

To prevent infection with *M. tuberculosis*, patients with sputum smear-positive TB should be isolated regardless of drug resistance status. The risk of infection depends on proximity and duration of contact of an infectious TB patient.[25] Smear-positive patients bear the greatest risk of infection and isolation is recommended until sputum smear (drug-susceptible TB) or culture conversion (drug-resistant TB) is recorded.[26] Healthcare workers with contact to infectious patients should wear N95 or FFP2 respirator masks (European Norm (EN) 61010-1).

To avoid progression to disease, *M. bovis* bacille Calmette–Guérin (BCG) vaccination remains the only measure to prevent severe disease and TB meningitis in childhood, but does not serve to reduce overall burden of disease.[27] Since there is no effective preventive vaccine and with the latest vaccine trials not holding out much hope for improvement,[28] preventive chemotherapy is the only action to reduce risk to progression to disease after proven contact to *M. tuberculosis* and is mainly accomplished with isoniazid monotherapy for 6–12 months.[29] Recently, alternative short course preventive regimes have been presented.[30] In the field of drug-resistant TB, there is currently no reliable evidence for effective preventive therapy. A preventive therapy is useful in contacts to drug-resistant TB with high risk for progression to disease (i.e., immunocompromised individuals) with at least two susceptible drugs of the pathogen for at least 6 months and a clinical follow-up of 2 years.[31] However, the WHO has not yet adapted any recommendations for M/XDR preventive therapy and urges for more clinical trials in this field of interest.[26]

REFERENCES

1. Lim SS, Vos T, Flaxman AD, Danaei G, Shibuya K, Adair-Rohani H, et al. A comparative risk assessment of burden of disease and injury attributable to 67 risk factors and risk factor clusters in 21 regions, 1990–2010: a systematic analysis for the global burden of disease study 2010. *Lancet* 2012;**380**:2224–60.

2. World Health Organisation. WHO global tuberculosis report. 2013.
3. Falzon D, Gandhi N, Migliori GB, Sotgiu G, Cox H, Holtz TH, et al. Resistance to fluoroquinolones and second-line injectable drugs: impact on mdr-tb outcomes. *Eur Resp J* 2012;1–24.
4. Böttger EC. The ins and outs of mycobacterium tuberculosis drug susceptibility testing. *Clin Microbiol Infect* 2011;**17**:1128–34.
5. Skrahina A, Hurevich H, Zalutskaya A, Sahalchyk E, Astrauko A, Hoffner S, et al. Multidrug-resistant tuberculosis in Belarus: the size of the problem and associated risk factors. *Bull WHO* 2013;**91**:36–45.
6. Riley R. Aerial dissemination of pulmonary tuberculosis. *Am Rev Tuberc* 1957;931–41.
7. Rehm J, Samokhvalov AV, Neuman MG, Room R, Parry C, Lönnroth K, et al. The association between alcohol use, alcohol use disorders and tuberculosis (TB). A systematic review. *BMC Public Health* 2009;**9**:450.
8. Bothamley GH, Ditiu L, Migliori GB, Lange C. Active case finding of tuberculosis in Europe: a tuberculosis network European trials group (tbnet) survey. *Eur Resp J* 2008;**32**:1023–30.
9. Dooley KE, Chaisson RE. Tuberculosis and diabetes mellitus: convergence of two epidemics. *Lancet Infect Dis* 2009;**9**:737–46.
10. Philips JA, Ernst JD. Tuberculosis pathogenesis and immunity. *Ann Rev Pathol* 2012;**7**:353–84.
11. Adams DO. The granulomatous inflammatory response. A review. *Am J Pathol* 1976;**84**:164–91.
12. Lange C, Mori T. Advances in the diagnosis of tuberculosis. *Respirology (Carlton, Vic)* 2010;**15**:220–40.
13. Steingart KR, Sohn H. Xpert® mtb/rif assay for pulmonary tuberculosis and rifampicin resistance in adults. The Cochrane Library.
14. Jafari C, Kessler P, Sotgiu G, Ernst M, Lange C. Impact of a mycobacterium tuberculosis-specific interferon-γ release assay in bronchoalveolar lavage fluid for a rapid diagnosis of tuberculosis. *J Intern Med* 2011;**270**:254–62.
15. Hillemann D, Rüsch-Gerdes S, Richter E. Feasibility of the genotype MTBDRsl assay for fluoroquinolone, amikacin-capreomycin, and ethambutol resistance testing of mycobacterium tuberculosis strains and clinical specimens. *J Clin Microbiol* 2009;**47**:1767–72.
16. Orenstein EW, Basu S, Shah NS, Andrews JR, Friedland GH, Moll AP, et al. Treatment outcomes among patients with multidrug-resistant tuberculosis: systematic review and meta-analysis. *Lancet Infect Dis* 2009;**9**:153–61.
17. Johnston JC, Shahidi NC, Sadatsafavi M, Fitzgerald JM. Treatment outcomes of multidrug-resistant tuberculosis: a systematic review and meta-analysis. *PLoS One* 2009;**4**:e6914.
18. Ahuja SD, Ashkin D, Avendano M, Banerjee R, Bauer M, Bayona JN, et al. Multidrug resistant pulmonary tuberculosis treatment regimens and patient outcomes: an individual patient data meta-analysis of 9,153 patients. *PLoS Med* 2012;**9**:e1001300.
18a Lange C, Abubakar I, Alffenaar JW, Bothamley G, Caminero JA, Carvalho AC, et al. Management of patients with multidrug-resistant/extensively drug-resistant tuberculosis in Europe: a TBNET consensus statement. *Eur Respir J* 2014 Mar 23. [Epub ahead of print].
19. World Health Organisation Guidelines for the programmatic management of drug-resistant tuberculosis 2011 update. 2011.

20. Van Deun A, Maug AKJ, Salim MAH, Das PK, Sarker MR, Daru P, et al. Highly effective, and inexpensive standardized treatment of multidrug-resistant tuberculosis. *Am J Resp Crit Care Med* 2010;**182**:684–92.
21. Chang K-C, Yew W-W. Management of difficult multidrug-resistant tuberculosis and extensively drug-resistant tuberculosis: Update 2012. *Respirology (Carlton, Vic)* 2013;**18**: 8–21.
22. Diacon AH, Pym A, Grobusch M, Patientia R, Rustomjee R, Page-Shipp L, et al. The diarylquinoline tmc207 for multidrug-resistant tuberculosis. *N Engl J Med* 2009;**360**: 2397–405.
23. Gler MT. Delamanid for multidrug-resistant pulmonary tuberculosis. *N Engl J Med* 2012; **366**:2151–60.
24. Skripconoka V, Danilovits M, Pehme L, Tomson T, Skenders G, Kummik T, et al. Delamanid improves outcomes and reduces mortality in multidrug-resistant tuberculosis. *Eur Resp J* 2013;**41**:1393–400.
25. Marais BJ, Obihara CC, Warren RM, Schaaf HS, Gie RP, Donald PR. The burden of childhood tuberculosis: a public health perspective. *Int J Tuberc Lung Dis* 2005;**9**:1305–13.
26. World Health Organization. Guidelines for the programmatic management of drug-resistant tuberculosis—2011 update. Communication 2011.
27. Trunz BB, Fine P, Dye C. Effect of bcg vaccination on childhood tuberculous meningitis and miliary tuberculosis worldwide: a meta-analysis and assessment of cost-effectiveness. *Lancet* 2006;**367**:1173–80.
28. Tameris MD, Hatherill M, Landry BS, Scriba TJ, Snowden MA, Lockhart S, et al. Safety and efficacy of mva85a, a new tuberculosis vaccine, in infants previously vaccinated with bcg: a randomised, placebo-controlled phase 2b trial. *Lancet* 2013;1021–8.
29. Smieja M, Marchetti C, Cook D, Smaill F.M. Isoniazid for preventing tuberculosis in non-HIV infected persons (review). The Cochrane Collaboration 2010.
30. Sterling TR, Villarino ME, Borisov AS. Three months of rifapentine and isoniazid for latent tuberculosis infection. *N Eng J Med* 2011;2155–66.
31. Schaaf HS, Marais BJ. Management of multidrug-resistant tuberculosis in children: a survival guide for paediatricians. *Paediatric Resp Rev* 2011;**12**:31–8.

Chapter 20

Acinetobacter Infections

Nicola Petrosillo,[1] Cecilia Melina Drapeau[2] and Stefano Di Bella[1]
[1]*National Institute for Infectious Diseases "Lazzaro Spallanzani," Rome, Italy,*
[2]*Department of Medical Microbiology, King's College Hospital, London, UK*

CASE PRESENTATION

A 70-year-old man was admitted to hospital because of high grade temperature, confusion, and pain in hypogastrium. His past medical history included diabetes mellitus, hypertension, and hyperlipidemia. During the previous 5 years, the patient had experienced three hospital admissions with acute urinary retention episodes, which eventually led to indwelling of a long-term urinary catheter. At the present admission the patient was found to be confused, his blood pressure was 100/60 mmHg, his pulse 125 beats per minute, his respiratory rate 25 breaths per minute, and his oxygen saturation was 97% in room air. The white cell count was 14,000 per cubic millimeter; serum creatinine was 1.8 mg per deciliter (creatinine clearance calculated according to the Cockroft–Gault formula was 37 ml per minute), and C-reactive protein was 12 mg per deciliter. The working diagnosis was a catheter-associated urinary tract infection. Blood cultures and urine cultures were obtained and empirical antimicrobial therapy with intravenous piperacillin/tazobactam adjusted for his renal function (2.25 g every 6 hours) was started. Both blood cultures and urine cultures grew extended spectrum beta-lactamase (ESBL)-producing *Escherichia coli*. The piperacillin/tazobactam was therefore switched to ertapenem (1 g every 24 hours) and the urinary catheter was changed.

The patient experienced progressive clinical improvement; however, after 8 days the patient became febrile again and developed dyspnea with episodes of drowsiness. His oxygen saturation dropped to 82% despite receiving a fraction of inspired oxygen of 50%, his respiratory rate increased up to 40 per minute, and the patient had to be mechanically ventilated through orotracheal intubation. A chest X-ray and a computed tomography (CT) showed a pulmonary infiltrate in the right middle lobe. Empirical antibiotic therapy including intravenous vancomycin (500 mg every 12 hours) and meropenem

Emerging Infectious Diseases. DOI: http://dx.doi.org/10.1016/B978-0-12-416975-3.00020-0

(1 g every 12 hours) was started. After an initial clinical improvement with defervescence, and at day 10 of mechanical ventilation the patient started spiking temperature up to 39°C, and became hypotensive (100/70 mmHg), and tachycardic (110 beats/minute); partial oxygen pressure was 90 mmHg with a fraction of inspired oxygen of 90%. A bronchial aspirate was performed and a full septic screen including blood culture was repeated. *Acinetobacter baumannii* resistant to carbapenems and susceptible only to colistin (MIC <1 μg/ml), and tigecycline (MIC ≤ 2 μg/ml) was isolated both from bronchial aspirate (10^5 colony-forming units per milliliter) and from blood cultures. The patient was then started on intravenous and aerosolized colistin and intravenous tigecycline. Intravenous colistin was administered with a loading dose of 6,000,000 IU followed by 3,000,000 IU every 12 hours. Tygecycline was given with a loading dose of 100 mg followed by 50 mg every 12 hours. A CT scan of the chest showed a worsening of the parenchymal infiltrate; blood culture was repeated after 7 days of treatment, *A. baumannii* was again isolated from blood, and the colistin MIC was 4 μg/ml. His respiratory function further deteriorated and the patient died.

1. WHY THIS CASE WAS SIGNIFICANTLY IMPORTANT AS AN EMERGING INFECTION

Infections by *Acinetobacter* species, mainly *baumannii* complex, were not considered a common healthcare-associated threat up to 20 years ago. In the last decades, its clinical significance has been promoted by its remarkable ability to up-regulate or acquire resistance determinants, contributing to making it a growing public health problem. Multidrug-resistant strains of *A. baumannii* continue to spread in the hospital setting worldwide, in both an endemic and epidemic manner, with a mortality rate ranging from 40 to 70% in ventilator-associated pneumonia (VAP),[1,2] 25–30% in meningitis,[3,4] and 34–49% in bacteremia.[5,6] Despite the pathogenicity of *Acinetobacter* being questioned for a long time, during the last decade some important studies have demonstrated that *A. baumannii* remains significantly associated with mortality after adjustment for other important risk factors.[7] Moreover, another study showed how inappropriate use of antibiotics caused 7.6% excess in mortality compared with patients treated with appropriate antibiotics, again supporting the thesis of an important attributable mortality due to *Acinetobacter*.[8] More recently, a further threat has been represented by the spread of carbapenem.[9] and colistin-resistant *A. baumannii* infections (Figure 20.1).

2. WHAT IS THE CAUSATIVE AGENT?

The genus *Acinetobacter* comprises Gram-negative, strictly aerobic, non-fermenting, non-motile (akinetos: non-motile), catalase-positive, oxidase-negative bacteria. *Acinetobacter* are rod-shaped during rapid growth and

FIGURE 20.1 Worldwide reports of colistin heteroresistance and resistance of *Acinetobacter baumannii*. Updated to July 2013.

coccobacillary in the stationary phase. Members of the genus *Acinetobacter* are considered ubiquitous organisms; in fact they can be recovered from virtually any sample obtained from soil or surface water.[9] *Acinetobacter* species of human origin grows well on solid media that are routinely used in clinical microbiology laboratories, at a 37°C incubation temperature. Colonies are 1 to 2 mm in diameter, non-pigmented, domed, and mucoid, with smooth to pitted surfaces.[10] The inability of *Acinetobacter* spp. to reduce nitrate or grow anaerobically distinguishes these organisms from Enterobacteriaceae.[10]

Currently, the *Acinetobacter* genus comprises 27 species with valid names and several other putative species with provisional designations.[11] From the clinical point of view, the *Acinetobacter baumannii* group constituted by *A. baumannii*, *Acinetobacter* genomic species 3 (now named *A. pittii*), and *Acinetobacter* genomic species 13TU (now named *A. nosocomialis*)[12] are important nosocomial pathogens causing both sporadic infections and outbreaks, mainly in the intensive care units. These three clinically important species are phylogenetically related and cannot be differentiated by currently available phenotypic identification systems.

3. WHAT IS THE FREQUENCY OF THE DISEASE?

In the early 1980s, hospital outbreaks of *Acinetobacter baumannii* infections in Europe had been investigated epidemiologically using molecular typing methods.[13,14] In the majority of cases, one or two epidemic strains were detected in a given epidemiological setting. Transmission of such strains has been observed between hospitals, most probably via transfer of colonized patients. Spread of multidrug-resistant *A. baumannii* is not confined to hospitals within a city but also occurs on a national, international, and intercontinental scale.[15,16]

Epidemiological data on *A. baumannii* derive from large national and international studies on nosocomial pathogens and/or healthcare-acquired infection.

In Italy, in a large study on 125 ICUs recording 4489 infections, *Acinetobacter* accounted for 7.1% of infections with positive microbiological findings.[17] Of 275 *A. baumannii* strains, 108 (39.2%) were carbapenem resistant.

A wide prevalence survey conducted in 2007 [Extended Prevalence of Infection in Intensive Care (EPIC II) study] on 1265 ICUs in 75 countries worldwide collected data from 13,796 adult inpatients of which 7087 (51%) were classified as infected. *Acinetobacter* was involved in 9% of all infections.[18] In this study, rates of infection with *Acinetobacter* differed most markedly, ranging from 3.7% in North America to 19.2% in Asia.

In USA, the National Healthcare Safety Network (NHSN), which collects surveillance healthcare data, assessed the antimicrobial resistance patterns for healthcare-associated infections reported during 2009–2010.[19] According to their results, *Acinetobacter* accounted for 1.8% of 81,139 pathogens reported from 69,475 healthcare-associated infections. Thirty-seven percent of the *Acinetobacter* strains collected in this study were resistant to carbapenems.

In South America, the SENTRY Antimicrobial Surveillance Program reported data of 12,811 bacterial organisms collected between 2008 and 2010 from 10 Latin American medical centers.[20] The bacterial isolates were collected from hospitalized patients with serious community-acquired or nosocomial-acquired infections. According to this study *Acinetobacter* caused 7% of BSI, 18% of pneumonias, and 10% of skin and soft tissue infections. A significant trend for imipenem resistance was noticed: 68% of *Acinetobacter* isolates were imipenem resistant, compared to less than 10% in the period 1997–1999.

4. HOW ARE THE BACTERIA TRANSMITTED?

The most common modality of transmission of *A. baumannii* is contact, mainly from the hands of hospital staff; however, the bacteria can spread through the air over short distances in water droplets and can be found in skin scales from colonized patients.[21] Moreover, the hospital environment is a reservoir for *A. baumannii*[22] as this pathogen can survive in the environment for extended periods because of its innate resistance to desiccation[23,24] (Figure 20.2). *Acinetobacter* has been isolated during outbreaks from various sites in the patients' environment, including bed curtains, furniture, and hospital equipment[25] (Table 20.1). Outbreaks of *A. baumannii* associated with contaminated inanimate fomites are reported to resolve once the common source is identified and removed, replaced, or adequately disinfected.[26] Evidence is accumulating that contaminated surfaces make an important contribution to the epidemic and endemic transmission of *A. baumannii*.[27]

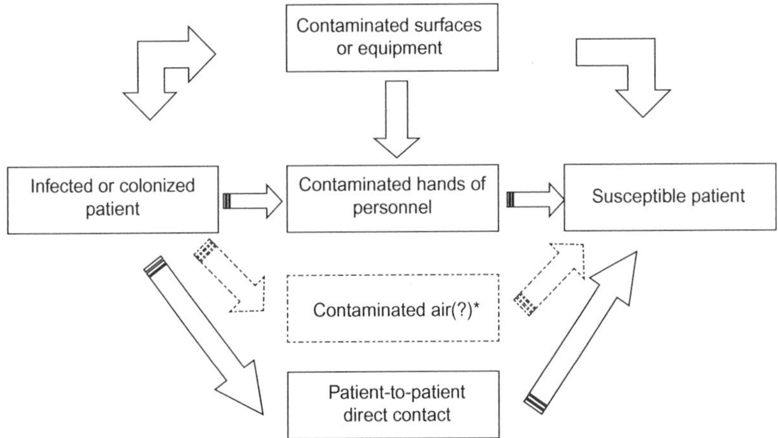

FIGURE 20.2 Common transmission routes of *Acinetobacter* in the healthcare facilities.

TABLE 20.1 ***Acinetobacter baumannii*** **Outbreaks and Items Involved in Environmental Contamination**

Suctioning equipment	Chair/recliner
Washbasin	Television
Bedrail	Paper towel dispenser
Bedside	Door
Table	Light switch
Ventilator	Blood pressure equipment
Infusion pump	Telephone
Sink	Television remote control
Hygroscopic bandage	Dresser
Shower trolley	Glucose meter
Pillow	Wheel chair
Mattress	Gait belt
Resuscitation equipment	Oxygen flow meter
Stainless steel trolley	Cardiac monitor
Cabinet	Closet
Curtain	Dressing cart
Enteral feeding pump	Medication drawer

From Manian et al.[28] and Fournier et al.,[29] modified.

5. WHICH FACTORS ARE INVOLVED IN DISEASE PATHOGENESIS? WHAT ARE THE PATHOGENIC MECHANISMS?

Adherence to host cells, as demonstrated *in vitro* using bronchial epithelial cells,[28] is considered to be the first step in the colonization process. Outgrowth on mucosal surfaces and medical devices, such as intravascular catheters and endotracheal tubes,[29] can lead to biofilm formation, which enhances the risk of infection of the bloodstream and airways; a regulatory role in biofilm formation is played by quorum sensing.[30] Among the several factors that can be involved in *A. baumannii* infection in experimental studies on animal models, the lipopolysaccharide has been shown to elicit a pro-inflammatory response.[31] *In vitro*, the *A. baumannii* outer membrane protein A has been demonstrated to cause cell death;[32] moreover, iron-acquisition mechanisms and resistance to the bactericidal activity of human serum are considered to be important for survival in the blood during bloodstream infections.[33]

The main risk factors for multidrug-resistant *A. baumannii* colonization or infection are listed in Table 20.2. *Acinetobacter* resistance mechanisms have an important role in the clinical course of this infection. Resistance mechanisms frequently expressed in nosocomial strains of *Acinetobacter*

TABLE 20.2 Risk Factors for Multidrug-Resistant *Acinetobacter baumannii*

Prior use of antibiotics, especially carbapenems and 3rd generation cephalosporins
Stay in intensive care
Intubation for mechanical ventilation
Length of ICU and hospital stay
Severity of illness
Sex
Tracheostomy
Hydrotherapy
Transfusions
Arterial and central venous catheter indwelling
Foley catheters
Exposure to patients infected with *A. baumannii*
Prior carriage of *A. baumannii*
Environmental contamination with *A. baumannii*

From Falagas et al.,[38] modified.

include production of β-lactamases, alterations in cell-wall channels (porins), and efflux pumps. *A. baumannii* can become resistant to quinolones through mutations in the genes gyrA and parC and can become resistant to aminoglycosides by expressing aminoglycoside-modifying enzymes. AmpC β-lactamases cause resistance to cephalosporins. Porin channels in *A. baumannii* are thought to be involved in β-lactam resistance. Overexpression of bacterial efflux pumps can decrease the concentration of β-lactam antibiotics in the periplasmic space and can actively expel quinolones, tetracyclines, chloramphenicol, disinfectants, and tigecycline.[34]

Carbapenem resistance in *A. baumannii* is mainly mediated by the production of carbapenem-hydrolyzing enzymes.[13] Genes encoding a variety of carbapenem-hydrolyzing-β-lactamases, belonging to classes A, B, and D, have been detected in *A. baumannii*. The class D OXA-type carbapenemases are by far the most prevalent carbapenemases in *A. baumannii*.[35]

6. WHAT ARE THE CLINICAL MANIFESTATIONS?

6.1 Nosocomial Infections

6.1.1 Ventilator-Associated Pneumonia

The greatest impact of *Acinetobacter* has been as a causative agent of nosocomial pneumonia, particularly ventilator-associated cases, because of its high rate of tracheostomy colonization.[36] In fact, in most institutions, the majority of *A. baumannii* isolates are from the respiratory tracts of hospitalized patients. In many circumstances, it is very difficult to distinguish upper airway colonization from true pneumonia. There is no doubt, however, that true VAP due to *A. baumannii* occurs. In large surveillance studies from the United States, between 5 and 10% of cases of ICU-acquired pneumonia were due to *A. baumannii*.[37]

6.1.2 Bloodstream Infection (BSI)

In a large study on 20,978 nosocomial BSIs in the United States (1995–2002), *A. baumannii* was the 10th most common etiologic agent, being responsible for 1.3% of all monomicrobial nosocomial bloodstream infections and for 1.6% of bloodstream infections among ICU patients.[38] In this study, *A. baumannii* infections were the latest of all bloodstream infections to occur during hospitalization, after a mean of 26 days from the time of hospital admission. The crude mortality was 34% surpassed only by *Candida* and *Pseudomonas* (39% each).

6.1.3 Meningitis

Nosocomial, post-neurosurgical *A. baumannii* meningitis is an increasingly important entity. Typical patients have undergone neurosurgery and have an external ventricular drain.[39] In a review of four studies comprising 281 adult patients with hospital-acquired bacterial meningitis, 3.6% (10 patients) had meningitis due to *Acinetobacter*.[40] The median time to develop *Acinetobacter* meningitis after a neurosurgical procedure is 12 days (range 1–40 days).[41] Mortality may be as high as 70%, although the cause of mortality is often difficult to discern.[39] In a study in which just one out of 13 patients had a carbapenem-resistant isolate, mortality from *Acinetobacter* meningitis was 30%[3] whereas all nine patients with meningitis due to carbapenem-resistant *A. baumannii* in a hospital where carbapenem-resistant *A. baumannii* was endemic died.[39]

6.1.4 Urinary Tract Infection

Urinary tract infections caused by *Acinetobacter* are uncommon. These infections are almost always associated with urinary catheters. In a large study on healthcare-associated infections conducted by the Centers for Disease Control and Prevention, 0.9% of catheter-associated urinary tract infections were caused by *Acinetobacter*.[19] Urinary tract infections related to indwelling urinary tract catheters usually run a more benign clinical course and are more frequent in rehabilitation centers than in ICUs.[42]

6.1.5 Traumatic Battlefield and Other Wounds

Nosocomial *A. baumannii* wound infections have been reported in natural or man-made disasters (earthquakes, bombing, military operations).[43–45] During the Iraq conflict, several deep wound and burn wound infections and bone infections have been reported in repatriated casualties, often caused by multidrug-resistant *A. baumannii* strains.[43] Recent data indicate that contamination of the environment of field hospitals and transmission of the organism within the healthcare premises have had a major role in the acquisition of *A. baumannii* infections among these patients.[46]

6.1.6 Other Manifestations

Acinetobacter endocarditis is an uncommon clinical manifestation, more often associated with prosthetic valves.[47,48] *Acinetobacter* may also cause endophthalmitis or keratitis, sometimes related to contact lens use or following eye surgery,[49,50] and peritonitis in patients undergoing peritoneal dialysis.[51]

6.2 Community-Acquired Infections

Community-acquired infections caused by *Acinetobacter* are quite uncommon. Most community-acquired *Acinetobacter* infections have been reported from countries with tropical or subtropical climates (e.g., China, Taiwan, and tropical Australia), and mainly affect patients with comorbidities, such as chronic obstructive pulmonary disease, renal disease, and diabetes mellitus, and/or are associated with heavy smoking and excessive alcohol consumption.[52] Multidrug resistance in these organisms is uncommon.[53] In a review including 80 patients with community-acquired *A. baumannii* infections, 51 had pneumonia and bacteremia occurred in the remaining 29 patients. The mortality of patients with community-acquired *Acinetobacter* pneumonia and/or bacteremia is considerable (56%);[52] however, a publication bias is possible. The primary portal of entry may be throat carriage, which occurs in up to 10% of community residents with excessive alcohol consumption.[54]

7. HOW DO YOU DIAGNOSE?

Diagnosis of infections caused by *Acinetobacter* usually relies on classical microbiological cultures from biological specimens (sputum, bronchoaspirate, bronchoalveolar lavage fluid, urine, blood, pus, etc.). Polymerase chain reaction for specific genes, including resistance genes, has been used especially for characterizing nosocomial *Acinetobacter* outbreaks.[55] *Acinetobacter* species grow well on solid media that are routinely used in clinical microbiology laboratories, at a 37°C incubation temperature. Colonies are 1 to 2 mm, non-pigmented, domed, and mucoid, with smooth to pitted surfaces.[10] Until today, strain isolation has been of fundamental importance to obtain antimicrobial susceptibilities. However, recently, several authors have validated the matrix-assisted laser desorption ionization-time of flight mass spectrometry (MALDI-TOF MS) to detect antibiotic resistance.[56] MALDI-TOF MS has recently demonstrated to be able not only to detect carbapenem resistance in *Acinetobacter* strains but also to distinguish between metallo-β-lactamases and oxicillinases.

8. HOW DO YOU DIFFERENTIATE THIS DISEASE FROM SIMILAR ENTITIES?

Infections caused by *A. baumannii* do not differ for clinical manifestation from other entities. However, there are some specific features that can help clinicians to strongly suspect *Acinetobacter*. These features have been evaluated mainly for ventilator-associated pneumonia (VAP). Most of VAP are caused by bacterial pathogens that normally colonize the oropharynx and gut, or that are acquired via transmission by healthcare workers from environmental surfaces or from other patients. In *A. baumannii* VAP, time to infection from the beginning of mechanical ventilation can be helpful to differentiate this entity from other etiologies. In a study of 267 VAP with at least one positive microbiological finding, *Acinetobacter* was the second most frequent pathogen (after *Staphylococcus aureus*) and more than 90% of *Acinetobacter* were isolated from patients with "late VAP," i.e., a VAP occurring more than 4 days after mechanical ventilation.[57,58] Moreover, recent studies show that the previous colonization and carbapenem exposure are risk factors for developing *Acinetobacter* infection in ICU.[59] In summary, broad-spectrum antibiotic exposure and late onset VAP should lead to suspect *Acinetobacter* as the likely pathogen.

9. WHAT IS THE THERAPEUTIC APPROACH?

Acinetobacter is generally intrinsically resistant to a number of commonly used antibiotics, including aminopenicillins, first- and second-generation cephalosporins and chloramphenicol.[60,61] It has also a remarkable capacity to acquire mechanisms that confer resistance to broad-spectrum β-lactams, aminoglycosides, fluoroquinolones, and tetracyclines.[62] Below are described the main antimicrobial options for *A. baumannii* infections.

9.1 Carbapenems

Carbapenems remain the most important therapeutic option for serious infections caused by carbapenem-susceptible *A. baumannii* strains. However, increasing carbapenem resistance is creating therapeutic challenge, especially considering that most *A. baumannii* strains that are resistant to carbapenems are also resistant to the majority of other antibiotics (except the polymyxins or tigecycline).

9.2 Sulbactam

Several studies have evaluated the efficacy of sulbactam in different infections by *A. baumannii*. Of the β-lactamase inhibitors, sulbactam possesses the greatest intrinsic bactericidal activity against *A. baumannii* isolates.[13] Sulbactam has been used successfully for the treatment of MDR *A. baumanni* infections, such as meningitis, VAP, and catheter-related bacteremia.[63,64] In a study on BSI, seven (87.5%) out of eight cases of *A. baumannii* bacteremia treated with sulbactam were cured, and results were similar to those obtained in 42 cases treated with imipenem with a cure rate of 83%.[5]

9.3 Colistin

Polymyxin E (colistin) and polymyxin B are commonly used in the treatment of multidrug-resistant *A. baumannii* infections,[65] because of their *in vitro* activity against most strains. Colistin is the most used polymyxin in Europe. In a comparison between colistin and imipenem for patients with VAP caused by *A. baumannii*, both antibiotics resulted in a clinical cure of 57%.[66] Pharmacokinetics of colistin is still controversial. Dalfino et al. prospectively evaluated a colistin regimen based on a loading dose of 9 MU and a 9-MU twice-daily fractionated maintenance dose (titrated on renal function) in patients with sepsis due to Gram-negative bacteria susceptible only to colistin.[67] The authors analyzed 28 infectious episodes, mainly bloodstream infections (64%) and VAPs (36%). Twenty-eight infectious episodes were due to

Acinetobacter (46.4%). Clinical cure was observed in 82% of patients, which supported the hypothesis that a high-dose, extended-interval colistin administration has a high efficacy in critically ill patients.

Despite its large use, in a worldwide study including 4686 *Acinetobacter*, colistin resistance of the collected strains (2006–2009) was less than 2%.[68] It is likely that the use of combined therapy in the treatment of *A. baumannii* infections may reduce the probability of selection of resistant subpopulations.[69]

For the treatment of MDR *A. baumannii* central nervous system infections, intraventricular administration of colistin has been demonstrated to be effective.[70,71]

Finally, the main collateral effect of colistin is nephrotoxocity that occurs in a dose-dependent manner.[72] A less frequent adverse event is represented by neurotoxicity.

9.4 Tigecycline

Tigecycline is active against *A. baumannii* strains, including imipenem-resistant and MDR strains,[73] although resistant strains and the increase in MIC during the last years have been described.[74,75] From 2009 to 2011, several reports of retrospective analysis of 181 *A. baumannii* infections, most commonly VAP (142 cases) and bacteremia (20 cases), showed clinical cure in 65.4% and 65%, respectively (65.2% of all cases).[76–80] However, it should be kept in mind that, in spite of *in vitro* tigecycline activity, cases of breakthrough bacteremia by *A. baumannii* in patients on tigecycline have been reported.[81] Other limitations of this drug are the development of resistance (24.6% in one study)[79] and poor concentration in urine and cerebrospinal fluid.[82,83] The bacteriostatic effect of tigecycline limits its use in bloodstream infections.

9.5 Aminoglycosides

Amikacin and tobramycin are the two agents that appear to retain activity against many *A. baumannii* isolates.[84] A retrospective study comparing colistin and tobramycin in 64 ICU patients (32 patients in each group) with *Acinetobacter* infections (mostly pneumonias) found no significant difference both in mortality and time to death between the two groups.[85] Since scarce data are available on the clinical efficacy of aminoglycosides monotherapy, these antimicrobials are usually used as part of combination therapy.

9.6 Combination Therapy

Several combination therapies have been proposed. Colistin is no longer used as monotherapy but is mostly combined with another antibiotic to prevent the possibility of resistance development.[86] For carbapenem-resistant strains, colistin is usually given co-administered with sulbactam or aminoglycosides or tigecycine. A widely investigated combination therapy is that including colistin and rifampicin. Despite the evidence of synergy between colistin and rifampicin in a proportion of *in vitro* and animal studies, the evidence from the limited clinical studies available suggests that colistin combination therapy is not superior to monotherapy.[87] Recently, the only existing randomized controlled trial comparing colistin and rifampicin with colistin alone was published. The study enrolled 210 ICU patients with life-threatening infections due to extensively drug-resistant *A. baumannii*. There was no difference in 30-day mortality between the two arms; however, microbiologic eradication rate was higher in the combination therapy group.[88]

Similarly, the colistin/sulbactam combination therapy has been evaluated in patients with severe MDR *A. baumannii* VAP. Of 89 patients, 52 received colistin and 37 received colistin/sulbactam therapy. Clinical response rates were 29.8% and 40% in colistin and colistin/sulbactam groups, respectively. The bacteriological response rates were 72.3% and 85.7% in colistin and colistin/sulbactam groups, respectively. The authors conclude that although the difference was not statistically significant, clinical cure rates or bacteriological clearance rates were better in the combination group than colistin monotherapy.[89]

10. WHAT ARE THE PREVENTIVE AND INFECTION CONTROL MEASURES?

Prevention of *A. baumannii* infection requires knowledge about the epidemiology of this infection. The various types of interventions used to control *Acinetobacter* spreading may be grouped into seven categories, including administrative support, judicious use of antimicrobials, surveillance (routine and enhanced), standard and contact precautions, environmental measures, education and decolonization.[90] Some of these measures are difficult to administer, and some are still controversial (i.e., decolonization). However, a multifaceted intervention comprising all these points is the cornerstone to prevent *A. baumannii* diffusion in a healthcare setting.

When *A. baumannii* enters a facility through a colonized/infected patient, contact precautions (CP) represent the best way for the control of its spread. There are several data on the effectiveness of CP in the control of multidrug-resistant Gram-negative organisms during epidemic conditions; limited evidence derives from the use of CP in endemic situations. However, Rodriguez-Bano et al. reported successful hospital-wide control of multidrug-resistant *A. baumannii* by using a bundle strategy that included contact precautions, active surveillance screening, education on hand hygiene, environmental cultures, strict environmental sanitation, and regular meetings of the personnel with discussion and feedback of the data.[91]

Most international guidelines recommend isolating patients with infection/colonization by *A. baumannii* in order to prevent hospital spread. However, there are multiple ways to organize such isolation measures. Transfer of patients to

TABLE 20.3 Preventive and Infection Control Measures Against *Acinetobacter* Infection

Identification and control of a source of infection in case of outbreak
Standard precautions (hand hygiene, proper glove use, barrier protections, compliance among personnel) as a preventive measure
Contact precautions (dedicated patient care equipment, contact isolation measures)
Physical separation of colonized/infected patients in single room to reduce the risk of acquisition of *Acinetobacter*, when possible
Surveillance (passive or active), including lab alerts, active screening
Continuous education programs
Environmental cleaning, including procedures for the proper use of detergents and disinfectants
Closure of hospital units only in limited circumstances or outbreaks to interrupt the spread of the microorganism and to permit a thorough cleaning/disinfection or a point source control

special isolation wards or housed in nursing cohorts, i.e., in separate rooms on general wards with designated nursing staff exclusively responsible for the cohort, could be an option. Another alternative is to isolate them in single or cohort rooms on general wards without designated personnel; a further strategy is to house the colonized/infected patients in rooms with non-infected/colonized patients but strictly applying barrier precautions, i.e., gloves, gowns, hand washing, etc., when caring for them ("barrier nursing"). Table 20.3 shows the most important preventive and infection control measures.

In conclusion, among Gram-negative organisms playing a significant role in nosocomial infections, *Acinetobacter* spp. (mainly *A. baumannii*) are now known to have major importance in healthcare settings and particularly in critical areas. The main factors that have contributed to confer such importance to these organisms include their current relatively high incidence in most hospitals and their multidrug resistance, which is related to misuse/overuse of wide-spectrum antibiotics. Both factors represent a serious challenge for epidemiologists, infection control professionals, and clinicians.

REFERENCES

1. Fagon JY, Chastre J, Domart Y, Trouillet JL, Gibert C. Mortality due to ventilator-associated pneumonia or colonization with Pseudomonas or Acinetobacter species: assessment by quantitative culture of samples obtained by a protected specimen brush. *Clin Infect Dis* 1996;**23**:538–42.

2. Garnacho J, Sole-Violan J, Sa-Borges M, Diaz E, Rello J. Clinical impact of pneumonia caused by Acinetobacter baumannii in intubated patients: a matched cohort study. *Crit Care Med* 2003;**31**:2478–82.
3. Chen SF, Chang WN, Lu CH, et al. Adult Acinetobacter meningitis and its comparison with non-Acinetobacter gram-negative bacterial meningitis. *Acta Neurol Taiwan* 2005;**14**:131–7.
4. Jimenez-Mejias ME, Pachon J, Becerril B, Palomino-Nicas J, Rodriguez-Cobacho A, Revuelta M. Treatment of multidrug-resistant Acinetobacter baumannii meningitis with ampicillin/sulbactam. *Clin Infect Dis* 1997;**24**:932–5.
5. Cisneros JM, Reyes MJ, Pachon J, et al. Bacteremia due to Acinetobacter baumannii: epidemiology, clinical findings, and prognostic features. *Clin Infect Dis* 1996;**22**:1026–32.
6. Kuo LC, Lai CC, Liao CH, et al. Multidrug-resistant Acinetobacter baumannii bacteraemia: clinical features, antimicrobial therapy and outcome. *Clin Microbiol Infect* 2007;**13**:196–8.
7. Robenshtok E, Paul M, Leibovici L, et al. The significance of Acinetobacter baumannii bacteraemia compared with Klebsiella pneumoniae bacteraemia: risk factors and outcomes. *J Hosp Infect* 2006;**64**:282–7.
8. Zaragoza R, Artero A, Camarena JJ, Sancho S, Gonzalez R, Nogueira JM. The influence of inadequate empirical antimicrobial treatment on patients with bloodstream infections in an intensive care unit. *Clin Microbiol Infect* 2003;**9**:412–8.
9. Baumann P, Doudoroff M, Stanier RY. A study of the Moraxella group. II. Oxidative-negative species (genus Acinetobacter). *J Bacteriol* 1968;**95**:1520–41.
10. Mandell. *Principles and practice of infectious diseases.* 7th ed Elsevier Churchill Livingstone; 2013.
11. Espinal P, Roca I, Vila J. Clinical impact and molecular basis of antimicrobial resistance in non-baumannii Acinetobacter. *Future Microbiol* 2011;**6**:495–511.
12. Nemec A, Krizova L, Maixnerova M, et al. Genotypic and phenotypic characterization of the Acinetobacter calcoaceticus-Acinetobacter baumannii complex with the proposal of Acinetobacter pittii sp. nov. (formerly Acinetobacter genomic species 3) and Acinetobacter nosocomialis sp. nov. (formerly Acinetobacter genomic species 13TU). *Res Microbiol* 2011;**162**:393–404.
13. Peleg AY, Seifert H, Paterson DL. Acinetobacter baumannii: emergence of a successful pathogen. *Clin Microbiol Rev* 2008;**21**:538–82.
14. Villegas MV, Hartstein AI. Acinetobacter outbreaks, 1977–2000. *Infect Control Hosp Epidemiol* 2003;**24**:284–95.
15. Naas T, Kernbaum S, Allali S, Nordmann P. Multidrug-resistant Acinetobacter baumannii, Russia. *Emerg Infect Dis* 2007;**13**:669–71.
16. Peleg AY, Bell JM, Hofmeyr A, Wiese P. Inter-country transfer of Gram-negative organisms carrying the VIM-4 and OXA-58 carbapenem-hydrolysing enzymes. *J Antimicrob Chemother* 2006;**57**:794–5.
17. Malacarne P, Boccalatte D, Acquarolo A, et al. Epidemiology of nosocomial infection in 125 Italian intensive care units. *Minerva Anestesiol* 2010;**76**:13–23.
18. Vincent JL, Rello J, Marshall J, et al. International study of the prevalence and outcomes of infection in intensive care units. *JAMA* 2009;**302**:2323–9.
19. Sievert DM, Ricks P, Edwards JR, et al. Antimicrobial-resistant pathogens associated with healthcare-associated infections: summary of data reported to the National Healthcare Safety Network at the Centers for Disease Control and Prevention, 2009–2010. *Infect Control Hosp Epidemiol* 2013;**34**:1–14.
20. Gales AC, Castanheira M, Jones RN, Sader HS. Antimicrobial resistance among Gram-negative bacilli isolated from Latin America: results from SENTRY Antimicrobial

Surveillance Program (Latin America, 2008–2010). *Diagn Microbiol Infect Dis* 2012;**73**:354–60.

21. Bernards AT, Frenay HM, Lim BT, Hendriks WD, Dijkshoorn L, van Boven CP. Methicillin-resistant Staphylococcus aureus and Acinetobacter baumannii: an unexpected difference in epidemiologic behavior. *Am J Infect Control* 1998;**26**:544–51.
22. Silvia Munoz-Price L, Namias N, Cleary T, et al. Acinetobacter baumannii: association between environmental contamination of patient rooms and occupant status. *Infect Control Hosp Epidemiol* 2013;**34**:517–20.
23. Wendt C, Dietze B, Dietz E, Ruden H. Survival of Acinetobacter baumannii on dry surfaces. *J Clin Microbiol* 1997;**35**:1394–7.
24. Jawad A, Seifert H, Snelling AM, Heritage J, Hawkey PM. Survival of Acinetobacter baumannii on dry surfaces: comparison of outbreak and sporadic isolates. *J Clin Microbiol* 1998;**36**:1938–41.
25. van den Broek PJ, Arends J, Bernards AT, et al. Epidemiology of multiple Acinetobacter outbreaks in The Netherlands during the period 1999–2001. *Clin Microbiol Infect* 2006;**12**:837–43.
26. Dancer SJ. The role of environmental cleaning in the control of hospital-acquired infection. *J Hosp Infect* 2009;**73**:378–85.
27. Otter JA, Yezli S, French GL. The role played by contaminated surfaces in the transmission of nosocomial pathogens. *Infect Control Hosp Epidemiol* 2011;**32**:687–99.
28. Lee JC, Koerten H, van den Broek P, et al. Adherence of Acinetobacter baumannii strains to human bronchial epithelial cells. *Res Microbiol* 2006;**157**:360–6.
29. Tomaras AP, Dorsey CW, Edelmann RE, Actis LA. Attachment to and biofilm formation on abiotic surfaces by Acinetobacter baumannii: involvement of a novel chaperone-usher pili assembly system. *Microbiology* 2003;**149**:3473–84.
30. Smith MG, Gianoulis TA, Pukatzki S, et al. New insights into Acinetobacter baumannii pathogenesis revealed by high-density pyrosequencing and transposon mutagenesis. *Genes Dev* 2007;**21**:601–14.
31. Brade H, Galanos C. Biological activities of the lipopolysaccharide and lipid A from Acinetobacter calcoaceticus. *J Med Microbiol* 1983;**16**:211–4.
32. Choi CH, Hyun SH, Lee JY, et al. Acinetobacter baumannii outer membrane protein A targets the nucleus and induces cytotoxicity. *Cell Microbiol* 2008;**10**:309–19.
33. Dorsey CW, Beglin MS, Actis LA. Detection and analysis of iron uptake components expressed by Acinetobacter baumannii clinical isolates. *J Clin Microbiol* 2003;**41**:4188–93.
34. Munoz-Price LS, Weinstein RA. Acinetobacter infection. *N Engl J Med* 2008;**358**: 1271–81.
35. Poirel L, Nordmann P. Carbapenem resistance in Acinetobacter baumannii: mechanisms and epidemiology. *Clin Microbiol Infect* 2006;**12**:826–36.
36. Glew RH, Moellering Jr. RC, Kunz LJ. Infections with Acinetobacter calcoaceticus (Herellea vaginicola): clinical and laboratory studies. *Medicine (Baltimore)* 1977;**56**:79–97.
37. Gaynes R, Edwards JR. Overview of nosocomial infections caused by gram-negative bacilli. *Clin Infect Dis* 2005;**41**:848–54.
38. Wisplinghoff H, Bischoff T, Tallent SM, Seifert H, Wenzel RP, Edmond MB. Nosocomial bloodstream infections in US hospitals: analysis of 24,179 cases from a prospective nationwide surveillance study. *Clin Infect Dis* 2004;**39**:309–17.
39. Metan G, Alp E, Aygen B, Sumerkan B. Acinetobacter baumannii meningitis in postneurosurgical patients: clinical outcome and impact of carbapenem resistance. *J Antimicrob Chemother* 2007;**60**:197–9.

40. Kim BN, Peleg AY, Lodise TP, et al. Management of meningitis due to antibiotic-resistant Acinetobacter species. *Lancet Infect Dis* 2009;**9**:245–55.
41. Siegman-Igra Y, Bar-Yosef S, Gorea A, Avram J. Nosocomial acinetobacter meningitis secondary to invasive procedures: report of 25 cases and review. *Clin Infect Dis* 1993;**17**:843–9.
42. Wise KA, Tosolini FA. Epidemiological surveillance of Acinetobacter species. *J Hosp Infect* 1990;**16**:319–29.
43. Davis KA, Moran KA, McAllister CK, Gray PJ. Multidrug-resistant Acinetobacter extremity infections in soldiers. *Emerg Infect Dis* 2005;**11**:1218–24.
44. Oncul O, Keskin O, Acar HV, et al. Hospital-acquired infections following the 1999 Marmara earthquake. *J Hosp Infect* 2002;**51**:47–51.
45. Kennedy PJ, Haertsch PA, Maitz PK. The Bali burn disaster: implications and lessons learned. *J Burn Care Rehabil* 2005;**26**:125–31.
46. Scott P, Deye G, Srinivasan A, et al. An outbreak of multidrug-resistant Acinetobacter baumannii-calcoaceticus complex infection in the US military health care system associated with military operations in Iraq. *Clin Infect Dis* 2007;**44**:1577–84.
47. Menon T, Shanmugasundaram S, Nandhakumar B, Nalina K, Balasubramaniam. Infective endocarditis due to Acinetobacter baumannii complex—a case report. *Indian J Pathol Microbiol* 2006;**49**:576–8.
48. Olut AI, Erkek E. Early prosthetic valve endocarditis due to Acinetobacter baumannii: a case report and brief review of the literature. *Scand J Infect Dis* 2005;**37**:919–21.
49. Kau HC, Tsai CC, Kao SC, Hsu WM, Liu JH. Corneal ulcer of the side port after phacoemulsification induced by Acinetobacter baumannii. *J Cataract Refract Surg* 2002;**28**:895–7.
50. Levy J, Oshry T, Rabinowitz R, Lifshitz T. Acinetobacter corneal graft ulcer and endophthalmitis: report of two cases. *Can J Ophthalmol* 2005;**40**:79–82.
51. Bergogne-Berezin E, Towner KJ. Acinetobacter spp. as nosocomial pathogens: microbiological, clinical, and epidemiological features. *Clin Microbiol Rev* 1996;**9**:148–65.
52. Falagas ME, Karveli EA, Kelesidis I, Kelesidis T. Community-acquired Acinetobacter infections. *Eur J Clin Microbiol Infect Dis* 2007;**26**:857–68.
53. Chen MZ, Hsueh PR, Lee LN, Yu CJ, Yang PC, Luh KT. Severe community-acquired pneumonia due to Acinetobacter baumannii. *Chest* 2001;**120**:1072–7.
54. Anstey NM, Currie BJ, Hassell M, Palmer D, Dwyer B, Seifert H. Community-acquired bacteremic Acinetobacter pneumonia in tropical Australia is caused by diverse strains of Acinetobacter baumannii, with carriage in the throat in at-risk groups. *J Clin Microbiol* 2002;**40**:685–6.
55. D'Andrea MM, Giani T, D'Arezzo S, et al. Characterization of pABVA01, a plasmid encoding the OXA-24 carbapenemase from Italian isolates of Acinetobacter baumannii. *Antimicrob Agents Chemother* 2009;**53**:3528–33.
56. Alvarez-Buylla A, Picazo JJ, Culebras E. Optimized method for Acinetobacter species carbapenemase detection and identification by matrix-assisted laser desorption ionization-time of flight mass spectrometry. *J Clin Microbiol* 2013;**51**:1589–92.
57. Fagon JY, Chastre J, Wolff M, et al. Invasive and noninvasive strategies for management of suspected ventilator-associated pneumonia. A randomized trial. *Ann Intern Med* 2000;**132**:621–30.
58. Park DR. The microbiology of ventilator-associated pneumonia. *Respir Care* 2005;**50**: 742–63 discussion 63–5

59. Giannella M, Di Bella S, D'Este G, et al. Risk factors for Acinetobacter baumannii colonisation and infection among patients admitted to intensive care units. ECCMID 2013, Berlin, Germany; 2013. p. 2054.
60. Vila J, Marcos A, Marco F, et al. In vitro antimicrobial production of beta-lactamases, aminoglycoside-modifying enzymes, and chloramphenicol acetyltransferase by and susceptibility of clinical isolates of Acinetobacter baumannii. *Antimicrob Agents Chemother* 1993;**37**:138–41.
61. Seifert H, Baginski R, Schulze A, Pulverer G. Antimicrobial susceptibility of Acinetobacter species. *Antimicrob Agents Chemother* 1993;**37**:750–3.
62. Dijkshoorn L, Nemec A, Seifert H. An increasing threat in hospitals: multidrug-resistant Acinetobacter baumannii. *Nat Rev Microbiol* 2007;**5**:939–51.
63. Corbella X, Ariza J, Ardanuy C, et al. Efficacy of sulbactam alone and in combination with ampicillin in nosocomial infections caused by multiresistant Acinetobacter baumannii. *J Antimicrob Chemother* 1998;**42**:793–802.
64. Wood GC, Hanes SD, Croce MA, Fabian TC, Boucher BA. Comparison of ampicillin-sulbactam and imipenem-cilastatin for the treatment of acinetobacter ventilator-associated pneumonia. *Clin Infect Dis* 2002;**34**:1425–30.
65. Falagas ME, Kasiakou SK. Colistin: the revival of polymyxins for the management of multidrug-resistant gram-negative bacterial infections. *Clin Infect Dis* 2005;**40**:1333–41.
66. Garnacho-Montero J, Ortiz-Leyba C, Jimenez-Jimenez FJ, et al. Treatment of multidrug-resistant Acinetobacter baumannii ventilator-associated pneumonia (VAP) with intravenous colistin: a comparison with imipenem-susceptible VAP. *Clin Infect Dis* 2003;**36**:1111–8.
67. Dalfino L, Puntillo F, Mosca A, et al. High-dose, extended-interval colistin administration in critically ill patients: is this the right dosing strategy? A preliminary study. *Clin Infect Dis* 2012;**54**:1720–6.
68. Gales AC, Jones RN, Sader HS. Contemporary activity of colistin and polymyxin B against a worldwide collection of Gram-negative pathogens: results from the SENTRY Antimicrobial Surveillance Program (2006–09). *J Antimicrob Chemother* 2011;**66**:2070–4.
69. Vila J, Pachon J. Therapeutic options for Acinetobacter baumannii infections: an update. *Expert Opin Pharmacother* 2012;**13**:2319–36.
70. Lopez-Alvarez B, Martin-Laez R, Farinas MC, Paternina-Vidal B, Garcia-Palomo JD, Vazquez-Barquero A. Multidrug-resistant Acinetobacter baumannii ventriculitis: successful treatment with intraventricular colistin. *Acta Neurochir (Wien)* 2009;**151**:1465–72.
71. Dalgic N, Ceylan Y, Sancar M, et al. Successful treatment of multidrug-resistant Acinetobacter baumannii ventriculitis with intravenous and intraventricular colistin. *Ann Trop Paediatr* 2009;**29**:141–7.
72. Pogue JM, Lee J, Marchaim D, et al. Incidence of and risk factors for colistin-associated nephrotoxicity in a large academic health system. *Clin Infect Dis* 2011;**53**:879–84.
73. Pachon-Ibanez ME, Jimenez-Mejias ME, Pichardo C, Llanos AC, Pachon J. Activity of tigecycline (GAR-936) against Acinetobacter baumannii strains, including those resistant to imipenem. *Antimicrob Agents Chemother* 2004;**48**:4479–81.
74. Betriu C, Rodriguez-Avial I, Gomez M, et al. Antimicrobial activity of tigecycline against clinical isolates from Spanish medical centers. Second multicenter study. *Diagn Microbiol Infect Dis* 2006;**56**:437–44.
75. Morfin-Otero R, Dowzicky MJ. Changes in MIC within a global collection of Acinetobacter baumannii collected as part of the Tigecycline Evaluation and Surveillance Trial, 2004 to 2009. *Clin Ther* 2012;**34**:101–12.

76. Gordon NC, Wareham DW. A review of clinical and microbiological outcomes following treatment of infections involving multidrug-resistant Acinetobacter baumannii with tigecycline. *J Antimicrob Chemother* 2009;**63**:775–80.
77. Chan JD, Graves JA, Dellit TH. Antimicrobial treatment and clinical outcomes of carbapenem-resistant Acinetobacter baumannii ventilator-associated pneumonia. *J Intensive Care Med* 2010;**25**:343–8.
78. Guner R, Hasanoglu I, Keske S, Kalem AK, Tasyaran MA. Outcomes in patients infected with carbapenem-resistant Acinetobacter baumannii and treated with tigecycline alone or in combination therapy. *Infection* 2011;**39**:515–8.
79. Ye JJ, Lin HS, Kuo AJ, et al. The clinical implication and prognostic predictors of tigecycline treatment for pneumonia involving multidrug-resistant Acinetobacter baumannii. *J Infect* 2011;**63**:351–61.
80. Curcio D, Fernandez F, Vergara J, Vazquez W, Luna CM. Late onset ventilator-associated pneumonia due to multidrug-resistant Acinetobacter spp.: experience with tigecycline. *J Chemother* 2009;**21**:58–62.
81. Peleg AY, Potoski BA, Rea R, et al. Acinetobacter baumannii bloodstream infection while receiving tigecycline: a cautionary report. *J Antimicrob Chemother* 2007;**59**:128–31.
82. Reid GE, Grim SA, Aldeza CA, Janda WM, Clark NM. Rapid development of Acinetobacter baumannii resistance to tigecycline. *Pharmacotherapy* 2007;**27**:1198–201.
83. Ray L, Levasseur K, Nicolau DP, Scheetz MH. Cerebral spinal fluid penetration of tigecycline in a patient with Acinetobacter baumannii cerebritis. *Ann Pharmacother* 2010;**44**:582–6.
84. Fishbain J, Peleg AY. Treatment of Acinetobacter infections. *Clin Infect Dis* 2010;**51**: 79–84.
85. Gounden R, Bamford C, van Zyl-Smit R, Cohen K, Maartens G. Safety and effectiveness of colistin compared with tobramycin for multi-drug resistant Acinetobacter baumannii infections. *BMC Infect Dis* 2009;**9**:26.
86. Li J, Rayner CR, Nation RL, et al. Heteroresistance to colistin in multidrug-resistant Acinetobacter baumannii. *Antimicrob Agents Chemother* 2006;**50**:2946–50.
87. Petrosillo N, Ioannidou E, Falagas ME. Colistin monotherapy vs. combination therapy: evidence from microbiological, animal and clinical studies. *Clin Microbiol Infect* 2008;**14**:816–27.
88. Durante-Mangoni E, Signoriello G, Andini R, et al. Colistin and rifampicin compared with colistin alone for the treatment of serious infections due to extensively drug-resistant Acinetobacter baumannii: a multicenter, randomized clinical trial. *Clin Infect Dis* 2013;**57**:349–58.
89. Kalin G, Alp E, Akin A, Coskun R, Doganay M. Comparison of colistin and colistin/sulbactam for the treatment of multidrug resistant Acinetobacter baumannii ventilator-associated pneumonia. *Infection* 2013;**42**:37–42.
90. Siegel JD, Rhinehart E, Jackson M, Chiarello L. Management of multidrug-resistant organisms in health care settings, 2006. *Am J Infect Control* 2007;**35**:S165–93.
91. Simor AE, Lee M, Vearncombe M, et al. An outbreak due to multiresistant *Acinetobacter baumannii* in a burn unit: risk factors for acquisition and management. *Infect Control Hosp Epidemiol* 2002;**23**:261–7.

Chapter 21

Infections Due to NDM-1 Producers

Patrice Nordmann,[1,2,3] Laurent Dortet[1,3] and Laurent Poirel[1,2,3]

[1]*Medical and Molecular Microbiology Unit, Department of Medicine, Faculty of Science, University of Fribourg, Switzerland,* [2]*INSERM U914, South-Paris Medical School, K.-Bicêtre, France,* [3]*Associated National Reference Center for Antibiotic Resistance, K.-Bicêtre, France*

INTRODUCTION

Infections with multidrug-resistant Gram-negative bacteria are of major global concern given limited therapeutic options. Among those bacteria, carbapemase producers in Enterobacteriaceae are playing a growing role. Currently, one of the most clinically significant carbapenemases is the recently described NDM-1 (New Delhi metallo-β-lactamase). The epidemiology of NDM-1, its interspecies dispersion, and recognized implications for patient care and public health deserve our utmost attention.

CASE PRESENTATION

A 67-year-old man was admitted to the emergency unit of a Paris hospital in January 2013. He had been hospitalized for 1 week in India where a urinary catheter was implemented prior to his transfer to Paris. He had a history of Hodgkin's lymphoma treated by six cures of chemotherapy based on bleomycin 1 year prior to his hospitalization. His complaint had started a few days ago, which comprised fever and a tired state. At the patient admission, the urinary catheter was withdrawn, blood cultures were sampled, and the patient was empirically treated with imipenem and amikacin.

Twelve hours after the patient admission, a *Klebsiella pneumoniae* isolate was recovered from blood cultures taken at the admission. It was resistant to all β-lactams including imipenem, meropenem, ertapenem, and doripenem. This isolate was, in addition, resistant to cotrimoxazole, nalidixic acid, and aminoglycosides except gentamicin. The strain was susceptible only to colistin, chloramphenicol, fosfomycin, tigecycline, fluoroquinolones, and

Emerging Infectious Diseases. DOI: http://dx.doi.org/10.1016/B978-0-12-416975-3.00021-2

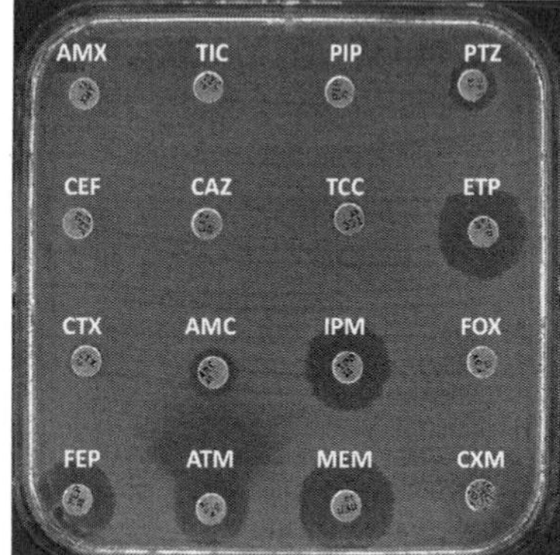

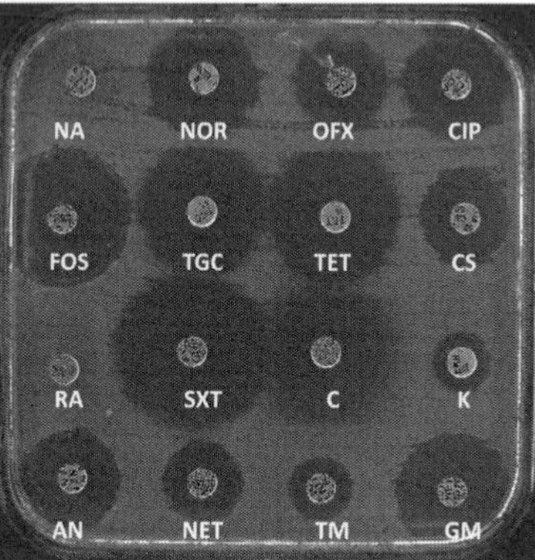

FIGURE 21.1 Disk diffusion antibiogram of the NDM-1-producing *K. pneumoniae* isolate from the reported case. The $bla_{\text{NDM-1}}$ gene was located onto an IncHI plasmid of ≈120 kb co-bearing the $bla_{\text{CTX-M-15}}$, $bla_{\text{TEM-1}}$, $bla_{\text{OXA-9}}$, *qnrS1*, and *aacA4* resistance genes. PTZ, piperacillin + tazobactam; PIP, piperacillin; TIC, ticarcillin; AMX, amoxicillin; ETP, ertapenem; TCC, ticarcillin + clavulanic acid; CAZ, ceftazidime; CF, cefalotin; FOX, cefoxitin; IMP, imipenem; AMC, amoxicillin + clavulanic acid; CTX, cefotaxime; CMX, cefuroxime; MEM, meropenem; ATM, aztreonam; FEP, cefepime; FT, nitrofurantoin; NOR, norfloxacin; OFX, ofloxacin; CIP, ciprofloxacin; FOS, fosfomycin; TGC, tigecycline; TE, tetracycline; CS, colistin; SSS, sulfonamide; SXT, sulfamethoxazole + trimethoprim; C, chloramphenicol; NET, netilmicin; GM, gentamicin; AN, amikacin; TM, tobramycin; RA, rifampicin.

nitrofurantoin. Biochemical and molecular tests identified this *K. pneumoniae* isolate as producing the NDM-1 metallo-β-lactamase. Carbapenemase production by the *K. pneumoniae* isolate was assessed using the biochemical Carba NP test. PCR amplification followed by sequencing of several β-lactamase genes and of plasmid-mediated quinolone resistance *qnr* and of aminoglycoside resistance *aacA4* encoding genes revealed the presence of the β-lactamase genes $bla_{\text{NDM-1}}$ together with $bla_{\text{TEM-1}}$, $bla_{\text{SHV-11}}$, $bla_{\text{CTX-M-15}}$, and $bla_{\text{OXA-9}}$ genes. This *K. pneumoniae* isolate also harbored the *aacA4* gene encoding the AAC(6′)-Ib acetyltransferase that confers high-level resistance to aminoglycosides (except to gentamicin), and the *qnrS1* gene conferring increased MICs to quinolones (Figure 21.1). The $bla_{\text{NDM-1}}$ gene together with $bla_{\text{CTX-M-15}}$, $bla_{\text{TEM-1}}$, $bla_{\text{OXA-9}}$, *qnrS1*, and *aacA4* genes were found to be located on a 120-kb plasmid belonging to the IncHI incompatibility group.

Antibiotic treatment was switched to levofloxacin and gentamicin for 6 days. Since the patient still had fever, antibiotic treatment was changed to fosfomycin, colistin, and gentamicin for 8 additional days. After 28 days of antibiotic treatment, the withdrawn urinary catheter was no longer colonized with the carbapenem-resistant *K. pneumoniae*. The antibiotic treatment failure with levofloxacin had been further explained by the expression of a plasmid-mediated QnrS1 resistance determinant conferring at least *in vitro* increased MICs to quinolones.

1. WHY WAS THIS CASE IDENTIFIED AS EMERGENT?

The multi-drug resistance pattern of the *K. pneumoniae* isolate was quite intriguing. Detailed analysis of the resistance mechanisms identified that isolate as an NDM-1 producer.

One of the most recently and most clinically significant carbapenemases is NDM-1. This carbapenemase belongs to the class B of Ambler classification of β-lactamases that includes the metallo-β-lactamases (MBLs). NDM-1 was first identified in 2008 in a *K. pneumoniae* isolate recovered from a Swedish patient who has been previously hospitalized in New Delhi, India.[1] Since then, NDM carbapenemases have been the focus of worldwide attention due to their rapid dissemination among Enterobacteriaceae, *Acinetobacter* spp., and *Pseudomonas aeruginosa*.[2,3] Within less than 8 years, NDM producers have been identified worldwide with a rapid dissemination from the main reservoir, namely, the Indian subcontinent and a secondary reservoir the Balkan states.[4,5] Therefore, this epidemiological pattern seems to be an emerging phenomenon.

NDM-1, as observed for all MBLs, confers resistance to a broad range of β-lactams including penicillins, cephalosporins, and carbapenems, but spares aztreonam. However, almost all NDM producers are resistant to nearly all antibiotic families including all β-lactams (aztreonam included), all aminoglycosides, fluoroquinolones, nitrofurantoin, and sulfonamides (Figure 21.1). This resistance pattern is explained by the quite systematic association of the $bla_{\text{NDM-1}}$ expression with other antibiotic resistance determinants, such as those encoding clavulanic acid-inhibited expanded-spectrum β-lactamases (ESBLs), AmpC cephalosporinases, other types of carbapenemases (OXA-48-, VIM-, KPC-types), resistance to aminoglycosides (16S RNA methylases), macrolides (esterases), rifampicin (rifampicin-modifying enzymes), quinolones (Qnr proteins), chloramphenicol, and sulfamethoxazole.[6–11]

2. WHAT IS THE CAUSATIVE AGENT?

NDM producers were mainly described in Enterobacteriaceae that have been recovered from many clinical settings. Enterobacteriaceae represents a large family of Gram-negative bacteria that includes, along with many harmless bacteria, many of the more common pathogens, such as *Salmonella*, *Shigella*, *Escherichia coli*, *Yersinia pestis*, *Klebsiella*, *Shigella*, *Proteus*, *Enterobacter*, *Serratia*, and *Citrobacter*. This family is the only representative in the order Enterobacteriales of the class Gammaproteobacteria in the phylum Proteobacteria. Among the NDM-1-producing Enterobacteriaceae, *K. pneumoniae* and *E. coli* are the most often reported species. Those two enterobacterial species are part of the normal gut flora. However, the NDM-1 carbapenemase is also frequently reported from other enterobacterial species that could be transiently found in the gut flora, but mainly in the environmental sample including *Klebsiella oxytoca*, *Enterobacter cloacae*, *Enterobacter aerogenes*, *Citrobacter freundii*, *Proteus mirabilis*, *Serratia marcescens*, and *Providencia* spp. Until now, NDM-1 has rarely been identified by highly virulent enterobacterial isolates such as *Salmonella typhi* or *Shigella* sp.[12–14] In addition, although most of the NDM-producing bacteria are Enterobacteriaceae, this carbapenemase has been also identified in *Acinetobacter* spp., and in rare cases in *P. aeruginosa*.

3. WHAT IS THE FREQUENCY OF THE DISEASE? (PREVALENCE, INCIDENCE, BURDEN, AND IMPACT OF THE DISEASE)

After the initial descriptions of NDM producers, an extended survey was performed in India, Pakistan, Bangladesh, and the UK during 2008–2009, and allowed to identify 180 NDM-1-producing enterobacterial isolates,[15] being mainly recovered from Indian patients (143/180 strains, 79.4%) (Figure 21.2). The obvious link with the Indian subcontinent was then confirmed in most of the following reports on NDM-1-related infections. Accordingly, among the 235 enterobacterial isolates with reduced susceptibility to ertapenem and collected from intra-abdominal infections as part of the SMART study, the most commonly identified carbapenemase was NDM-1 (50%), and all the $bla_{\text{NDM-1}}$-positive isolates were from India.[16] Since then, the high local prevalence of NDM-1 producers was pointed out by several reports from Indian studies.[17–22] A study from a main hospital in Mumbai reported a 5–8% prevalence rate of NDM producers among Enterobacteriaceae.[23,24] This range of prevalence

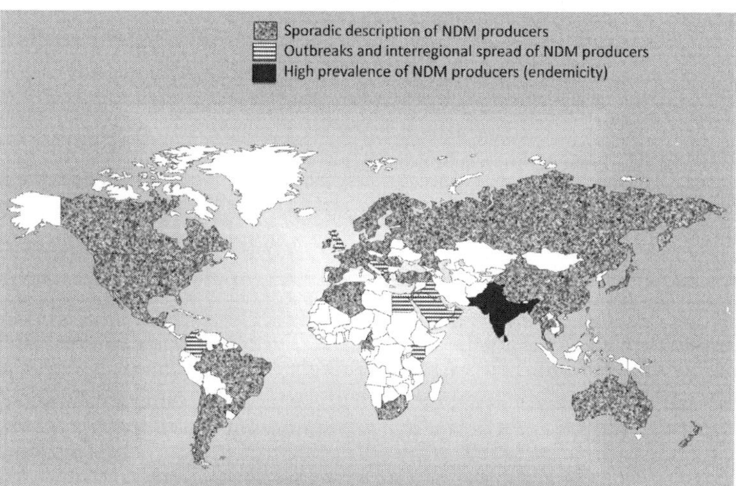

FIGURE 21.2 Geographical spread of NDM-producing Enterobacteriaceae as updated in February 2014.

(6.9%) was then confirmed by investigating 780 consecutive and non-duplicate enterobacterial isolates recovered from hospitalized and outpatients from February to July 2010 in Varanasi, Northern India.[25] Finally, a 7% prevalence rate was observed among 885 Gram-negative bacteria isolated from clinical samples (pus, blood, sputum, fluids) received at the Microbiology Department of Sassoon Hospital (Pune, India) from August to December 2010.[26] Also, during the screening of 200 hospitalized and non-hospitalized patients, a higher prevalence of fecal carriage (18.5%) was reported in two military hospitals in Rawalpindi, Pakistan.[27] In addition, and more surprisingly, the $bla_{\text{NDM-1}}$ gene was detected from drinking water and seepage samples recovered in the New Delhi area, India.[28]

Apart from this recognized Indian main reservoir, several studies pointed out the Balkan states as a potential secondary reservoir for NDM-producing Enterobacteriaceae[4,5,29] (Figure 21.2). More recently, the Middle East has been reported to be another reservoir of NDM producers, with scattered reports identified in Kuwait, Sultanate of Oman, Saudi Arabia, and United Arab Emirates.[30–35] Considering the important population exchange between the Middle East and the Indian subcontinent, this might explain this dissemination of NDM producers in the Middle East region.

Out of these three main geographical reservoirs, NDM-producing Enterobacteriaceae have been identified worldwide (Figure 21.2). As an example, NDM has been reported to be the third most commonly identified carbapenemase (25/341) in France, after OXA-48 (257/341) and KPC (39/341); data collected by the Associated National Reference Center corresponding to the period 2012.[36] In 2013, the number of NDM producers has actually doubled, leading to push it as the second most commonly identified carbapenemase in France (personal data).

K. pneumoniae and *E. coli* are the most often described species harboring bla_{NDM} genes. In hospital settings, NDM-1-producing *K. pneumoniae* are reported quite often. In some cases, those *K. pneumoniae* strains have been the source of nosocomial outbreaks, such as in Italy,[37] the UK,[38] Turkey,[39] Kenya,[40] China,[41] Colombia,[42] Guatemala,[43] Canada,[44] and also in Chicago, IL.[45] Two outbreaks (Turkey and Colombia) occurred in neonatal intensive care units, causing several deaths.[42,46]

In addition, a nosocomial dissemination of NDM-1-producing *E. coli* has been reported in Bulgaria.[46] However, *E. coli* being less ready than *K. pneumoniae* to dissiminate in nosocomial settings, fortunately those situations remain rare.

An interesting report corresponds to a French woman who had lived in India for several years and had never been hospitalized there. At admission for a breast tumor in a French hospital, bacterial cultures from the surface of this tumor grew an NDM-1-producing *E. coli*.[47] She was therefore colonized with that strain, which was then identified by performing a rectal swab screening. It was demonstrated that this NDM-1-producing *E. coli* persisted in the gastrointestinal tract of the patient for over a year, despite the lack of specific antibiotic selective pressure during that period.[48]

More surprisingly, two French community patients without a history of foreign travel developed urinary tract infections (UTIs) caused by NDM-1-producing *K. pneumonia*; those cases therefore being considered as autochtonous.[49,50]

4. WHAT ARE THE TRANSMISSION ROUTES?

The transmission routes of NDM producers are those usually reported for Enterobacteriaceae. They are transmitted mostly by contaminated hands, water, and the environment. An environmental surveillance study conducted in New Delhi detected NDM-1-producing bacteria in 4% ($n = 2/50$) of drinking water samples and 30% ($n = 51/171$) of seepage samples.[28] In addition, environmental contamination with carbapenemase producers (including NDM producers) has already been described out of India.[51,52]

Therefore, the contaminated environment may play a significant role as a source of hospital-acquired infections. NDM transmission has been linked to the endoscopic camera head used for urologic procedures, where camera sheathing was not routinely used, although the camera head was regularly cleaned with detergent wipes.[38] We have identified that enterobacterial isolates producing NDM-1 may persist in the environment for up to at least 3 months (P. Nordmann, personal data).

5. WHICH FACTORS ARE INVOLVED IN DISEASE PATHOGENESIS? WHAT ARE THE PATHOGENIC MECHANISMS?

As stated above, the NDM-1 enzyme may be identified in many Gram-negative species. There is no specific pathogenic factor known to be associated with the NDM-1 gene.[53–55] The enterobacterial isolate itself may be pathogenic. This NDM-1 determinant may be identified in common enterobacterial species but also in spontaneous virulent Enterobacteriaceae such as in *Shigella boydii*.[28]

Genetic investigations and *in vivo* infection models revealed neither any virulence factor associated with the bla_{NDM-1} gene nor an increased mortality caused by the infectious agent *per se*.[8,56–60] Nevertheless, the broad-spectrum hydrolysis profile of NDM-1, including all β-lactams with the exception of aztreonam, complicates the treatment of the infection. NDM-1 compromises the efficacy of carbapenems (ertapenem, imipenem, meropenem, doripenem), which are often last-resort antibiotics. In addition, the systematic genetic association of the bla_{NDM-1} gene to other resistance traits, such as genes encoding resistance to aminoglycosides, tetracycline, chloramphenicol, sulfonamides, rifampicin, but also extended-spectrum β-lactamases (hydrolyzing aztreonam), often leads to multidrug-resistant (MDR) isolates. This quite systematic association between the $bla_{\text{NDM-1}}$ gene and other resistance genes results from two key elements: (1) the $bla_{\text{NDM-1}}$ gene is often located on plasmids which co-harbor other resistance genes, therefore leading to co-associated resistance through plasmid acquisition, and (2) the $bla_{\text{NDM-1}}$-bearing plasmids are often identified in isolates possessing chromosomal determinants co-harboring other resistance plasmids. Those MDR strains, also commonly named "superbugs," may cause lethal infections, not due to a peculiar virulence trait, but as a consequence of lack of efficient antibacterial treatments.

6. WHAT ARE THE CLINICAL MANIFESTATIONS?

Initial case detection of NDM-1 production was characterized from a *K. pneumoniae* isolate from urinary tract infection of a 59-year-old man who returned to Sweden after hospitalization in India in January 2008.[1] This patient was concurrently colonized with an NDM-1-producing *E. coli* in the stool. This first case underlines the infectious and carriage state with NDM-1 producers. Subsequently, two NDM-1-producing *E. coli* strains from 2006 were retrospectively identified in stored clinical isolates from healthcare facilities in New Delhi, India, via the SENTRY Antimicrobial Surveillance Program.[7] Infections caused by NDM producers include urinary tract infections, peritonitis, septicemia, pulmonary infections, soft tissue infections, and device-associated infections. There is no gender preference and most of the cases are adults. The vast majority of infections are urinary tract infections as observed for any enterobacterial infections. Both hospital- and community-acquired infections have been reported. No special clinical manifestations have been associated with NDM producers as compared to wild-type susceptible strains. As observed for other multidrug-resistant bacteria (KPC, OXA-48 producers), it is highly probable that colonization of the gut flora precedes in most of the cases the infection by NDM producers and oro-fecal transmission in the community might occur mostly through hand contamination, food, and water. Indeed, colonization with NDM-1 producers in the gut flora has been extensively reported.[61]

7. HOW DO YOU DIAGNOSE?

Detection of NDM producers requires three consecutive steps: detection of the enterobacterial isolate itself, detection of the carbapenemase production, and finally detection of the $bla_{\text{NDM-1}}$ gene. Two situations may be observed: detection of NDM-1 producers among infected patients and detection of NDM-1 producers among carriers.

7.1 Diagnosis of Infected Patients

Identification of NDM producers relies first on a preliminary analysis of susceptibility testing. The US guidelines (CLSI) (updated in 2013) retained as breakpoints for Enterobacteriaceae susceptibility (S) ≤ 1 and resistance (R) ≥ 4 mg/L for imipenem and meropenem, and S ≤ 0.5 and R ≥ 2 mg/L for ertapenem. The European guidelines (EUCAST) (updated in 2013) are slightly different and propose breakpoints for imipenem and meropenem as follows: susceptible (S) ≤ 2 and resistant (R) ≥ 8 mg/L, and for ertapenem S ≤ 0.5 and R ≥ 1 mg/L. Although some discrepancies might exist for several isolates depending on the reference used to interpret the antibiogram, MIC values of ertapenem are often higher than those of other carbapenems with NDM producers. Consequently, ertapenem seems to be the best molecule for detecting carbapenemase producers, including NDM producers. We believe that, whatever the EUCAST and CLSI guidelines are (which may change overtime), a decreased susceptibility to any carbapenems shall prompt a search for any carbapenemase activity.

Detection methods based on the inhibitory properties of several divalent ion chelators (e.g., EDTA, dipicolinic acid) may identify metallo-β-lactamase (MBL) producers (but not NDM producers specifically). A disk-diffusion test based on the detection of a synergy between a carbapenem-containing disk (imipenem or meropenem) and a disk containing an MBL inhibitor (EDTA or mercaptopropionic acid or dipicolinic acid) has been proposed.[62] A combined disk technique using a carbapenem disk and the same carbapenem disk supplemented with EDTA (10 μL of a 0.1 M solution at pH 8) has been also proposed.[63] Using this test, a 5 mm increase of the inhibition diameter around the disk containing imipenem plus EDTA compared to imipenem alone likely indicates the production of an MBL. However, those two phenotypic methods are time-consuming and false-negative results often arise, in particular when low level of resistance is observed.[63] Among those phenotypic methods, the Etest MBL strip (bioMérieux, La Balmes-les-Grottes, France), a two-sided strip containing gradients of imipenem alone on one side, and imipenem supplemented with EDTA on the other side, is also commonly used for the detection of MBL producers. Using this test, at least three doubling dilutions of the MIC in the presence of EDTA is considered as a positive result.[64] However, several NDM-producing isolates exhibit low MIC of carbapenems, leading to interpretable results using the Etest MBL strip.

Detection of carbapenemase activity may be done by using the modified Hodge test (MHT). The MHT lacks sensitivity (50%) for detecting NDM-1 producers since production of carbapenemase may be low in several NDM-1 producers. Addition of a $ZnSO_4$ (100 μg/ml) supplementation in the culture medium increases the sensitivity of detection to 85.7%. However, this test which is time-consuming (at least 24 h) has a low specificity with *Enterobacter* spp. due to a frequent overexpression of a chromosomal cephalosporinase.

UV spectrophotometry analysis has been developed to detect carbapenem hydrolysis. This method is based on the detection of imipenem absorbance decrease in the presence of a crude extract of bacterial enzymes. This crude extract can be obtained from an overnight culture of the tested strain by mechanical lysis using sonication. This UV spectrophotometry-based technique is not expensive and has a 100% sensitivity and a 98.5% specificity for detecting carbapenemase activity.[65] However, it is time-consuming and requires trained microbiologists and expensive equipment.

Analysis of carbapenem hydrolysis by using the MALDI-TOF technology has been shown to be a useful technique to detect carbapenemase production in a few hours. This technique was based on detection of a carbapenem (imipenem, meropenem, or ertapenem) spectrum and its main derivatives resulting from carbapenem hydrolysis. After 3 to 4 hours of incubation of the isolate to be screened with a carbapenem, the bacteria were pelleted by centrifugation, and the supernatant containing the carbapenem and its metabolites was tested by MALDI-TOF mass spectrophotometry. Disappearance of the peak corresponding to the native carbapenem and appearance of peak(s) corresponding to the metabolite(s) as a result of the carbapenem hydrolysis indicate carbapenemase activity.[66–70] This test has excellent sensitivity and specificity. However, it requires trained microbiologists to implement the technique and fancy laboratory equipment.

The most interesting test is the rapid Carba NP test (Figure 21.3). It is based on the detection of the hydrolysis of imipenem by a color change of a pH indicator. This test is 100% sensitive and 100% specific for the detection of any type of carbapenemase produced by Enterobacteriaceae, including NDM producers.[71–73] A second version of the Carba NP test (the Carba NP test II) has been developed to rapidly differentiate between the diverse carbapenemase types found in Enterobacteriaceae (metallo- or non-metallo-enzymes). The CarbaNP test II combines the inhibition properties of EDTA as a chelating agent with the high efficiency of the Carba NP test for identification of any type of MBL producer, including all NDM producers.[74] The Carba NP test has been evaluated to detect carbapenemase-producing Enterobacteriaceae directly from spiked blood cultures.[75] The proposed strategy allows detection of carbapenemase producers in less than 2 hours and NDM producers in less than 5 hours, with sensitivity and specificity of 100%, respectively[75] (Figure 21.4).

Identification of the NDM-1 producer at the genetic level is currently required mostly for the purpose of hygiene. A number of genotypic approaches have been reported, based on PCR techniques, including real-time PCR methods able to detect bla_{NDM}-positive isolates directly from clinical samples.[76] Those methods, however, have the disadvantages of being time-consuming, being focused on NDM producers being therefore unable to identify any novel carbapenemase gene, and are quite expensive. Commercial DNA microarray methods are marketed and increase the convenience of these tests.[77] Although

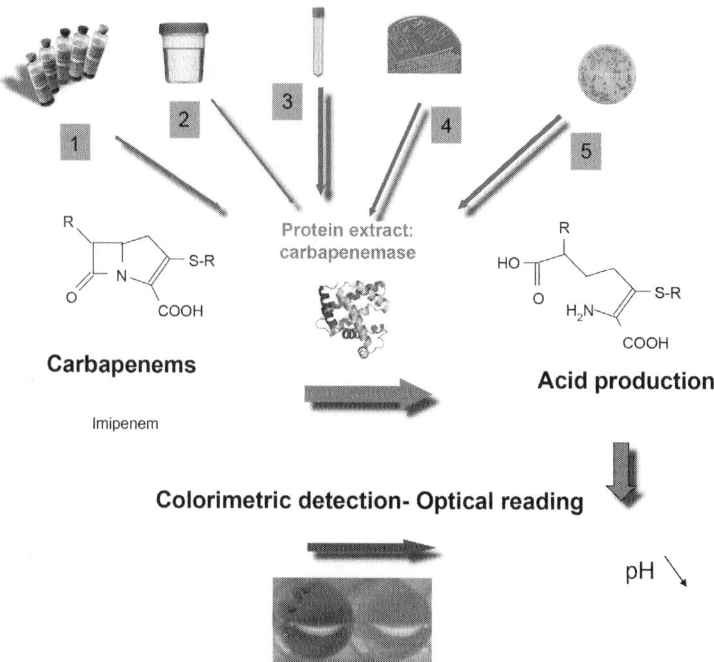

FIGURE 21.3 Principle of the detection of carbapenemase activity using the Carba NP test. The following starting samples may be used: blood cultures (1), urines (2), other clinical samples (3), isolated bacteria (4), or bacteria grown on selective culture media (5).

they cannot overcome general limitations of genotypic techniques, these DNA microarray methods are able to identify the carbapenemase genes, and the main extended-spectrum β-lactamase and acquired cephalosporinases genes in on run, leading to a better understanding of the β-lactamase content of the tested isolate. Finally, molecular amplification of the *bla*$_{NDM}$ gene followed by sequencing makes it possible to identify the exact nature of the NDM variant. It usually required at least an additional period of time of 48 h.

7.2 Identification of Carriers

Since the prevention of dissemination of carbapenemase producers partially relies on an early and accurate detection of carriers, recommendations for the screening of colonized patients have been introduced in several countries. Commonly, "at-risk" patients, meaning those being colonized with carbapenemase producers, are patients transferred from a foreign hospital and those hospitalized in intensive care units, in transplantation units, and immune compromised patients.

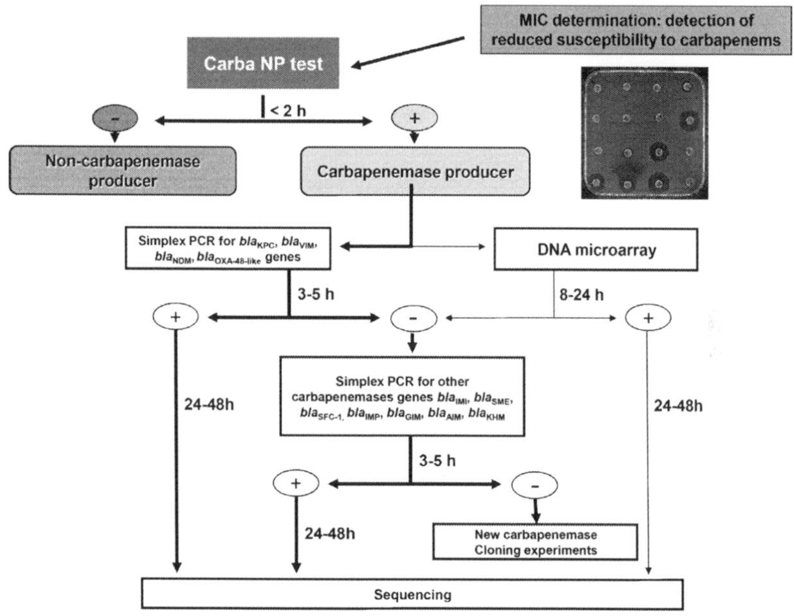

FIGURE 21.4 Diagnostic flow chart for identification of carbapenemase-producing enterobacterial isolate including the NDM-1 producer. The Carba NP test is used for rapid identification of carbapenemase producers. Then, molecular techniques (PCRs or DNA microarray) are required for a precise identification of the carbapenemase gene. Bold arrows indicate the most efficient way for identification of the carbapenemase activity and then carbapenemase genes.

Since the intestinal flora is the main reservoir of Enterobacteriaceae, rectal swabs and stools are the most suitable clinical samples for performing this screening. These specimens may be plated on screening medium, either directly or after an enrichment step in broth containing imipenem 0.5–1 μg/ml or ertapenem 0.5 μg/ml.[78,79] In outbreak situations, this enrichment step might increase the sensitivity of the screening, and consequently reduce the number of potential false-negative results by increasing the inoculum of the targeted strain. However, its disadvantage is the induced delay (12–24 h) needed to confirm or reject carbapenemase detection. Although the efficiency of this enrichment step has not been evaluated for NDM producers, it has already been shown to improve the detection of KPC producers.[78,79]

Regardless of the enrichment step, the specimens have to be plated on selective media. For that purpose, several screening media have been evaluated and compared to the screening of carriers of NDM producers. One of the first tested media was the ChromID ESBL culture medium (bioMérieux) containing cefpodoxime as a selector and which is routinely used to screen ESBL producers. Since NDM enzymes have a broad-spectrum activity, they hydrolyze not only carbapenems, but also expanded-spectrum cephalosporins. Therefore, detection of NDM-producing isolates using ChromID ESBL (aimed do detect ESBL producers) is possible but with a low specificity since the selective agent is a cephalosporin and not a specific carbapenemase substrate (e.g., a carbapenem).

Several media supplemented with a carbapenem have been evaluated for the screening of carbapenemase producers. The first marketed screening medium targeting KPC producers was the CHROMagar KPC medium that contains meropenem (CHROMagar, Paris, France).[80] Using this medium, carbapenem-resistant bacteria are well detected when they exhibit relatively high-level resistance to carbapenems. Its main disadvantage is its lack of sensitivity, since it does not detect carbapenemase producers with low-level carbapenem resistance. Indeed, although NDM producers have often high-level resistance to carbapenems, several isolates exhibited MICs between 0.5 and 1 μg/ml, making their detection difficult on screening media containing high concentrations of carbapenems.[63,81,82]

Colorex KPC (E&O Laboratories, Bonnybridge, UK), another screening medium for carbapenemase producers, also contains meropenem. Since the content of this medium is reported to be identical to that of CHROMagar KPC, only NDM producers with high-level resistance to carbapenems may be detected, leading to an accurate detection of 57 to 64% of NDM-producing Enterobacteriaceae using this medium.[27,83]

A third commercially available screening medium also contains a carbapenem (CRE Brilliance, Thermo Fisher Scientific, UK). According to the study, sensitivities for detection of patients colonized with NDM producers were reported to be 63 to 85% using this medium.[83,84]

Another screening medium also containing a carbapenem is ChromID CARBA (bioMérieux, La Balmes-les-Grottes, France). This commercially available medium has been reported to be more sensitive (87.5 to 94%) than the others for the detection of NDM-producing Enterobacteriaceae.[27,83,84] Recently, ChromID OXA-48 has been developed for identification of carbapenemases of the OXA-48 type but not of the NDM type.

Finally, a home-made screening medium containing ertapenem, cloxacillin, and zinc, namely, the SUPERCARBA medium, has been shown to have excellent sensitivity and specificity for the detection of carbapenemase producers, including NDM producers. The zinc supplementation and the low ertapenem concentration allow the efficient detection of all NDM producers regardless of their level of resistance to carbapenems.[82,85] Consequently, using the SUPERCARBA medium and then performing the Carba NP test on isolated colonies might be proposed as the recommended strategy for screening of carbapenemase producers.[86]

In order to avoid the additional 24 to 48 h before the carriage status of the patient can be established using the above screening media, an in-house quantitative real-time PCR assay using the TaqMan chemistry has been developed to detect the NDM-encoding genes directly from spiked stool samples. The bacterial extraction from stool samples was performed manually or adapted to a fully automated extraction system. This assay was found to be 100% specific and sensitive with detection limits reproducible below 1×10^1 CFU/100 mg of feces.[76] However, this technology remains expensive and is thus considered to be a valuable tool in the follow-up of an outbreak and cohorting of colonized patients.

The proposed strategy for detection of carbapenemase producers at the infection stage or at the carrier level has been proposed recently.[87] This strategy may be implemented in any lab in the world, even those that are not specialized for detection of multidrug-resistant isolates.

8. HOW DO YOU DIFFERENTIATE THIS DISEASE FROM SIMILAR ENTITIES?

As mentioned above, NDM producers are neither more virulent nor the source of different clinical entities as compared to any other enterobacterial isolate whatever their susceptibility is. Differentiation of NDM producers from producers of other types of carbapenemases relies on molecular biology (Figure 21.4). This identification is currently only needed for the purposes of epidemiology rather than for management of the patient or her/his treatment.

9. WHAT IS THE THERAPEUTIC APPROACH?

The NDM producers are multidrug-resistant organisms that cause infections associated with substantial morbidity and mortality. Most of the NDM-1 producers remain susceptible only to two bactericidal antibiotics (colistin and fosfomycin) and a single bacteriostatic antibiotic (tigecycline). *In vitro* synergy combination assays performed with NDM-1 producers with those three antibiotic molecules showed a synergistic activity of colistin and fosfomycin, of colistin and tigecycline in rare cases, while most of the associations remain neutral for most of the tested isolates.[88] Successful treatments have been obtained by using colistin, fosfomycin, or tigecycline for treating infections mostly due to other types of carbapenemase producers.[89–91] The main problem with colistin is its potential nephrotoxicity, and with tigecycline is its lack of activity for treating urinary tract infections and septicemia originating from urinary tract infections. In addition, acquired colistin resistance among producers of other types of carbapenemase has been reported already.[92–94] It is clear that we do not know currently the best therapeutic approach for treating infections due to NDM-1 producers.

Since NDM-1 does not hydrolyze aztreonam, a combination therapy including aztreonam and avibactam (also named NXL-104), a novel serine β-lactamase inhibiting the most frequent broad-spectrum hydrolyzing-β-lactamases hydrolyzing aztreonam, has been suggested as a possible strategy against NDM-1 producers. This therapeutic option has been demonstrated to be an efficient combination therapy *in vitro* but is not available on the market yet.[95,96]

10. WHAT ARE THE PREVENTIVE AND INFECTION CONTROL MEASURES?

The implementation of screening and isolation measures is more effective if the diagnosis of colonization is made early.[97] CurrentCenters for Disease Control and Prevention recommendations for preventing carbapenemase producer transmission in healthcare facilities have been published and mostly written according to experience drawn from the management of KPC outbreaks rather than NDM producers.[98] We believe, however, that those recommendations may also be followed to prevent the spread of NDM producers. Core prevention measures are well supported by evidence and should be utilized by all facilities regardless of the prevalence of carbapenemase producers in the facility or region. These are based on standard precautions as well as contact precautions that apply to any multidrug-resistant bacteria. The intent of contact precautions is to prevent transmission by minimizing the contamination of healthcare personnel who have contact with the patient or the patient's environment. Adherence to contact precautions requires the appropriate use of gowns and gloves by healthcare personnel for all interactions that may involve contact with the patient or the patient's environment. As part of contact precautions, patients should be placed in single-patient rooms, or if not available, then patients with the same carbapenemase producers who are suitable roommates can be placed together in the same room. In addition, non-critical medical equipment or disposable medical items (e.g., blood pressure cuffs, disposable stethoscopes) should be dedicated to individual patient use. Patients colonized or infected with NDM producers who are in short-stay acute care hospitals or long-term hospitalization units should be placed on contact precautions. The use of contact precautions for residents in long-term care settings (e.g., skilled nursing facilities, nursing homes) is more complex and must include consideration of the potential impact of these interventions on their well-being and rehabilitation potential. To facilitate prompt implementation of contact precautions, both acute and long-term care facilities should have systems in place to identify patients with a history of colonization or infection by NDM producers when they are readmitted. In addition to placing NDM producers-colonized or -infected patients in single-patient rooms, acute and long-term care facilities should consider cohorting patients together in the same ward. If feasible, there should be designated staffing to care exclusively for patients with NDM producers to minimize the risk of transmission to other patients.

Use of chlorhexidine bathing has been demonstrated to successfully reduce bloodstream infections and colonization with MRSA and VRE primarily in intensive care unit settings, but its role in reducing carbapenemase transmission is less clear.[99] Similarly, decontamination of the gut flora for carbapenemase producers remains debatable.[100]

11. CONCLUSION

The issue of NDM-1 producers is one of the most important challenges of modern medicine for the following reasons:

- The size of the uncontrolled reservoir, mostly South-East Asia, India, and Pakistan, where policies for antibiotic stewardship and detection of antibiotic-resistant bacteria have not been implemented.
- The rapidity with which NDM-1 producers may spread with their following emergence in clinical settings.
- The spread of NDM-1 producers not only in hospitals but also in the community, in particular *E. coli* in South-East Asia where transmission control remains impossible.
- One of the main vectors is *E. coli*, which remains the most important source of infections for humankind.
- The multiple vectors of the $bla_{\text{NDM-1}}$ gene—plasmids, genomes, bacterial species—which implies that those multiple outbreaks will be difficult to control.
- The multidrug resistance pattern of the NDM-1 producers, which may have a tremendous impact for development of advanced medical procedures such as organ transplantation, bone marrow transplantation, and cancer therapy.
- The possibility of evolution of those multidrug-resistant NDM-1 producers toward pan-drug resistance.

Hope remains that novel marketed antibiotics will soon limit the clinical impact of those resistant bacteria representing currently one of the most serious threats to mankind.

REFERENCES

1. Yong D, Toleman MA, Giske CG, et al. Characterization of a new metallo-beta-lactamase gene, bla(NDM-1), and a novel erythromycin esterase gene carried on a unique genetic structure in Klebsiella pneumoniae sequence type 14 from India. *Antimicrob Agents Chemother* 2009;**53**:5046–54.
2. Leverstein-Van Hall MA, Stuart JC, Voets GM, Versteeg D, Tersmette T, Fluit AC. Global spread of New Delhi metallo-beta-lactamase 1. *Lancet Infect Dis* 2010;**10**:830–1.
3. Nordmann P, Dortet L, Poirel L. Carbapenem resistance in Enterobacteriaceae: here is the storm!. *Trends Mol Med* 2012;**18**:263–72.

4. Gecaj-Gashi A, Hasani A, Bruqi B, Mulliqi-Osmani G. Balkan NDM-1: escape or transplant? *Lancet Infect Dis* 2011;**11**:586.
5. Livermore DM, Walsh TR, Toleman M, Woodford N. Balkan NDM-1: escape or transplant? *Lancet Infect Dis* 2011;**11**:164.
6. Barguigua A, El Otmani F, Lakbakbi El Yaagoubi F, Talmi M, Zerouali K, Timinouni M. First report of a Klebsiella pneumoniae strain coproducing NDM-1, VIM-1 and OXA-48 carbapenemases isolated in Morocco. *APMIS: Acta Path Micro Imm Scand* 2013;**121**:675–7.
7. Castanheira M, Deshpande LM, Mathai D, Bell JM, Jones RN, Mendes RE. Early dissemination of NDM-1- and OXA-181-producing Enterobacteriaceae in Indian hospitals: report from the SENTRY Antimicrobial Surveillance Program, 2006–2007. *Antimicrob Agents Chemother* 2011;**55**:1274–8.
8. Dolejska M, Villa L, Poirel L, Nordmann P, Carattoli A. Complete sequencing of an IncHI1 plasmid encoding the carbapenemase NDM-1, the ArmA 16S RNA methylase and a resistance-nodulation-cell division/multidrug efflux pump. *J Antimicrob Chemother* 2013;**68**:34–9.
9. Poirel L, Dortet L, Bernabeu S, Nordmann P. Genetic features of blaNDM-1-positive Enterobacteriaceae. *Antimicrob Agents Chemother* 2011;**55**:5403–7.
10. Poirel L, Ros A, Carricajo A, et al. Extremely drug-resistant Citrobacter freundii isolate producing NDM-1 and other carbapenemases identified in a patient returning from India. *Antimicrob Agents Chemother* 2011;**55**:447–8.
11. Samuelsen O, Naseer U, Karah N, et al. Identification of Enterobacteriaceae isolates with OXA-48 and coproduction of OXA-181 and NDM-1 in Norway. *J Antimicrob Chemother* 2013;**68**:1682–5.
12. Cabanes F, Lemant J, Picot S, et al. Emergence of Klebsiella pneumoniae and Salmonella metallo-beta-lactamase (NDM-1) producers on reunion island. *J Clin Microbiol* 2012;**50**:3812.
13. Fischer J, Schmoger S, Jahn S, Helmuth R, Guerra B. NDM-1 carbapenemase-producing Salmonella enterica subsp. enterica serovar Corvallis isolated from a wild bird in Germany. *J Antimicrob Chemother* 2013;**68**:2954–6.
14. Savard P, Gopinath R, Zhu W, et al. First NDM-positive Salmonella sp. strain identified in the United States. *Antimicrob Agents Chemother* 2011;**55**:5957–8.
15. Kumarasamy KK, Toleman MA, Walsh TR, et al. Emergence of a new antibiotic resistance mechanism in India, Pakistan, and the UK: a molecular, biological, and epidemiological study. *Lancet Infect Dis* 2010;**10**:597–602.
16. Lascols C, Hackel M, Marshall SH, et al. Increasing prevalence and dissemination of NDM-1 metallo-beta-lactamase in India: data from the SMART study (2009). *J Antimicrob Chemother* 2011;**66**:1992–7.
17. Krishna BV. New Delhi metallo-beta-lactamases: a wake-up call for microbiologists. *Indian J Med Microbiol* 2010;**28**:265–6.
18. Mochon AB, Garner OB, Hindler JA, et al. New Delhi metallo-beta-lactamase (NDM-1)-producing Klebsiella pneumoniae: case report and laboratory detection strategies. *J Clin Microbiol* 2011;**49**:1667–70.
19. Muir A, Weinbren MJ. New Delhi metallo-beta-lactamase: a cautionary tale. *J Hosp Infect* 2010;**75**:239–40.
20. Raghunath D. New metallo beta-lactamase NDM-1. *Indian J Med Res* 2010;**132**:478–81.
21. Roy S, Singh AK, Viswanathan R, Nandy RK, Basu S. Transmission of imipenem resistance determinants during the course of an outbreak of NDM-1 Escherichia coli in a sick newborn care unit. *J Antimicrob Chemother* 2011;**66**:2773–80.

22. Roy S, Viswanathan R, Singh AK, Das P, Basu S. Sepsis in neonates due to imipenem-resistant Klebsiella pneumoniae producing NDM-1 in India. *J Antimicrob Chemother* 2011;**66**:1411–3.
23. Deshpande P, Rodrigues C, Shetty A, Kapadia F, Hedge A, Soman R. New Delhi Metallo-beta lactamase (NDM-1) in Enterobacteriaceae: treatment options with carbapenems compromised. *J Assoc Phys India* 2010;**58**:147–9.
24. Deshpande P, Shetty A, Kapadia F, Hedge A, Soman R, Rodrigues C. New Delhi metallo 1: have carbapenems met their doom? *Clin Infect Dis* 2010;**51**:1222.
25. Seema K, Ranjan Sen M, Upadhyay S, Bhattacharjee A. Dissemination of the New Delhi metallo-beta-lactamase-1 (NDM-1) among Enterobacteriaceae in a tertiary referral hospital in north India. *J Antimicrob Chemother* 2011;**66**:1646–7.
26. Bharadwaj R, Joshi S, Dohe V, Gaikwad V, Kulkarni G, Shouche Y. Prevalence of New Delhi metallo-beta-lactamase (NDM-1)-positive bacteria in a tertiary care centre in Pune, India. *Int J Antimicrob Agents* 2012;**39**:265–6.
27. Perry JD, Naqvi SH, Mirza IA, et al. Prevalence of faecal carriage of Enterobacteriaceae with NDM-1 carbapenemase at military hospitals in Pakistan, and evaluation of two chromogenic media. *J Antimicrob Chemother* 2011;**66**:2288–94.
28. Walsh TR, Weeks J, Livermore DM, Toleman MA. Dissemination of NDM-1 positive bacteria in the New Delhi environment and its implications for human health: an environmental point prevalence study. *Lancet Infect Dis* 2011;**11**:355–62.
29. Struelens MJ, Monnet DL, Magiorakos AP, Santos O'Connor F, Giesecke JEuropean NDMSP. New Delhi metallo-beta-lactamase 1-producing Enterobacteriaceae: emergence and response in Europe. *Euro Surveill* 2010;15.
30. Zowawi HM, Balkhy HH, Walsh TR, Paterson DL. beta-Lactamase production in key gram-negative pathogen isolates from the Arabian Peninsula. *Clin Microbiol Rev* 2013;**26**:361–80.
31. Sonnevend A, Al Baloushi A, Ghazawi A, et al. Emergence and spread of NDM-1 producer Enterobacteriaceae with contribution of IncX3 plasmids in the United Arab Emirates. *J Med Microbiol* 2013;**62**:1044–50.
32. Poirel L, Al Maskari Z, Al Rashdi F, Bernabeu S, Nordmann P. NDM-1-producing Klebsiella pneumoniae isolated in the Sultanate of Oman. *J Antimicrob Chemother* 2011;**66**:304–6.
33. Jamal W, Rotimi VO, Albert MJ, Khodakhast F, Udo EE, Poirel L. Emergence of nosocomial New Delhi metallo-beta-lactamase-1 (NDM-1)-producing Klebsiella pneumoniae in patients admitted to a tertiary care hospital in Kuwait. *Int J Antimicrob Agents* 2012;**39**:183–4.
34. Dortet L, Poirel L, Al Yaqoubi F, Nordmann P. NDM-1, OXA-48 and OXA-181 carbapenemase-producing Enterobacteriaceae in Sultanate of Oman. *Clin Microbiol Infect* 2012;**18**:E144–8.
35. Shibl A, Al-Agamy M, Memish Z, Senok A, Khader SA, Assiri A. The emergence of OXA-48- and NDM-1-positive Klebsiella pneumoniae in Riyadh, Saudi Arabia. *Int J Infect Dis* 2013;**17**:e1130–3.
36. Dortet L, Cuzon G, Nordmann P. Dissemination of carbapenemase-producing Enterobacteriaceae in France, 2012. *J Antimicrob Chemother* 2013.
37. Gaibani P, Ambretti S, Berlingeri A, et al. Outbreak of NDM-1-producing Enterobacteriaceae in northern Italy, July to August 2011. *Euro Surveill* 2011;**16**:20027.
38. Koo VS, O'Neill P, Elves A. Multidrug-resistant NDM-1 Klebsiella outbreak and infection control in endoscopic urology. *BJU Int* 2012;**110**:E922–6.

39. Poirel L, Ozdamar M, Ocampo-Sosa AA, Turkoglu S, Ozer UG, Nordmann P. NDM-1-producing Klebsiella pneumoniae now in Turkey. *Antimicrob Agents Chemother* 2012;**56**:2784–5.
40. Poirel L, Revathi G, Bernabeu S, Nordmann P. Detection of NDM-1-producing Klebsiella pneumoniae in Kenya. *Antimicrob Agents Chemother* 2011;**55**:934–6.
41. Wang X, Xu X, Li Z, et al. An outbreak of a nosocomial NDM-1-producing Klebsiella pneumoniae ST147 at a teaching hospital in mainland China. *Microb Drug Resist* 2013.
42. Escobar Perez JA, Olarte Escobar NM, Castro-Cardozo B, et al. Outbreak of NDM-1-producing Klebsiella pneumoniae in a neonatal unit in Colombia. *Antimicrob Agents Chemother* 2013;**57**:1957–60.
43. Pasteran F, Albornoz E, Faccone D, et al. Emergence of NDM-1-producing Klebsiella pneumoniae in Guatemala. *J Antimicrob Chemother* 2012;**67**:1795–7.
44. Lowe CF, Kus JV, Salt N, et al. Nosocomial transmission of New Delhi metallo-beta-lactamase-1-producing Klebsiella pneumoniae in Toronto, Canada. *Infect Control Hosp Epidemiol* 2013;**34**:49–55.
45. Centers for Disease Control and Prevention. Notes from the field: New Delhi metallo-β-lactamase-producing Escherichia coli associated with endoscopic retrograde cholangio-pancreatography—Illinois, 2013. *MMWR Morb Mortal Wkly Rep* 2014;**62**:1051.
46. Poirel L, Savov E, Nazli A, Trifinova I, Todovora I, Gergova I, et al. Outbreak caused by NDM-1- and RmtB-producing Escherichia coli in Bulgaria. *Antimicrob Agents Chemother* 2014. in press.
47. Poirel L, Hombrouck-Alet C, Freneaux C, Bernabeu S, Nordmann P. Global spread of New Delhi metallo-beta-lactamase 1. *Lancet Infect Dis* 2010;**10**:832.
48. Poirel L, Herve V, Hombrouck-Alet C, Nordmann P. Long-term carriage of NDM-1-producing Escherichia coli. *J Antimicrob Chemother* 2011;**66**:2185–6.
49. Arpin C, Noury P, Boraud D, et al. NDM-1-producing Klebsiella pneumoniae resistant to colistin in a French community patient without history of foreign travel. *Antimicrob Agents Chemother* 2012;**56**:3432–4.
50. Nordmann P, Couard JP, Sansot D, Poirel L. Emergence of an autochthonous and community-acquired NDM-1-producing Klebsiella pneumoniae in Europe. *Clin Infect Dis* 2012;**54**:150–1.
51. Isozumi R, Yoshimatsu K, Yamashiro T, et al. bla(NDM-1)-positive Klebsiella pneumoniae from environment, Vietnam. *Emerg Infect Dis* 2012;**18**:1383–5.
52. Lerner A, Adler A, Abu-Hanna J, Meitus I, Navon-Venezia S, Carmeli Y. Environmental contamination by carbapenem-resistant Enterobacteriaceae. *J Clin Microbiol* 2013;**51**:177–81.
53. Fuursted K, Scholer L, Hansen F, et al. Virulence of a Klebsiella pneumoniae strain carrying the New Delhi metallo-beta-lactamase-1 (NDM-1). *Microb Infect/Institut Pasteur* 2012;**14**:155–8.
54. Peirano G, Mulvey GL, Armstrong GD, Pitout JD. Virulence potential and adherence properties of Escherichia coli that produce CTX-M and NDM beta-lactamases. *J Med Microbiol* 2013;**62**:525–30.
55. Peirano G, Schreckenberger PC, Pitout JD. Characteristics of NDM-1-producing Escherichia coli isolates that belong to the successful and virulent clone ST131. *Antimicrob Agents Chemother* 2011;**55**:2986–8.
56. Bonnin RA, Poirel L, Carattoli A, Nordmann P. Characterization of an IncFII plasmid encoding NDM-1 from Escherichia coli ST131. *PLoS One* 2012;**7**:e34752.

57. Hishinuma A, Yoshida A, Suzuki H, Okuzumi K, Ishida T. Complete sequencing of an IncFII NDM-1 plasmid in Klebsiella pneumoniae shows structural features shared with other multidrug resistance plasmids. *J Antimicrob Chemother* 2013;**68**:2415–7.
58. McGann P, Hang J, Clifford RJ, et al. Complete sequence of a novel 178-kilobase plasmid carrying bla(NDM-1) in a Providencia stuartii strain isolated in Afghanistan. *Antimicrob Agents Chemother* 2012;**56**:1673–9.
59. Sekizuka T, Matsui M, Yamane K, et al. Complete sequencing of the bla(NDM-1)-positive IncA/C plasmid from Escherichia coli ST38 isolate suggests a possible origin from plant pathogens. *PLoS One* 2011;**6**:e25334.
60. Yamamoto T, Takano T, Fusegawa T, et al. Electron microscopic structures, serum resistance, and plasmid restructuring of New Delhi metallo-beta-lactamase-1 (NDM-1)-producing ST42 Klebsiella pneumoniae emerging in Japan. *J Infect Chemother* 2013;**19**:118–27.
61. Bushnell G, Mitrani-Gold F, Mundy LM. Emergence of New Delhi metallo-beta-lactamase type 1-producing enterobacteriaceae and non-enterobacteriaceae: global case detection and bacterial surveillance. *Int J Infect Dis* 2013;**17**:e325–33.
62. Arakawa Y, Shibata N, Shibayama K, et al. Convenient test for screening metallo-beta-lactamase-producing gram-negative bacteria by using thiol compounds. *J Clin Microbiol* 2000;**38**:40–3.
63. Nordmann P, Poirel L, Carrer A, Toleman MA, Walsh TR. How to detect NDM-1 producers. *J Clin Microbiol* 2011;**49**:718–21.
64. Walsh TR, Bolmstrom A, Qwarnstrom A, Gales A. Evaluation of a new Etest for detecting metallo-beta-lactamases in routine clinical testing. *J Clin Microbiol* 2002;**40**:2755–9.
65. Bernabeu S, Poirel L, Nordmann P. Spectrophotometry-based detection of carbapenemase producers among Enterobacteriaceae. *Diag Microbiol Infect Dis* 2012;**74**:88–90.
66. Burckhardt I, Zimmermann S. Using matrix-assisted laser desorption ionization-time of flight mass spectrometry to detect carbapenem resistance within 1 to 2.5 hours. *J Clin Microbiol* 2011;**49**:3321–4.
67. Hrabak J, Chudackova E, Walkova R. Matrix-assisted laser desorption ionization-time of flight (maldi-tof) mass spectrometry for detection of antibiotic resistance mechanisms: from research to routine diagnosis. *Clin Microbiol Rev* 2013;**26**:103–14.
68. Hrabak J, Studentova V, Walkova R, et al. Detection of NDM-1, VIM-1, KPC, OXA-48, and OXA-162 carbapenemases by matrix-assisted laser desorption ionization-time of flight mass spectrometry. *J Clin Microbiol* 2012;**50**:2441–3.
69. Hrabak J, Walkova R, Studentova V, Chudackova E, Bergerova T. Carbapenemase activity detection by matrix-assisted laser desorption ionization-time of flight mass spectrometry. *J Clin Microbiol* 2011;**49**:3222–7.
70. Kempf M, Bakour S, Flaudrops C, et al. Rapid detection of carbapenem resistance in Acinetobacter baumannii using matrix-assisted laser desorption ionization-time of flight mass spectrometry. *PLoS One* 2012;**7**:e31676.
71. Dortet L, Bréchard L, Poirel L, Nordmann P. Comparison of diverse growing media for further detection of carbapenemase production using the Carba NP test. *J Med Microbiol* 2013.
72. Nordmann P, Poirel L, Dortet L. Rapid detection of carbapenemase-producing Enterobacteriaceae. *Emerg Infect Dis* 2012;**18**:1503–7.
73. Vasoo S, Cunningham SA, Kohner PC, et al. Comparison of a novel, rapid chromogenic biochemical assay, the Carba NP test with the modified hodge test for detection of carbapenemase producing Gram-negative bacilli. *J Clin Microbiol* 2013;**51**:3097–101.

74. Dortet L, Poirel L, Nordmann P. Rapid identification of carbapenemase types in Enterobacteriaceae and Pseudomonas spp. by using a biochemical test. *Antimicrob Agents Chemother* 2012;**56**:6437–40.
75. Dortet L, Brechard L, Poirel L, Nordmann P. Rapid detection of carbapenemase-producing enterobacteriaceae from blood cultures. *Clin Microbiol Infect* 2013.
76. Naas T, Ergani A, Carrer A, Nordmann P. Real-time PCR for detection of NDM-1 carbapenemase genes from spiked stool samples. *Antimicrob Agents Chemother* 2011;**55**:4038–43.
77. Cuzon G, Naas T, Bogaerts P, Glupczynski Y, Nordmann P. Evaluation of a DNA microarray for the rapid detection of extended-spectrum beta-lactamases (TEM, SHV and CTX-M), plasmid-mediated cephalosporinases (CMY-2-like, DHA, FOX, ACC-1, ACT/MIR and CMY-1-like/MOX) and carbapenemases (KPC, OXA-48, VIM, IMP and NDM). *J Antimicrob Chemother* 2012;**67**:1865–9.
78. Adler A, Navon-Venezia S, Moran-Gilad J, Marcos E, Schwartz D, Carmeli Y. Laboratory and clinical evaluation of screening agar plates for detection of carbapenem-resistant Enterobacteriaceae from surveillance rectal swabs. *J Clin Microbiol* 2011;**49**:2239–42.
79. Landman D, Salvani JK, Bratu S, Quale J. Evaluation of techniques for detection of carbapenem-resistant Klebsiella pneumoniae in stool surveillance cultures. *J Clin Microbiol* 2005;**43**:5639–41.
80. Moran Gilad J, Carmeli Y, Schwartz D, Navon-Venezia S. Laboratory evaluation of the CHROMagar KPC medium for identification of carbapenem-nonsusceptible Enterobacteriaceae. *Diag Microbiol Infect Dis* 2011;**70**:565–7.
81. Girlich D, Poirel L, Nordmann P. Value of the modified Hodge test for detection of emerging carbapenemases in Enterobacteriaceae. *J Clin Microbiol* 2012;**50**:477–9.
82. Girlich D, Poirel L, Nordmann P. Comparison of the SUPERCARBA, CHROMagar KPC, and Brilliance CRE screening media for detection of Enterobacteriaceae with reduced susceptibility to carbapenems. *Diag Microbiol Infect Dis* 2013;**75**:214–7.
83. Wilkinson KM, Winstanley TG, Lanyon C, Cummings SP, Raza MW, Perry JD. Comparison of four chromogenic culture media for carbapenemase-producing Enterobacteriaceae. *J Clin Microbiol* 2012;**50**:3102–4.
84. Day KM, Salman M, Kazi B, et al. Prevalence of NDM-1 carbapenemase in patients with diarrhoea in Pakistan and evaluation of two chromogenic culture media. *J Appl Microbiol* 2013;**114**:1810–6.
85. Nordmann P, Girlich D, Poirel L. Detection of carbapenemase producers in Enterobacteriaceae by use of a novel screening medium. *J Clin Microbiol* 2012;**50**:2761–6.
86. Nordmann P, Poirel L. Strategies for identification of carbapenemase-producing Enterobacteriaceae. *J Antimicrob Chemother* 2013;**68**:487–9.
87. Dortet L, Brechard L, Cuzon G, Poirel L, Nordmann P. Strategy for rapid detection of carbapenemase-producing Enterobacteriaceae. *Antimicrob Agents Chemother* 2014. in press.
88. Bercot B, Poirel L, Dortet L, Nordmann P. In vitro evaluation of antibiotic synergy for NDM-1-producing Enterobacteriaceae. *J Antimicrob Chemother* 2011;**66**:2295–7.
89. Cobo J, Morosini MI, Pintado V, et al. Use of tigecycline for the treatment of prolonged bacteremia due to a multiresistant VIM-1 and SHV-12 beta-lactamase-producing Klebsiella pneumoniae epidemic clone. *Diag Microbiol Infect Dis* 2008;**60**:319–22.
90. Humphries RM, Kelesidis T, Dien Bard J, Ward KW, Bhattacharya D, Lewinski MA. Successful treatment of pan-resistant Klebsiella pneumoniae pneumonia and bacteraemia with a combination of high-dose tigecycline and colistin. *J Med Microbiol* 2010;**59**:1383–6.
91. Pournaras S, Vrioni G, Neou E, et al. Activity of tigecycline alone and in combination with colistin and meropenem against Klebsiella pneumoniae carbapenemase

(KPC)-producing Enterobacteriaceae strains by time-kill assay. *Int J Antimicrob Agents* 2011;**37**:244–7.

92. Bogdanovich T, Adams-Haduch JM, Tian GB, et al. Colistin-resistant, Klebsiella pneumoniae carbapenemase (KPC)-producing Klebsiella pneumoniae belonging to the international epidemic clone ST258. *Clin Infect Dis* 2011;**53**:373–6.
93. Cannatelli A, D'Andrea MM, Giani T, et al. In vivo emergence of colistin resistance in Klebsiella pneumoniae producing KPC-type carbapenemases mediated by insertional inactivation of the PhoQ/PhoP mgrB regulator. *Antimicrob Agents Chemother* 2013;**57**:5521–6.
94. Kontopoulou K, Protonotariou E, Vasilakos K, et al. Hospital outbreak caused by Klebsiella pneumoniae producing KPC-2 beta-lactamase resistant to colistin. *J Hosp Infect* 2010;**76**:70–3.
95. Livermore DM, Mushtaq S, Warner M, et al. Activities of NXL104 combinations with ceftazidime and aztreonam against carbapenemase-producing Enterobacteriaceae. *Antimicrob Agents Chemother* 2011;**55**:390–4.
96. Shakil S, Azhar EI, Tabrez S, et al. New Delhi metallo-beta-lactamase (NDM-1): an update. *J Chemother* 2011;**23**:263–5.
97. Palmore TN, Henderson DK. Managing transmission of carbapenem-resistant enterobacteriaceae in healthcare settings: a view from the trenches. *Clin Infect Dis* 2013;**57**:1593–9.
98. Centers for Disease Control and Prevention. Guidance for control of infections with carbapenem-resistant or carbapenemase-producing Enterobacteriaceae in acute care facilities. *MMWR Morb Mortal Wkly Rep* 2009;**58**:256–60.
99. Naparstek L, Carmeli Y, Chmelnitsky I, Banin E, Navon-Venezia S. Reduced susceptibility to chlorhexidine among extremely-drug-resistant strains of Klebsiella pneumoniae. *J Hosp Infect* 2012;**81**:15–9.
100. Oren I, Sprecher H, Finkelstein R, et al. Eradication of carbapenem-resistant Enterobacteriaceae gastrointestinal colonization with nonabsorbable oral antibiotic treatment: a prospective controlled trial. *Am J Infect Control* 2013;**41**:1167–72.

Chapter 22

The *Exserohilum rostratum* Incident: The Compounding Pharmacy as a Source of Emerging Infections

Larry Lutwick
Division of Infectious Diseases, Departments of Medicine and Biomedical Sciences, Western Michigan University Homer Stryker MD School of Medicine, Kalamazoo, MI, USA

CASE PRESENTATION[1]

A 51-year-old woman was seen in an emergency department 1 week after a cervical epidural steroid injection, complaining of new occipital headache. She had not had cervical injections before, was not on any long-term medication, and had no history of any immunocompromising condition. No findings were found on examination, a non-contrast head CT was normal, no lumbar puncture was performed, and she was sent home.

The following day, she returned and was hospitalized after complaining of nausea, vertigo, ataxia, and diplopia. The examination again was essentially normal, routine blood tests were unremarkable, brain magnetic resonance imaging (MRI) was normal, and there was no fever. A repeat MRI was repeated 2 days later when she developed slurred speech, right hemiparesis, a left facial droop and anisocoria, and showed a pontine diffusion abnormality. Cerebrospinal fluid (CSF) exam revealed a glucose of 36 mg/dL (serum 105), total protein 153 mg/dL, cell count 850/mm^3 with 84% neutrophils, and Gram-negative stain and subsequent bacterial culture. Treated broadly for bacterial and HSV infection, she deteriorated, required endotracheal intubation on day 4 and was transferred to a tertiary care facility. There, an MRI revealed pontile, midbrain, and cerebellar changes and diffuse meningeal enhancement. Repeat CSF exam was not significantly changed and CSF testing for a variety of viruses and cryptococcal and histoplasmal antigens were all negative.

Emerging Infectious Diseases. DOI: http://dx.doi.org/10.1016/B978-0-12-416975-3.00022-4

Progressive decreased diffusion defects were found on repeat MRI with evidence of brainstem infarction and ventriculomegaly. An externalized ventricular drain was placed. MRI of the neck showed inflammation and possible fluid collection at the injection site. On day 9 of illness, she no longer had papillary, corneal, and gag reflexes and amphotericin B was added empirically. The following day, death from neurologic criteria was pronounced. A mold, subsequently identified as *Exserohilum* species, was isolated from her CSF on the day of death.

Autopsy revealed a grossly necrotic brainstem with microscopic evidence of angioinvasion of fungal hyphae. Hemorrhagic infarctions in the brain and spinal cord were also found.

1. WHY THIS CASE WAS SIGNIFICANTLY IMPORTANT AS AN EMERGING INFECTION

Several cases of fungal meningitis in patients who had received epidural injections of a corticosteroid were reported to the Centers for Disease Control and Prevention (CDC) by the Tennessee Department of Health in September 2012. By early October 2012, there were 11 such cases in two states and within a month, 404 cases in 19 states with 29 deaths. Although the initial report of this outbreak was due to *Aspergillus*, almost all other cases were due to a mold rarely implicated in human infection, *Exserohilum rostratum*.

The injected corticosteroid was implicated and contamination was traced to three lots of a preservative-free methylprednisolone produced at the New England Compounding Center in Framingham, MA. By the time the affected lots were recalled on October 8, 2012, more than 13,000 individuals had been injected with the drug from these lots during the previous 4 months.[2] A rapid response was needed. As commented on by Feldmesser,[3] "the high mortality associated with this multistate outbreak necessitated very rapid mobilization of the CDC with the organization of committees to define case definitions, optimal approaches to therapy, and the establishment of research priorities. The response on the part of public health officials was nothing short of heroic."

Subsequently, other manifestations related to the tainted steroid injections were described including paravertebral and epidural abscesses, sacroilitis, and peripheral joint infections. The outbreak slowed over time but by October 2013, there had been 751 total cases with 64 deaths involving 20 states.[4] Furthermore, clinicians began to notice relapses of *Exserohilum* CNS infection after recommended treatment.[5]

Although not the first time that a compounding pharmacy-produced drug was implicated as a vehicle for human infection, this unprecedented outbreak has underscored the emerging issue of infections of various etiologies from pharmaceuticals manufactured by compounding pharmacies.

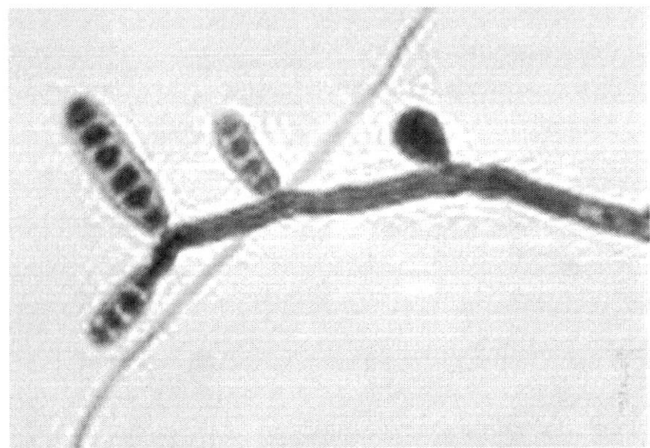

FIGURE 22.1 *E. rostratum* conidia. *From CDC (public domain).*

2. WHAT IS THE CAUSATIVE AGENT?

In this outbreak, the organism that was implicated in a majority of the cases where an etiology was proven was *Exserohilum rostratum* (Figure 22.1). The genus *Exserohilum* can be lumped into the phaeohyphomycoses, meaning "fungi with dark hyphae" diseases. Under the umbrella of phaeohyphomycosis are more than 100 fungal species in 60 genera over several orders,[6] all of which contain melanin. They are also referred to as dematiaceous fungi. Three of the 35 species of *Exserohilum* have been reported as causes of human disease (species *rostratum*, *longirostratum*, and *mcginnisi*). *E. rostratum* is not uncommonly isolated from soil and marine environments and recognized as a phytopathogen, particularly for some grasses, causing leaf spot and crown and root rot. Indeed, it has been investigated for the biocontrol of weeds.[7]

The unusual circumstance of fungal contamination of an immunosuppressive medication injected into the human body allowed *E. rostratum* to be a crossover fungal pathogen; here, one that can cause disease in plants and humans. Of the perhaps 5 million fungal species, about 270,000 can cause plant diseases and 325 are known to infect humans.[8] Many of these crossover fungi are dematiaceous soil saphrophytes that are quite weak human pathogens, causing infection in the immunocompromised or in those who have sustained penetrating superficial or deep trauma; the latter includes this iatrogenic outbreak.

A previous, small cluster was reported in 2002.[9] In that cluster of five patients, the mold isolated was *Exophilala dermatitidis*. More recently, in the wake of the *E. rostratum* fiasco, two other small fungal-associated clusters were reported. In one, eight patients developed *Bipolaris hawaiiensis* (another dematiaceous fungus) endophthalmitis following bevacizumab (an antivascular endothelial growth factor, anti-VEGF)/triamcinolone (a different corticosteroid) compounded medication injection.[10] In the other,[11] subcutaneous abscesses occurred in at least 24 individuals in four states with mixed fungal and bacterial isolates linked to contaminated compounded methylprednisolone intramuscular injections.

3. WHAT IS THE FREQUENCY OF THE DISEASE?

Staes and colleagues[12] reviewed iatrogenic infections related to compounding pharmacies over 13 years ending in 2012. Prior to the *E. rostratum* incident, they noted 11 outbreaks, involving 207 patients and 17 deaths. Bacteria, primarily Gram-negative bacilli, were involved in a majority of them linked to intravenous and intravitreal injections. Among such bacterial events are: *Pseudomonas fluorescens* bacteremia from heparinized saline flushes;[13] *Sphingomonas paucimobilis* bacteremia from intravenous fentanyl;[14] several *Serrratia marcescens* clusters—meningitis, epidural abscess and joint infections from bethamethasone,[15] bacteremia from prefilled heparin and saline syringes[16] or from intravenous magnesium solution,[17] and endophthalmitis from intravitreal bevacizumab;[18] alpha-hemolytic streptococcal endophthalmitis from bevacizumab;[19] and mixed *Pseudomonas aeruginosa*/*Burkholderia cepacia* endophthalmitis from a Trypan blue ophthalmic solution.[20]

Even viral infections have been implicated with hepatitis C virus being transmitted by intravenous injection of a tainted radiopharmaceutical.[21]

Usually in these outbreaks, the infected organism is tracked back to a compounding pharmacy product manufactured in both inadequate sterility and poor quality control and the unopened product was found "smoking," that is, the same organism was isolated or identified from the product, analogous to a "smoking gun."

The scope of each outbreak depends on the size of the area of distribution of the given compounded pharmaceutical. In the *E. rostratum* outbreak, the contaminated lots were distributed to 20 states, but only to certain medical facilities in each state and depending on the facility, the steroid might be injected into the epidural space or into a joint cavity.

4. HOW ARE THE FUNGI TRANSMITTED?

The commonality in all the outbreaks is injection of the material into a usual sterile site, whether it be the bloodstream, the vitreous humor of the eye, the epidural space, the joint cavity, or muscle.

5. WHICH FACTORS ARE INVOLVED IN DISEASE PATHOGENESIS? WHAT ARE THE PATHOGENIC MECHANISMS?

As in the *Exserohilum* outbreak described, many microorganisms involved with compounding pharmacy-related infections are uncommon causes of human disease, uncommon pathogens of so-called low virulence potential. As pointed out by Casadevall and Pirofski,[22] virulence is derived from the Latin *virulentus* meaning "full of poison" and they are quick to point out that virulence has been used to characterize the ability of a microbe to cause disease (literally, to deliver poison), but examining the organism alone is not enough. Rather, however, virulence depends upon the presence of a susceptible host and the host–microbe interaction.[22,23] Therefore, a virulent pathogen in one host may not be virulent in a different species or in a host given immunity by vaccination. This has been proposed as the "damage-response framework"[24] and has been used to further the understanding why infection occurs in clinical practice.[25,26] The increasing number of individuals with factors impacting on this framework such as immunosuppression from HIV infection, from treatment of malignancies or transplanted tissue rejection prevention, and from the presence of implanted devices such as joints, heart valves, and pacemakers continues to affect the host–organism relationship.

In the *E. rostratum* outbreak, this "non-pathogen" became highly virulent. The inoculation of a fungus into a site with limited immunological responsiveness allowed the organism to produce progressive necrotizing fatal disease in the central nervous system (CNS), helped by the co-administered steroid. When inoculated into a joint space, the manifestations were slower to occur and easier to control with antifungal intervention. The lots were also contaminated to some degree with other microorganisms, both fungal and bacterial, but few of the cases were caused by other organisms for which the damage-response framework produced a different risk of disease. Indeed, some of the other fungi were unable to grow at body temperature as the thermophilic *E. rostratum* could.

The pathogenic factors of *E. rostratum* are not clearly established. Some microorganisms clearly have such factors including the antiphagocytic capsules in *Cryptococcus neoformans* and *Neisseria meningitidis* or the toxins in *Vibrio cholerae* and *Clostridium tetani*. In part, the ability of *E. rostratum* to be vasculotropic, that is, to invade arteries causing thrombosis and necrosis, may be important in the CNS presentations. Similar tropisms exist in aspergillosis and zygomycosis, fungi well known to cause human disease especially in immunocompromised hosts, but also in a third, an oomycetous organism called *Pythium insidiosum*. Pythiosis[27] is an emerging, life-threatening human infection found primarily in Thailand that can cause large vessel arteritis requiring amputation, especially in those with hemoglobinopathies.

6. WHAT ARE THE CLINICAL MANIFESTATIONS?

The CDC case definitions for the *E. rostratum* outbreak[28] are as follows.

6.1 Probable Case

A person who received a preservative-free methylprednisolone acetate injection that definitely or likely came from one of the following three lots produced by the New England Compounding Center, and subsequently developed any of the following:

- Meningitis (clinically diagnosed meningitis with one or more of the following symptoms: headache, fever, stiff neck, or photophobia, in addition to a CSF profile showing pleocytosis regardless of glucose or protein levels) of unknown etiology following epidural or paraspinal injection;
- Posterior circulation stroke without an embolic source and without documentation of a normal CSF profile, following epidural or paraspinal injection;
- Osteomyelitis, abscess, or other infection (e.g., soft tissue infection) of unknown etiology, in the spinal or paraspinal structures at or near the site of injection following epidural or paraspinal injection; or
- Osteomyelitis or worsening inflammatory arthritis of a peripheral joint of unknown etiology diagnosed following joint injection.

6.2 Confirmed Case

A probable case with evidence of a fungal pathogen associated with the clinical syndrome.

As of October 23, 2013, there were 751 total infected individuals,[4] 233 with meningitis only, 151 with meningitis and spinal or paraspinal infection, seven with stroke without lumbar puncture, 325 with only spinal or paraspinal infection, 33 with only peripheral joint infection, and two with peripheral joint infection and spinal or paraspinal infection. There had been 64 deaths. Of the total, five of the 20 affected states comprised almost 82% (Michigan 264, Tennessee 153, Indiana 93, Virginia 54, New Jersey 51) of cases.

7. HOW DO YOU DIAGNOSE?

7.1 Imaging Studies

The case report illustrates that multiple imaging studies can be required to demonstrate the progressive abnormalities of *Exserohilum rostratum* meningitis.

The majority of cases in this outbreak, especially later in the outbreak, are localized spinal or paraspinal infections such as epidural abscess, discitis, vertebral osteomyelitis, or other complications at or near the injection site. These infections may occur on their own or in patients previously diagnosed with fungal meningitis. Although patients with these infections frequently have new or worsening back pain, symptoms may be mild and difficult to distinguish from the baseline chronic pain. Cases have occurred in patients with pain at or near baseline. Based on current information, CDC[29] recommends that with patients with new or worsening symptoms at or near the injection site, an MRI with contrast should be obtained. A low threshold must exist for repeat MRI studies in patients who continue to have symptoms localizing to the site of injection, even after a normal study. With patients being treated for fungal meningitis who had no previous evidence of localized infection at the site of injection, even in the absence of new or worsening symptoms at or near the injection site, an MRI of the injection site should be done about 2–3 weeks after diagnosis of meningitis.

7.2 Culture

Since fungal hyphae are primarily in the CNS tissues or joint synovium and in lower amounts in the CSF or joint fluid, the yield of fluid fungal culture is low. Fluid volume cultured should be large, a minimum of 10 mL. Synovial tissue exam will have a higher yield in fungal joint infection and the tissue can be examined with staining to visualize the hyphae (Figure 22.2). Tissue obtained from vertebral biopsy or epidural/paraspinal collections should be cultured and examined histopathologically.

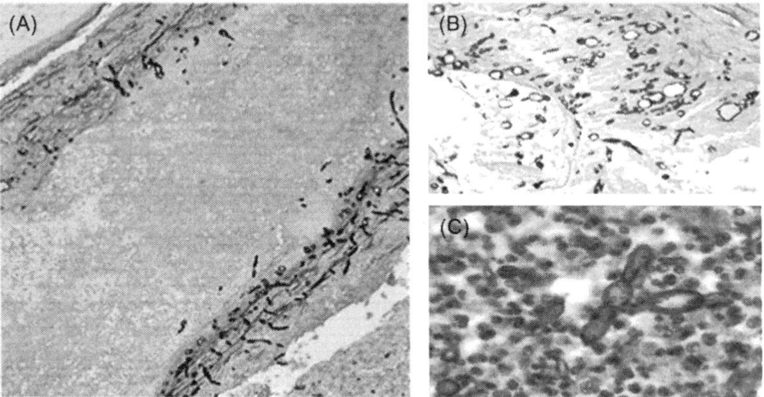

FIGURE 22.2 Fungal hyphae can be visualized with silver stain within vessel walls (A) and in area of necrosis in basilar artery (B). Using a polyfungal immunohistochemistry reagent, fungal hyphae is also seen in the purulent exudate in spinal meningitis (C). *From CDC (public domain).*

7.3 PCR

CDC has developed a research test using polymerase chain reaction (PCR) and DNA sequencing to detect fungal DNA in CSF, other body fluids, and tissues from patients from the outbreak. The test[30] detected *E. rostratum* DNA in 123 samples from 114 patients (28% of the 413 patients for whom samples were available). More sensitive than culture, of 139 patients who had a specimen tested by PCR and culture, *E. rostratum* DNA was detected by PCR in 41 (29%) but was only recovered from culture in 14%. In total, 33% of patients had fungi detected in specimens by either culture or PCR. Clearly, a negative result from culture and PCR did not rule out the infection. There were no false-positive PCR results (100% specificity) among 136 specimens from patients who did not meet the case definition.

7.4 Fungal Antigens

CSF (1,3)-β-D-glucan testing (Fungitell, Beacon Laboratories) has been reported positive in three of five outbreak patients.[31] More anecdotally, a serum galactomannan assay may have a role in the diagnosis of *Exserohilum* infection.[32] The case was disseminated *E. rostratum* infection in an immunosuppressed host, and not part of the tainted corticosteroid outbreak.

8. HOW DO YOU DIFFERENTIATE THIS DISEASE FROM SIMILAR ENTITIES?

In the investigation of compounding pharmacy-associated outbreaks, the first step is to identify the link. Suspicion should be aroused by the reports of, usually, unusual bacterial or fungal isolates in individuals who had received an injection or infusion of a medication into a sterile site such as the bloodstream or epidural space. Such reports are made to governmental agencies like the state health department or CDC that might receive other such reports. As with single suspected cases of foodborne illness, an isolated case without others or without the "smoking" pharmaceutical can be much more difficult to discern.

9. WHAT IS THE THERAPEUTIC APPROACH?

9.1 CNS[33]

The CDC suggested therapy is voriconazole, initially at 6 mg/kg every 12 hours. Serum trough levels should be done on the fifth day of treatment, and the dose adjusted, aiming for a trough level range of 2 to 5 μg/mL. Trough levels >5 μg/mL should be avoided because of the risk of neurotoxicity and other drug-related adverse events. Levels should be done weekly for the initial 4–6 weeks of treatment and when dose adjustments are made to maintain trough levels within the range of 2 to 5 μg/mL. Usually, therapy is started intravenously (IV). Changing from IV to oral therapy is done only after a patient is clinically stable or improving.

Patients with mild disease may be started on oral voriconazole at 6 mg/kg every 12 hours on an empty stomach, monitoring trough levels as with IV therapy. A slightly higher dose may be needed and longer time to achieve optimal levels may be noted if gastrointestinal intolerance or poor absorption is encountered. For toxicity and drug–drug interactions, the reader should consult the referenced citation.

Liposomal amphotericin B in addition to voriconazole may be used in those with severe disease, in those who do not improve or deteriorate, or who manifest new sites of disease activity on voriconazole. Liposomal amphotericin B may also be considered as an alternative to voriconazole in those who are unable to tolerate voriconazole. This drug should be given at a dose of 5 to 6 mg/kg IV daily. The liposomal preparation of amphotericin B is preferred over other lipid formulations because of better CNS penetration. Higher doses (7.5 mg/kg IV daily) may be considered for patients who are not improving, recognizing the potential for increased nephrotoxicity on this dose.

9.2 Bone and Joint[34]

For discitis, vertebral osteomyelitis, or epidural abscess, voriconazole at a dose of 6 mg/kg every 12 hours is recommended. For osteoarticular infections that do not involve the spine, voriconazole therapy can start with a loading dose of 6 mg/kg every 12 hours for two doses, followed by 4 mg/kg every 12 hours. Drug levels should be monitored as in CNS infection. In more severe osteoarticular infection, clinical instability, discitis, vertebral osteomyelitis, or epidural abscess, voriconazole IV is used. Transition from IV to oral voriconazole is done only after a patient is clinically stable or improving.

Patients with mild osteoarticular infection not involving the spine who are able to take oral voriconazole as prescribed and who are able to be monitored closely may be started on oral voriconazole with a loading dose of 6 mg/kg every 12 hours for two doses, followed by 4 mg/kg every 12 hours, monitoring trough levels as above and dose adjustment as necessary. The target range for serum voriconazole trough levels (2 to 5 μg/mL) is readily achievable using the oral drug.

A lipid formulation of amphotericin B at a dose of 5 mg/kg IV daily can also be considered in addition to voriconazole in patients with severe osteoarticular infection and/or patients with clinical instability.

9.3 Other Treatment Considerations

Surgical interventions may well be required in both the CNS and bone/joint manifestations of *E. rostratum*.

The length of therapy is not well defined and duration of therapy will likely vary substantially depending upon individual patient circumstances and disease severity. In general, patients will likely need a minimum of 3–6 months of antifungal treatment. Patients with more severe CNS disease including complications such as stroke, those with persistent CSF abnormalities, and those with underlying immunosuppression will likely need to continue antifungal treatment for 6 months to 1 year. Follow-up monitoring after completion of therapy is important to detect potential relapse of infection.

10. WHAT ARE THE PREVENTIVE AND INFECTION CONTROL MEASURES?

The hallmark of prevention of this group of emerging infections is breaking the chains. Not the chains of infection transmission *per se*, but rather the chains that prevent the US Food and Food Administration (FDA) from adequate regulation of the compounding pharmacy industry.

As stated by Outterson,[35] the FDA rules are often forged in crisis. Indeed, the Food, Drug, and Cosmetic Act was only adopted by the US Congress in the wake of another pharmaceutical fiasco. That episode resulted in the deaths of more than 100 adults and children related to a pharmaceutical company's production of an elixir of sulfanilamide (a sulfonamide antimicrobial) using diethylene glycol.[36] Subsequently, in response in part to the world thalidomide tragedy, 1962 legislation was added to ensure proof of drug efficacy and safety through appropriate and adequate controlled investigations. Approval of pharmaceuticals for use by the FDA requires review of the quality, efficacy, and safety of the drug prior to marketing, production of the drug under federal "Good Manufacturing Practice" (GMP), regulation of drug labeling for safe prescribing and use, measurement by FDA of a benefit/risk analysis, and adherence to GMP sterility requirements.[37]

Because of the lack, in some cases, of a specific form such as an oral suspension, ointment, or a preservative-free formulation of a drug, compounding pharmacies were called on to produce these customized medications. Prior to the changes in FDA regulations in the 1960s, which led to the dispensing of FDA-approved medications, it has been estimated that pharmacists compounded about 80% of prescriptions. In the 1980s, a growing call for these non-FDA approved formulations produced an upswing in the number of compounding pharmacies, which are regulated by state rules and regulations and not the FDA. Guharoy et al.[38] cite data suggesting that as many as 7500 pharmacies now exist in the USA that specialize in compounding and as many as 3% of total prescriptions on the USA are compounded ones.

As chronicled by Outterson,[35] the FDA had struggled to regulate "industrial scale" compounding for several decades and a compounding statute was to amend the law with a new section, 503A. Part of the section was to ban advertising and promotion of these compounded drugs since in its traditional form, compounding pharmacies respond to individual prescriptions so advertising was not needed. Eventually, the US Supreme Court, based on First Amendment issues, ruled that advertising was constitutional. New regulations were issued without advertising and interstate shipping sections, but requiring prescriptions and better manufacturing and testing procedures. Still, pharmacies, including compounding pharmacies, are under the control of state Boards of Pharmacy, and compliance with GMP is not assured.[38] Indeed, in addition to issues of sterility already discussed, Gunderman et al.[39] cite issues of deliberate dilution of about 98,000 prescriptions including oncology medications to increase profit margin, and substandard or super standard potencies ranging from 0 to 450% of target dose including deaths from compounded colchicine, which was almost 10 times the labeled concentration.

Not only are substantially improved FDA regulations and oversight of these compounding pharmacies vitally needed for sterility and standardization of dosing, but prescribing clinicians need to be very aware that use, particularly inappropriate use, of such compounded non-FDA approved drugs can produce the possibility of invalidation of malpractice insurance, personal liability, and even criminal prosecution.[37]

REFERENCES

1. Lyons JL, Gireesh ED, Trivedi JB, et al. Fatal Exserohilum meningitis and central nervous system vasculitis after cervical epidural methylprednisolone injection. *Ann Intern Med* 2012;**157**:835–6.
2. Kainer MA, Reagan DR, Nguyen DB, et al. Tennessee fungal meningitis investigative team: fungal infections associated with contaminated methyl prednisolone in Tennessee. *N Engl J Med* 2012;**367**:2194–203.
3. Feldmesser M. Fungal disease following contaminated steroid injections. Exserohilum is ready for its close-up. *Am J Path* 2013;**183**:661–4.
4. <http://www.cdc.gov/hai/outbreaks/meningitis-map-large.html>; 2013 [accessed 14.04.13].
5. Smith RM. Relapse of fungal meningitis associated with contaminated methylprednisolone. *N Engl J Med* 2013;**368**:26.
6. Revankar SG, Sutton DA. Melanized fungi in human disease. *Clin Microbiol Rev* 2010;**23**:884–928.
7. Chandramohan S, Charudattan R. Control of seven grasses with a mixture of three fungal pathogens with restricted host ranges. *Biol Control* 2001;**22**:246–55.
8. Gauthier GM, Keller NP. Crossover fungal pathogens: the biology and pathogenesis of fungi capable of crossing kingdoms to infect plants and humans. *Fungal Genet Biol* 2013;**61**:146–57.
9. CDC. Exophiala infection from contaminated injectable steroids prepared by a compounding pharmacy—United States, July—November 2002. *Morbid Mortal Week Rep* 2002; **51**:1109–12.
10. Sheyman AT, Cohen BZ, Friedman AH, Ackert JM. An outbreak of fungal endophthalmitis after intravitreal injection of compounded bevicizumab and triamcinolone. *JAMA Ophthalmol* 2013;**13**:864–9.

11. ProMED-mail. Fungal infection, contaminated drug—USA (11): new outbreak, request for information. ProMED-Mail 2013;7 Jun:20130607.176736.
12. Staes C, Jacobs J, Mayer J, Allen J. Description of outbreaks of health-care-associated infections related to compounding pharmacies, 2000–2012. *Am J Health Syst Pharm* 2013;**70**:1301–12.
13. Gershman MD, Kennedy DJ, Noble-Wang J, et al. Multistate outbreak of pseudomonas fluorescens bloodstream infection after exposure to contaminated heparinized saline flush prepared by a compounding pharmacy. *Clin Infect Dis* 2008;**47**:1372–9.
14. Maragakis LL, Chaiwarith R, Srinivasan A, et al. Sphingomonas paucimobilis bloodstream infections associated with contaminated fentanyl. *Emerg Infect Dis* 2009;**15**:12–18.
15. Civen R, Vugia DJ, Alexander R, et al. Outbreak of Serratia marcescens infections following injection of betamethasome compounded at a community pharmacy. *Clin Infect Dis* 2006;**43**:831–7.
16. Blossom D, Noble-Wang J, Su J, et al. Multistate outbreak of Serratia marcescens bloodstream infections caused by contamination of prefilled heparin and isotonic sodium chloride solution syringes. *Arch Intern Med* 2009;**169**:1705–11.
17. Sunenshine RH, Tan ET, Terashita DM, et al. A multistate outbreak of Serratia marcescens bloodstream infection associated with contaminated intravenous magnesium sulfate from a compounded pharmacy. *Clin Infect Dis* 2007;**45**:527–33.
18. Lee SH, Woo SJ, Park KH, et al. Serratia marcescens endophthalitis associated with intravitreal injections of bevacizumab. *Eye* 2010;**24**:226–32.
19. Goldberg RA, Flynn Jr HW, Isom RF, Miller D, Gonzalez S. An outbreak of Streptococcus endophthalmitis after intravitreal injection of bevacizumab. *Am J Ophthalmol* 2012;**153**:204–8.
20. Sunenshine R, Schultz M, Lawrence MG, et al. An outbreak of postoperative Gram-negative bacterial endophthalmitis associated with contaminated Trypan Blue ophthalmic solution. *Clin Infect Dis* 2009;**48**:1580–3.
21. Patel PR, Larson AK, Castel AD, et al. Hepatitis C virus infections from a contaminated radiopharmaceutical used in myocardial perfusion studies. *JAMA* 2006;**296**:2005–11.
22. Casadevall A, Profski L. Host-pathogen interactions: the attributes of virulence. *J Infect Dis* 2001;**184**:337–44.
23. Casadevall A, Pirofski L. Microbial virulence results from the interaction between host and microorganism. *Trends Microbiol* 2003;**11**:157–8.
24. Casadevall A, Pirofski L. Host-pathogen interactions: redefining the basic concepts of virulence and pathogenicity. *Infect Immun* 1999;**67**:3703–13.
25. Pirofski L, Casadevall A. The meaning of microbial exposure, infection, colonization, and disease in clinical practice. *Lancet Infect Dis* 2002;**2**:628–35.
26. Pirofski L, Casadevall A. Q&A: what is a pathogen? A question that begs the point. *BMC Biol* 2012;**10**:6.
27. Krajaejun T, Sathapatayavongs B, Pracharktam R, et al. Clinical and epidemiological analyses of human pythiosis in Thailand. *Clin Infect Dis* 2006;**43**:569–76.
28. <http://www.cdc.gov/hai/outbreaks/clinicians/casedef_multistate_outbreak.html>; 2013 [accessed 14.04.13].
29. <http://www.cdc.gov/hai/outbreaks/clinicians/diagnosis.html>; [accessed 14.04.13].
30. Gade L, Scheelite CM, Pham CD, et al. Detection of fungal DNA in human body fluids and tissues during a multistage outbreak of fungal meningitis and other infections. *Eukaryotic Cell* 2013;**12**:677–83.

31. Lyons JL, Roos KL, Marr KA, et al. Cerebrospinal fluid (1,3)-β-D-glucan detection as an aid for diagnosis of iatrogenic fungal meningitis. *J Clin Microbiol* 2013;**51**:1285–7.
32. Korem M, Polacheck I, Michael-Gayego A, Strahilevitz J. Galactomannan testing for early diagnosis of exserohilum rostratum infection. *J Clin Microbiol* 2013;**51**:2800–1.
33. <http://www.cdc.gov/hai/outbreaks/clinicians/guidance_cns.html>; [accessed 14.04.13].
34. <http://www.cdc.gov/hai/outbreaks/clinicians/treatment-joints.html>; [accessed 14.04.13].
35. Outterson K. Regulating compounding pharmacies after NECC. *N Engl J Med* 2012;**367**:1969–72.
36. Ballentine C. Taste of raspberries, taste of death; the 1937 elixir sulfanilamide incident. *FDA Consumer* 1981;**1**.
37. Sellers S, Utian WH. Pharmacy compounding primer for physicians. Prescriber beware. *Drugs* 2012;**72**:2043–50.
38. Guharoy R, Noviasky J, Haydar Z, Fakih MG, Hartman C. Compounding pharmacy conundrum. "We cannot live without them but we cannot love with them" according to the present paradigm. *Chest* 2013;**143**:896–900.
39. Gunderman J, Jozwiakowski M, Chollet J, Randell M. Potential risks of pharmacy compounding. *Drugs R D* 2013;**13**:1–8.

Chapter 23

Mucormycosis

M. Bulent Ertugrul[1] and Sevtap Arikan-Akdagli[2]

[1]*Adnan Menderes University Medical School, Department of Infectious Diseases and Clinical Microbiology, Aydin, Turkey,* [2]*Hacettepe University Medical School, Department of Medical Microbiology, Ankara, Turkey*

CASE PRESENTATION

A 65-year-old female patient was admitted to the emergency service complaining of nausea, vomiting, headache, and hypoesthesia affecting the left side of her face. She had been suffering from ulcerative colitis for 19 years and was taking methyl prednisolone (16 mg/day) and mesalazine (1000 mg/day).

On admission, her white blood cell (WBC) count was 28,600/mm^3, hemoglobin 12.6 g/L, platelet count 407,000/mm^3, and glucose concentration 316 mg/dL. Cranial computerized tomography was normal while magnetic resonance imaging yielded a soft tissue density at the level of ethmoid sinuses. Thereupon the patient was admitted to the neurology clinic. On the same day, she developed facial paralysis. As the laboratory assessment showed diabetic ketoacidosis, insulin infusion was initiated. Following insulin treatment, glucose concentration and ketonuria were under control. In addition to edema affecting the left periorbital site and left side of the face, left abducens nerve paralysis was added to the clinical features. She had oral mucosal aphthous lesions. Computerized tomography of paranasal sinuses showed soft tissue density at the level of the left ethmoid, maxillary, and sphenoid sinuses. Tissue samples of the left ethmoid and maxillary sinuses taken by the endoscopic method revealed mold hyphae. Tissue samples were cultured and yielded *Rhizopus* spp. This finding confirmed the diagnosis of rhino cerebral mucormycosis. Methyl prednisolone and mesalazine treatments were terminated and amphotericin B deoxycholate (0.5 mg/kg/day) was started. She underwent endoscopic sinus surgery and surgical debridement of ethmoid, sphenoid, and maxillary sinuses was performed. Subsequently, left optic, oculomotor, and trochlear nerve paralysis were observed. Amphotericin B deoxycholate dose was switched to 1 mg/kg/day.

Emerging Infectious Diseases. DOI: http://dx.doi.org/10.1016/B978-0-12-416975-3.00023-6

On the sixth day, she experienced paralysis of 8th, 9th, 10th, 11th, and 12th cranial nerves. At the same time, she complained of blurred vision in the left eye. When cranial MR was repeated, bone destruction affecting the left ethmoid, sphenoid, and maxillary sinuses and contrast enhancing lesion with mass affect were observed. She immediately underwent surgical debridement but unfortunately died following the operation.

(This is a published case report, *Klimik Journal*, 2011.[1])

1. WHY THIS CASE WAS SIGNIFICANTLY IMPORTANT AS AN EMERGING INFECTION

The increase in the incidence of cancer and the progress in intensive care treatment, including organ transplantations, advanced myelosuppressive cancer treatments, and liberal use of corticosteroids, have resulted in more and more immunocompromised patients being vulnerable to invasive fungal infections.[2] Aspergillosis and candidosis still remain the most prevalent opportunistic infections in such patients, but diseases caused by fungi classified in the order Mucorales have become increasingly important.[2] Infections due to Mucorales are more aggressive, producing acute onset, rapidly progressive, and commonly fatal angioinvasive fungal infection in immunocompromised patients.[3,4] The mortality rate varied with the site of infection and host: 96% of patients with disseminated infections, 85% of those with gastrointestinal infections, and 76% with pulmonary infections died.[5]

2. WHAT IS THE CAUSATIVE AGENT?

(See line 20 in the Case Presentation.)

According to the revised taxonomy, fungi causing mucormycosis are classified in the (new) phylum Glomeromycota, class Glomeromycetes, subphylum Mucoromycotina, order Mucorales.[6–8] The genera of *Rhizopus*, *Mucor*, *Lichtheimia* (formerly *Absidia*), *Cunninghamella*, *Rhizomucor*, *Apophysomyces*, and *Saksenaea* constitute those that are identified as causative agents of the majority of cases of mucormycosis.[9] Approximately half of all mucormycosis cases are caused by *Rhizopus* spp.[10] *Lichtheimia corymbifera* is the second and *Mucor* spp. are the third most common mucoralean fungi shown to be responsible for development of mucormycosis.[3] Other, less common, genera, including *Cunninghamella*, *Apophysomyces*, and *Saksenaea*, are increasingly being isolated and, for the most part, cause similar clinical syndromes.

The fungi classified in the order Mucorales are mainly saprophytic rapidly growing, and able to grow at temperatures higher than 37°C (except for *Mucor* spp.).[11] These molds are supposedly ubiquitous in nature and widely found on organic substrates, including bread, decaying fruits, vegetable matter, crop debris, soil between growing seasons, compost piles, and animal excreta.[12] The existence of non-septate (or pauciseptate) hyphae is typical and responsible for the rapid growth as well as the fragile structure.[11] While the terms mucormycosis, phycomycosis, and zygomycosis have so far been used to refer to the diseases caused by the order Mucorales, the recently revised and accepted term remains as mucormycosis.[5,7]

3. WHAT IS THE FREQUENCY OF THE DISEASE?

Cases with mucormycosis have been reported from all over the world.[5] Active population-based surveillance in San Francisco, USA, during 1992–1993 revealed that the incidence of mucormycosis was 1.7 cases per million persons per year.[13] A more recent study in a more general population in Spain found a lower incidence (0.43 cases/1 million inhabitants, or 0.62/100,000 hospital admissions).[14] A report of 391 cases by the GIMEMA Infection Program demonstrated that 11.5% (*n*:45) of patients with hematologic malignancies were infected by Mucorales.[15] A multicenter study in organ transplant recipients showed that mucormycosis accounted for 5.7% of all opportunistic mold infections in these patients.[16] Autopsy rates, the "gold standard" approach, have been in continuous decline globally during the last decades. Nevertheless, mucormycosis remains an uncommon disease, even in high-risk patients, and represents 8.3–13% of all fungal infections encountered at autopsy in such patients.[5] Postmortem prevalence evaluation shows that mucormycosis is 10–50-fold less frequent than candidiasis or aspergillosis with a frequency of one to five cases per 10,000 autopsies.[5]

The epidemiology of mucormycosis seems to be different between developed and developing countries.[2] In developed countries, the incidence of mucormycosis in patients with hematological malignancies has increased during the last decade, probably due to the more severe and prolonged postchemotherapy neutropenia.[17–20] Kontoyiannis et al.[21] noted that the incidence of mucormycosis increased from eight cases per 100,000 admissions in 1989–1993 to 20 cases per 100,000 admissions in 1994–1998 at the M.D. Anderson Cancer Center. Bitar et al. showed an increasing incidence from 0.7/million in 1997 to 1.2/million in 2006 in France.[17] However, in developing countries, the number of mucormycosis cases seems to be on the increase, occurring commonly in patients with uncontrolled diabetes mellitus.[2]

4. WHAT ARE THE TRANSMISSION ROUTES?

These molds can be widely recovered from environmental habitats such as soil and dust.[4] Sporangiospores released by Mucorales range from 3 to 11 μm in diameter, are easily aerosolized, and are readily dispersed throughout the environment.[12] This is the major mode of transmission. The portals of entry of Mucorales are usually the respiratory tract, the skin, and, less frequently, the gut.[22]

5. WHICH FACTORS ARE INVOLVED IN DISEASE PATHOGENESIS? WHAT ARE THE PATHOGENIC MECHANISMS?

(See lines 3–5, 11–13 in the Case Presentation.)

In immunocompetent hosts, both mononuclear and polymorphonuclear phagocytes kill Mucorales by the generation of oxidative metabolites and the cationic peptides, defensins.[23] Chamilos et al. showed that exposure of neutrophils to *Rhizopus oryzae* hyphae results in up-regulation in Toll-like receptor 2 expression and in a robust pro-inflammatory gene expression with rapid induction of NF-κB pathway-related genes.[24] Phagocytic cell-mediated inhibition of germination of spores is impaired by corticosteroids and diabetes mellitus. Therefore, deficiencies of circulating neutrophils (e.g., neutropenia) and impaired phagocyte function (as in diabetes mellitus or steroid therapy) are risk factors for this infection.[4] In the presence of hyperglycemia and low pH, which is found in patients with diabetic ketoacidosis, phagocytes are dysfunctional and have impaired chemotaxis and defective intracellular killing by both oxidative and non-oxidative mechanisms.[24]

Most persons who develop mucormycosis are immunocompromised or diabetic patients, although 15–20% of patients have no evidence of any underlying condition at the time of diagnosis.[10] Risk factors associated with mucormycosis include prolonged neutropenia and use of corticosteroids, hematological malignancies (leukemia, lymphoma, and multiple myeloma), aplastic anemia, myelodysplastic syndromes, solid organ or hematopoietic stem cell transplantation, human immunodeficiency virus infection, diabetic and metabolic acidosis, iron overload, deferoxamine use, burns, wounds, malnutrition, extremes of age, i.e., prematurity or advanced age, and intravenous drug abuse.[25] In the last decade, investigators suggested that widespread use of voriconazole (prophylactic or therapeutic) in patients with hematological malignancies or hematopoietic stem cell recipients might decrease the incidence of invasive aspergillosis while it may lead to an increase in the incidence of mucormycosis.[26,27] Kontoyiannis et al.[21] statistically showed that the increased incidence of mucormycosis reflects the increased and prolonged use of oral voriconazole. Lamaris et al. demonstrated that Mucorales gains increased and transient virulence in two infectious models when exposed to voriconazole.[28] In contrast, a recent study reported that the increase was not related to voriconazole use but most probably to an increase in high-risk patients, particularly those with underlying hematologic malignancies.[20]

6. WHAT ARE THE CLINICAL MANIFESTATIONS?

(See lines 1–15 in the Case Presentation.)

Based on its clinical presentation and anatomic site, invasive mucormycosis is classified as one of the following six major clinical forms: (1) rhinocerebral, (2) pulmonary, (3) cutaneous, (4) gastrointestinal, (5) disseminated, and (6) uncommon rare forms, such as endocarditis, osteomyelitis, peritonitis, and renal infection.[3,5,25] Pulmonary mucormycosis was most common in patients with hematological malignancies and rhinocerebral disease in patients with diabetes mellitus.[9]

6.1 Rhinocerebral Mucormycosis

Rhinocerebral mucormycosis is the most frequently encountered form of the disease and occurs in nearly half of all mucormycosis cases reported in the literature.[3,10,25] This form is seen particularly in diabetic patients.[9] The infection develops after inhalation of fungal sporangiospores into the paranasal sinuses. Upon germination, the invading fungus may spread inferiorly to invade the palate, posteriorly to invade the sphenoid sinus, laterally into the cavernous sinus to involve the orbits, or cranially to invade the brain.[29] The fungus invades the cranium through either the orbital apex or cribriform plate of the ethmoid bone and ultimately kills the host.[5] Patients with rhinocerebral mucormycosis typically present with a history of fever, unilateral facial pain or headaches, nasal congestion, epistaxis, visual disturbance, and lethargy.[3] Signs and symptoms that suggest this infection in susceptible individuals include multiple cranial nerve palsies, unilateral periorbital facial pain, orbital inflammation, eyelid edema, blepharoptosis, proptosis, acute ocular motility changes, internal or external ophthalmoplegia, headache, and acute vision loss.[5] These lesions are frequently accompanied by cranial nerve palsy of the III, IV, V, and VI nerves.[30] Although rhinocerebral mucormycosis occurs in immunocompromised patients, it is more common in patients with poorly controlled diabetes mellitus, and can be the first manifestation of the underlying metabolic abnormality.[25] In diabetic ketoacidosis, there is an increased availability of iron that facilitates fungal growth, while the iron chelator desferrioxamine can act as a siderophore to provide iron for fungal growth.[24]

6.2 Pulmonary Mucormycosis

Primary pulmonary mucormycosis is the second most common form of mucormycosis.[25] This is the most frequently seen form in patients with hematological malignancies.[9,10] Among the radiological parameters, the presence of multiple ($\geq$10) nodules and pleural effusion at the time of the initial chest computed tomography (CT) scan were also independently associated with pulmonary mucormycosis as compared with aspergillosis.[31] Wahba et al. observed that the CT finding of a reversed halo sign, a focal round area of groundglass attenuation surrounded by a ring of consolidation, is more common in patients with mucormycosis than in those with other invasive pulmonary fungal infections.[32] Patients with pulmonary mucormycosis typically present with a history of prolonged high-grade fever, non-productive cough, hemoptysis, pleuritic chest pain, and increasing shortness of breath.[5] Massive and potentially fatal hemoptysis can occur if the major pulmonary blood vessels are invaded.[3] Pulmonary mucormycosis may invade lung-adjacent organs, such as the mediastinum, pericardium, and chest wall.[5] The overall mortality rate in patients with pulmonary mucormycosis is 76%, but it is higher in severely immunosuppressed patients.[10]

6.3 Cutaneous Mucormycosis

Cutaneous mucormycosis is the third most common clinical presentation, after sinusitis and pulmonary disease, and it accounts for 19% of all mucormycosis cases.[10,33] The most common mucoralean fungi involved in cutaneous mucormycosis include *R. oryzae* and *Rhizopus microsporus* var. *rhizopodiformis*.[3] Primary cutaneous mucormycosis is generally due to local trauma or inoculation (surgery, burns, motor vehicle-related trauma, the use of needles, knife wounds, insect or spider bites, and other types of trauma), while secondary infection is due to hematogenous dissemination of the organisms to the skin.[3,25,33] Most patients with cutaneous mucormycosis have underlying diseases, such as hematological malignancies or diabetes mellitus, or solid organ transplantation, but a large proportion of them are immunocompetent.[33] Cutaneous mucormycosis has also been reported to occur as a result of injury in a natural disaster, such as the tsunami that struck South-East Asia in 2004.[34] This form can manifest as a superficial or deep infection and it can appear as pustules, blisters, nodules, necrotic eschar, echthyma gangrenosum-like lesions, or necrotizing cellulitis.[5,35] The disease may be of gradual onset and slowly progressive or it may be fulminant, leading to gangrene and hematogenous dissemination.[33] Mortality rates are 10–30%.[10,36]

6.4 Gastrointestinal Mucormycosis

Gastrointestinal zygomycosis is the least common form of the infection, accounting for less than 10% of all cases of mucormycosis.[10] Disease can be acquired by ingestion of the pathogens in contaminated foods such as fermented milk or bread products.[3] One-third of the cases of gastrointestinal mucormycosis occur in infants and children.[35] Patients with gastrointestinal mucormycosis typically present with a history of fever, abdominal pain, or distention, dyspepsia, nausea and vomiting, diarrhea, hematemesis, melena, and hematochezia.[3] The disease is characterized by fungal invasion into the gut mucosa, submucosa, and blood vessels.[25] This form of mucormycosis carries an extremely high mortality rate because of the high incidence of bowel perforation and the difficulty in establishing the diagnosis.[37]

6.5 Disseminated Mucormycosis

Disseminated mucormycosis involves two or more non-contiguous organs.[35] Disease generally arises from the lungs and spreads hematogenously to the central nervous system.[3] It can also spread to the liver, spleen, kidney, heart, and skin.[5] This form of mucormycosis is the second most common clinical presentation in patients with hematological malignancies.[9] The symptoms and evolution of disseminated mucormycosis vary widely, reflecting the host as well as the location and degree of vascular invasion and tissue infarction in the affected organs.[5] Immunosuppression and deferoxamine therapy appears to be the most significant risk factor for disseminated mucormycosis.[35,38] Other risk factors for dissemination include organ transplantation, chemotherapy, and corticosteroids therapy.[39–41] The death rate approaches 100%.[25]

7. HOW DO YOU DIAGNOSE?

(See lines 16–21 in the Case Presentation.)

Benefits of early diagnosis of mucormycosis include prevention of angioinvasion, direct tissue injury of lung, brain, and sinuses, extension into critical sites (eyes, brain, great vessels, etc.), progression to dissemination, reduced need for or extent of surgical resection, and improved outcome and survival.[30] Chamilos et al. showed that initiation of polyene therapy within 5 days after diagnosis of mucormycosis was associated with improvement in survival, compared with initiation of polyene therapy at 6 days or more after diagnosis (83% vs. 49% survival).[42] The reference standard for the definite diagnosis of mucormycosis concerns mycological (direct microscopy and culture), histopathological, and cytopathological examination from affected organs.[43] Direct microscopic examination is performed on all materials sent to the clinical laboratory. When possible, BAL fluid and sterile body fluids should be submitted for examination by clinical microbiology and cytopathology laboratories.[30] Every effort should be made to obtain tissue biopsies for histopathology and culture but this is often difficult in patients with hematologic malignancies because of severe thrombocytopenia.[44]

7.1 Histopathology

Histopathological examination may provide strong evidence for diagnosis of mucormycosis.[3] Histopathology may particularly be useful in differentiation of mucormycosis from aspergillosis, other hyalohyphomycoses, and phaeohyphomycosis. Hyphae usually vary from 6 to 16 μm in diameter, and are sparsely septate and irregularly branched.[30,45] The organism characteristically invades the walls of adjacent blood vessels, causing thrombosis and infarction.[43,44] Stains of fixed tissues with hematoxylin and eosin or

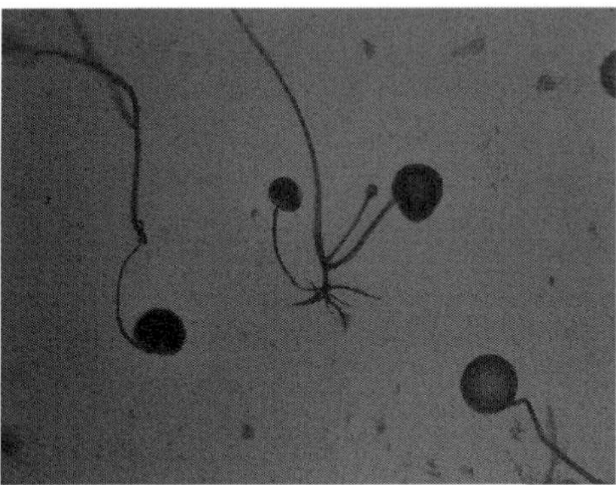

FIGURE 23.1 *Rhizopus* sp. on cellophane tape preparation (×10) stained by lactophenol cotton blue. *With the permission of Dr. Berna Gültekin.*

specialized fungal stains, such as Grocott methenamine-silver, periodic acid-Schiff, or lactophenol cotton blue stains, show broad-based, ribbon-like, non-septate hyphae (Figure 23.1) with wide-angle branching (approximately 90°; may vary from 45 to 90°).[43,45] The inflammatory responses can be neutrophilic, granulomatous, pyogranulomatous, or absent.[3,44] Additionally, immunohistochemistry with commercially available antizygomycete antibodies may help in the diagnosis.[44] Aspirated material from sinuses, sputum in pulmonary disease and biopsy material should be analyzed using 10% potassium hydroxide or optical brighteners such as Calcofluor White, Blankofluor, or Uvitex and finally examining the fungi by fluorescent microscope.[3,46]

7.2 Culture

Blood cultures are usually negative in all forms of mucormycosis.[3] Tissue specimens should be collected from sterile sites, if possible, with a strictly aseptic technique. During processing of the tissue specimens for cultivation, grinding should be avoided since aggressive processing may lead to the loss of the viability of the fragile hyphae of mucoralean fungi and false-negative culture results.[47] To optimize growth, clinical specimens should be inoculated onto appropriate media, such as Sabouraud glucose agar, and incubated at room temperature and 37°C.[3,30,43] The growth of Mucorales tends to be rapid, with mycelial elements expanding to cover the entire plate in only a few (1–7) days and it will demonstrate surface coloration varying from pure white to tan, brown, gray, or even black (Figure 23.2).[43] Malt extract agar is a useful alternative to broth media for the isolation of Mucorales. A positive culture result remains significant not only for definitive diagnosis but also for identification of the infecting mucoralean strain to genus and species level.

FIGURE 23.2 *Rhizopus* sp. on Sabouraud dextrose agar. *With the permission of Dr. Berna Gültekin.*

7.3 Molecular Diagnosis

Molecular techniques for detection of Mucorales by polymerase chain reaction (PCR) or other methods are not widely available and are reserved primarily for research purposes.[3] Different techniques have been reported: DNA probes targeting 18S subunit, ITS1 sequencing after PCR with pan-fungal primers, 18S-targeted semi-nested PCR, and real-time PCR targeting cytochrome b gene.[48] These techniques may be more rapid and more reliable than standard mycological identification. Further studies are needed to achieve better standardization of techniques and to improve sensitivity for identification in tissues.[3,48]

8. HOW DO YOU DIFFERENTIATE THIS DISEASE FROM SIMILAR ENTITIES?

No clinical history is completely specific for the diagnosis of invasive mucormycosis. The differential diagnosis includes infections caused by other angioinvasive pathogens, such as *Aspergillus*, *Fusarium*, *Pseudallescheria*, and *Scedosporium* species.[30] Clinical manifestations of pulmonary mucormycosis cannot be easily distinguished from those of pulmonary aspergillosis.[5] The differences in the frequency of masses, cavities, halo sign, and air crescent sign are insignificant between pulmonary mucormycosis and invasive pulmonary aspergillosis.[31] Galactomannan and 1,3-beta-D glucan detection tests may help to rule out invasive aspergillosis, the most frequent differential diagnosis, or combined *Aspergillus* and Mucorales infections.[44] Lass-Florl et al. showed a high efficiency of CT-guided percutaneous lung biopsy for differentiation of aspergillosis from mucormycosis in hematologic patients.[46]

9. WHAT IS THE THERAPEUTIC APPROACH?

(See lines 21–24 in the Case Presentation.)

Treatment of mucormycosis is through a multi-modality approach with an equally important three-point strategy that includes (1) antifungal therapy, (2) surgery, and (3) management of comorbid factors and adjunctive treatments for improving host response.[3] Among the more recent therapeutic developments in mucormycosis treatment are: the lipid formulations of amphotericin B (AmB), which are now the drugs of choice; the new triazole posaconazole, with promising efficacy as salvage treatment; the iron chelators deferasirox and deferiprone; and the echinocandins in combination with AmB and recombinant growth factors such as granulocyte colony stimulating factor (G-CSF) and granulocyte macrophage colony-stimulating factor (GMCSF).[44] Based on the published data, it seems reasonable to recommend either liposomal AmB (L-AMB) or AmB lipid complex (ABLC) as first-line treatment for mucormycosis. Starting dosages of 5–7.5 mg/kg/day for L-AMB and of 5 mg/kg/day for ABLC, respectively, are commonly used for adults and children.[49] Posaconazole monotherapy cannot be recommended as primary treatment of mucormycosis.[44] However, this drug offers an attractive oral treatment alternative for patients with mucormycosis who cannot tolerate or do not respond to intravenous AmB.[50–52] Posaconazole is recommended as an oral suspension 200 mg four times per day or 400 mg two times per day.[44] Although encouraging, these data are insufficient to support the recommendation for combination first-line therapy in mucormycosis. The use of a combination of a polyene and an echinocandin may, however, be an option in salvage therapy after failure of appropriate first-line therapy.[44] Recommendations regarding surgery in mucormycosis vary according to the site and extension of the disease. There is good evidence to recommend surgery for rhino-orbito-cerebral and soft tissue diseases and moderate evidence for pulmonary mucormycosis.[44]

10. WHAT ARE THE PREVENTIVE AND INFECTION CONTROL MEASURES?

Mucormycete sources in hospital environments are poorly understood, and most published outbreaks have involved cutaneous or gastrointestinal disease, associated with contaminated tape, elasticized dressing and bandage materials, wooden tongue depressors used as arm splints in neonates, prepared medications in adult intensive care unit patients, karaya gum adhesive, and contaminated cornstarch used in the manufacturing of medication tablets and ready-to-eat food products.[53] More work is needed to determine which healthcare facility environmental infection control measures are most effective in reducing patients' exposures to molds and to develop guidance for older facilities in which highly immunosuppressed patients are cared for.

Existing environmental infection control guidelines pertaining to mucormycosis prevention are based on limited data and are focused on aspergillosis. Environmental factors that place patients at risk for invasive aspergillosis may or may not be the same as those that place patients at risk for mucormycosis.[53] As there has not been any reported data on the preventive facilities for mucormycosis in the literature, general environmental infection control measures should be considered by healthcare professionals. Further clinical trials are needed to clarify the role of chemoprophylaxis in mucormycosis prevention.

REFERENCES

1. Ozkul A, Basak S, Ertugrul MB, et al. A case of rhinocerebral mucormycosis presenting with garcin syndrome. *J Klimik* 2011;**24**:187–90.
2. Meis JF, Chakrabarti A. Changing epidemiology of an emerging infection: zygomycosis. *Clin Microbiol Infect* 2009;**15**(Suppl. 5):10–14.
3. Chayakulkeeree M, Ghannoum MA, Perfect JR. Zygomycosis: the re-emerging fungal infection. *Eur J Clin Microbiol Infect Dis* 2006;**25**:215–29.
4. Rogers TR. Treatment of zygomycosis: current and new options. *J Antimicrob Chemother* 2008;**61**(Suppl. 1):i35–40.
5. Petrikkos G, Skiada A, Lortholary O, et al. Epidemiology and clinical manifestations of mucormycosis. *Clin Infect Dis* 2012;**54**(Suppl. 1):S23–34.
6. Hibbett DS, Bindera M, Bischoff JF, et al. A higher-level phylogenetic classification of the fungi. *Mycolog Res* 2007;**111**:509–47.
7. Kwon-Chung KJ. Taxonomy of fungi causing mucormycosis and entomophthoramycosis (zygomycosis) and nomenclature of the disease: molecular mycologic perspectives. *Clin Infect Dis* 2012;**54**:S8–15.
8. Walther G, Pawlowska J, Alastruey-Izquierdo A, et al. DNA barcoding in Mucorales: an inventory of biodiversity. *Persoonia* 2013;**30**:11–47.
9. Skiada A, Pagano L, Groll A, et al. Zygomycosis in Europe: analysis of 230 cases accrued by the registry of the European Confederation of Medical Mycology (ECMM) Working Group on Zygomycosis between 2005 and 2007. *Clin Microbiol Infect* 2011;**17**:1859–67.
10. Roden MM, Zaoutis TE, Buchanan WL, et al. Epidemiology and outcome of zygomycosis: a review of 929 reported cases. *Clin Infect Dis* 2005;**41**:634–53.
11. Larone DH. *Medically Important fungi: a guide to identification*. Washington DC: ASM Press; 2011.
12. Richardson M. The ecology of the Zygomycetes and its impact on environmental exposure. *Clin Microbiol Infect* 2009;**15**(Suppl. 5):2–9.
13. Rees JR, Pinner RW, Hajjeh RA, et al. The epidemiological features of invasive mycotic infections in the San Francisco Bay area, 1992–1993: results of population-based laboratory active surveillance. *Clin Infect Dis* 1998;**27**:1138–47.
14. Torres-Narbona M, Guinea J, Martinez-Alarcon J, et al. Impact of zygomycosis on microbiology workload: a survey study in Spain. *J Clin Microbiol* 2007;**45**:2051–3.
15. Pagano L, Girmenia C, Mele L, et al. Infections caused by filamentous fungi in patients with hematologic malignancies. A report of 391 cases by GIMEMA Infection Program. *Haematologica* 2001;**86**:862–70.

16. Husain S, Alexander BD, Munoz P, et al. Opportunistic mycelial fungal infections in organ transplant recipients: emerging importance of non-Aspergillus mycelial fungi. *Clin Infect Dis* 2003;**37**:221–9.
17. Bitar D, Van Cauteren D, Lanternier F, et al. Increasing incidence of zygomycosis (mucormycosis), France, 1997–2006. *Emerg Infect Dis* 2009;**15**:1395–401.
18. Brown J. Zygomycosis: an emerging fungal infection. *Am J Health Syst Pharm* 2005;**62**:2593–6.
19. Richardson M, Lass-Florl C. Changing epidemiology of systemic fungal infections. *Clin Microbiol Infect* 2008;**14**(Suppl. 4):5–24.
20. Saegeman V, Maertens J, Meersseman W, et al. Increasing incidence of mucormycosis in University Hospital, Belgium. *Emerg Infect Dis* 2010;**16**:1456–8.
21. Kontoyiannis DP, Lionakis MS, Lewis RE, et al. Zygomycosis in a tertiary-care cancer center in the era of Aspergillus-active antifungal therapy: a case-control observational study of 27 recent cases. *J Infect Dis* 2005;**191**:1350–60.
22. Ribes JA, Vanover-Sams CL, Baker DJ. Zygomycetes in human disease. *Clin Microbiol Rev* 2000;**13**:236–301.
23. Ibrahim AS, Spellberg B, Walsh TJ, et al. Pathogenesis of mucormycosis. *Clin Infect Dis* 2012;**54**(Suppl. 1):S16–22.
24. Chamilos G, Lewis RE, Lamaris G, et al. Zygomycetes hyphae trigger an early, robust proinflammatory response in human polymorphonuclear neutrophils through toll-like receptor 2 induction but display relative resistance to oxidative damage. *Antimicrob Agents Chemother* 2008;**52**:722–4.
25. Mantadakis E, Samonis G. Clinical presentation of zygomycosis. *Clin Microbiol Infect* 2009;**15**(Suppl. 5):15–20.
26. Marty FM, Cosimi LA, Baden LR. Breakthrough zygomycosis after voriconazole treatment in recipients of hematopoietic stem-cell transplants. *N Engl J Med* 2004;**350**:950–2.
27. Siwek GT, Dodgson KJ, de Magalhaes-Silverman M, et al. Invasive zygomycosis in hematopoietic stem cell transplant recipients receiving voriconazole prophylaxis. *Clin Infect Dis* 2004;**39**:584–7.
28. Lamaris GA, Ben-Ami R, Lewis RE, et al. Increased virulence of Zygomycetes organisms following exposure to voriconazole: a study involving fly and murine models of zygomycosis. *J Infect Dis* 2009;**199**:1399–406.
29. Hosseini SM, Borghei P. Rhinocerebral mucormycosis: pathways of spread. *Eur Arch Otorhinolaryngol* 2005;**262**:932–8.
30. Walsh TJ, Gamaletsou MN, McGinnis MR, et al. Early clinical and laboratory diagnosis of invasive pulmonary, extrapulmonary, and disseminated mucormycosis (zygomycosis). *Clin Infect Dis* 2012;**54**(Suppl. 1):S55–60.
31. Chamilos G, Marom EM, Lewis RE, et al. Predictors of pulmonary zygomycosis versus invasive pulmonary aspergillosis in patients with cancer. *Clin Infect Dis* 2005:**41**:60–6.
32. Wahba H, Truong MT, Lei X, et al. Reversed halo sign in invasive pulmonary fungal infections. *Clin Infect Dis* 2008;**46**:1733–7.
33. Skiada A, Rigopoulos D, Larios G, et al. Global epidemiology of cutaneous zygomycosis. *Clin Dermatol* 2012;**30**:628–32.
34. Andresen D, Donaldson A, Choo L, et al. Multifocal cutaneous mucormycosis complicating polymicrobial wound infections in a tsunami survivor from Sri Lanka. *Lancet* 2005;**365**:876–8.
35. Prabhu RM, Patel R. Mucormycosis and entomophthoramycosis: a review of the clinical manifestations, diagnosis and treatment. *Clin Microbiol Infect* 2004;**10**(Suppl. 1):31–47.

36. Skiada A, Petrikkos G. Cutaneous zygomycosis. *Clin Microbiol Infect* 2009;**15** (Suppl. 5):41–5.
37. Thomson SR, Bade PG, Taams M, et al. Gastrointestinal mucormycosis. *Br J Surg* 1991;**78**:952–4.
38. Gonzalez CE, Rinaldi MG, Sugar AM. Zygomycosis. *Infect Dis Clin North Am* 2002;**16**:895–914, vi.
39. Chakrabarti A, Das A, Sharma A, et al. Ten years' experience in zygomycosis at a tertiary care centre in India. *J Infect* 2001;**42**:261–6.
40. Kontoyiannis DP, Wessel VC, Bodey GP, et al. Zygomycosis in the 1990s in a tertiary-care cancer center. *Clin Infect Dis* 2000;**30**:851–6.
41. Nosari A, Oreste P, Montillo M, et al. Mucormycosis in hematologic malignancies: an emerging fungal infection. *Haematologica* 2000;**85**:1068–71.
42. Chamilos G, Lewis RE, Kontoyiannis DP. Delaying amphotericin B-based frontline therapy significantly increases mortality among patients with hematologic malignancy who have zygomycosis. *Clin Infect Dis* 2008;**47**:503–9.
43. Lass-Florl C. Zygomycosis: conventional laboratory diagnosis. *Clin Microbiol Infect* 2009;**15**(Suppl. 5):60–5.
44. Skiada A, Lanternier F, Groll AH, et al. Diagnosis and treatment of mucormycosis in patients with hematological malignancies: guidelines from the 3rd European Conference on Infections in Leukemia (ECIL 3). *Haematologica* 2013;**98**:492–504.
45. Frater JL, Hall GS, Procop GW. Histologic features of zygomycosis: emphasis on perineural invasion and fungal morphology. *Arch Pathol Lab Med* 2001;**125**:375–8.
46. Lass-Florl C, Resch G, Nachbaur D, et al. The value of computed tomography-guided percutaneous lung biopsy for diagnosis of invasive fungal infection in immunocompromised patients. *Clin Infect Dis* 2007;**45**:e101–4.
47. Ribes JA, Vanover-Sams CL, Baker DJ. Zygomycetes in human disease. *Clin Microbiol Rev* 2000;**13**:236–301.
48. Dannaoui E. Molecular tools for identification of Zygomycetes and the diagnosis of zygomycosis. *Clin Microbiol Infect* 2009;**15**(Suppl. 5):66–70.
49. Spellberg B, Walsh TJ, Kontoyiannis DP, et al. Recent advances in the management of mucormycosis: from bench to bedside. *Clin Infect Dis* 2009;**48**:1743–51.
50. Cornely OA, Vehreschild JJ, Ruping MJ. Current experience in treating invasive zygomycosis with posaconazole. *Clin Microbiol Infect* 2009;**15**(Suppl. 5):77–81.
51. Greenberg RN, Mullane K, van Burik JA, et al. Posaconazole as salvage therapy for zygomycosis. *Antimicrob Agents Chemother* 2006;**50**:126–33.
52. van Burik JA, Hare RS, Solomon HF, et al. Posaconazole is effective as salvage therapy in zygomycosis: a retrospective summary of 91 cases. *Clin Infect Dis* 2006;**42**:e61–5.
53. Llata E, Blossom DB, Khoury HJ, et al. A cluster of mucormycosis infections in hematology patients: challenges in investigation and control of invasive mold infections in high-risk patient populations. *Diagn Microbiol Infect Dis* 2011;**71**:72–80.

Chapter 24

Lyme Borreliosis

Daša Stupica[1], Gerold Stanek[2] and Franc Strle[1]
[1]*Department of Infectious Diseases, University Medical Centre Ljubljana, Ljubljana, Slovenia,*
[2]*Institute for Hygiene and Applied Immunology, Medical University of Vienna, Austria*

CASE PRESENTATION

A previously healthy 49-year-old male noticed a tick bite on his right pectoral region 2 days after hiking in the forests of Slovenia in June. Ten days later he noticed a round erythematous skin rash at the site of the tick bite. The next day, at the general physician's office, an erythema 6 × 5 cm was observed and blood for serological tests for Lyme borreliae was drawn. At the check-up visit 7 days later, the skin lesion was larger (18 × 9 cm), yet paler, otherwise the patient was asymptomatic. Serological test results for Lyme borreliae came back negative and the physician did not recommend antibiotics. In early August, the patient started to feel tightness and later on severe belt-like pain located in thoracic region, which was more pronounced at night and was resistant to analgesics. In early September, the patient was diagnosed with degree III atrioventricular block. For the preceding 5 days, his physical performance diminished, he was dyspneic and felt chest discomfort at physical activity. His right knee became painful, warm, and swollen but of normal color. At admission to hospital he was afebrile, without any skin rash; the right knee joint was swollen. C-reactive protein was 15 mg/L; white blood cell count, hemoglobin, platelet count, and blood biochemistry test results were normal; cerebrospinal fluid (CSF) analysis showed pleocytosis (lymphocytes 63×10^6/L, monocytes 3×10^6/L), proteins 1.2 g/l, and glucose 3.2 mmol/L (45% of serum glucose concentration). A transient heart pacemaker was inserted. High levels of serum and CSF borrelial IgG antibodies were found, and treatment with ceftriaxone 2 g OD intravenously for 14 days was started. The patient regained his physical capacity, radicular chest pain and knee swelling started to decline, and heart conduction abnormalities resolved. The pacemaker was removed during antibiotic treatment, and the patient's further course was uneventful. The patient had all major manifestations of LB which could have been prevented if EM had been

Emerging Infectious Diseases. DOI: http://dx.doi.org/10.1016/B978-0-12-416975-3.00024-8

recognised and properly treated. However, instead of antibiotic treatment, the physician who saw the patient at the time of EM, ordered tests for the presence of borrelial antibodies, and misinterpreted negative serology as an indication against antibiotic treatment.

1. WHY THIS CASE WAS SIGNIFICANTLY IMPORTANT AS AN EMERGING INFECTION

Lyme borreliosis (LB) is the most common tick-borne disease of pronounced public health importance in countries with moderate climate in the Northern Hemisphere.[1] Its increasing incidence[1,2] may be a consequence of a range of environmental factors, together with changing human behavior.[3] Unfortunately, LB has become a misdiagnosis also for non-specific chronic symptoms such as arthralgias, myalgias, headache, and fatigue—symptoms frequently present in the general population. The diagnosis of LB is too often based on erroneous interpretations of microbiological test results, whereas patients with typical clinical signs may stay undiagnosed and untreated.

2. WHAT IS THE CAUSATIVE AGENT?

The disease is caused by spirochetes of the *Borrelia burgdorferi* sensu lato complex (Lyme borreliae) that currently comprises a clade of 19 species.[1,4] *B. burgdorferi* sensu stricto (*B. burgdorferi*) is the sole species known to cause human infection in North America, whereas in Europe, *B. afzelii*, *B. garinii* (including recently designated *B. bavariensis*) and *B. burgdorferi* are responsible for most cases of LB. However, *B. spielmanii*, *B. valaisiana*, *B. bissettii*, and *B. lusitaniae* have been detected in human samples, suggesting that these species can also give rise to LB.[1] Borreliae consist of a protoplasmic cylinder surrounded by a cytoplasmic membrane, by the periplasm, which contains the flagella, and finally by an outer membrane. *B. burgdorferi* s.l. has an unusual genome: a small linear chromosome (about 950 kb) and numerous linear and circular plasmids. It has the largest number of plasmids known for any bacterium, comprising one-third of the entire genome.[1,4]

3. WHAT IS THE FREQUENCY OF THE DISEASE?

Information on the incidence rates of LB is rather limited. In the USA, the majority of cases are being reported from 12 states in the northeastern, middle, and south Atlantic, and north central regions of the country,[2] while in Europe, the incidence is the highest in central parts.[1] The incidence is increasing—e.g., in Slovenia, it increased from 169 cases per/100,000 in 2002 to 315/100,000 in 2009,[5] and in the USA from 8.2/100,000 in 2002 to 13.4/100,000 in 2009.[2]

Erythema migrans (EM) is by far the most common clinical presentation, representing about 90% of registered European LB cases, followed by neurological manifestations, acrodermatitis chronica atrophicans (ACA), arthritis, borrelial lymphocytoma, and heart involvement.[6] In the USA, the proportion of patients with arthritis is higher, but borrelial lymphocytoma and ACA are not observed.[7,8]

4. HOW ARE THESE BACTERIA TRANSMITTED?

The vectors of Lyme borreliae are hard ticks: *Ixodes ricinus* in Europe, *Ixodes persulcatus* in Asia, *Ixodes scapularis* in northeastern and midwestern USA, and *Ixodes pacificus* in western USA.[1,8,9] Transmission of Lyme borreliae occurs through injection of tick saliva during feeding. Attachment of about 36 hours was described for transmission of *B. burgdorferi* in North America, while attachment period of 24 hours was observed for transmission of *B. burgdorferi* s.l. by *I. ricinus*.[1,8–10]

Most transmissions to humans occur from May to September, coinciding with the activity of nymphs and with the increasing recreational use of tick habitats by the public. The main reservoirs for Lyme borreliae are small mammals, such as mice and voles, and some species of birds. In most tick habitats, deer are essential for the maintenance of tick populations because they can feed sufficient numbers of adult ticks.[9]

A typical habitat for the transmission of Lyme borreliae consists of deciduous or mixed forest, providing an appropriate environment for the development and survival of ticks and potential reservoir hosts.[1,9]

5. WHAT ARE THE PATHOGENIC MECHANISMS?

When an infected tick takes a blood meal, Lyme borreliae detach from the tick's midgut and increase in number. They undergo phenotypic changes, including switching from outer surface protein A (OspA) to OspC expression, which allows them to invade the tick's salivary glands and plays an essential role in the establishment of infection in a mammalian host.[1,7,8] During tick feeding, the borreliae are inserted into the skin from where they may disseminate to other locations. The risk of hematogenous dissemination of *B. burgdorferi* is strain dependent.[1,8,11]

Infection elicits innate and adaptive immune responses, resulting in both macrophage-mediated and antibody-mediated killing of borreliae. However, despite a robust immunological response, infection with Lyme borreliae can persist. Factors associated with persistence include down-regulation of expression of specific immunogenic surface-exposed proteins, including OspC, rapid and continuous alteration of the antigenic properties of a surface lipoprotein known as variable major protein-like sequence expressed (VlsE), and binding to various components of the extracellular matrix. Lyme borreliae do not produce toxins. Most tissue damage seems to result from host inflammatory reactions, which are partly influenced by *Borrelia* genospecies.[1,7,8,12,13] Host genetic factors have an important role in the expression and severity of infection in animals, but corresponding knowledge in humans is limited.[1]

6. WHAT ARE THE CLINICAL MANIFESTATIONS?

Complete presentation of LB, as depicted in the present case report, is unlikely, and would encompass a skin lesion (EM) developing after a tick bite, followed by heart and nervous system involvement, and then arthritis.[1,6–8,14] Months later, late involvement of nervous system, joints, and skin would arise (Table 24.1).

6.1 Skin

Skin is the most frequently affected tissue in LB and EM is by far the most common of skin manifestations.[1,6–8]

Erythema migrans affects people of all ages and both sexes. It begins days to weeks after tick bite with a small red macula or papule. The redness slowly enlarges and usually clears centrally, giving a ring-like appearance. Untreated lesions expand over days to months, ranging from a few centimeters to more than a meter in diameter and eventually disappear. EM is most often located on legs in adults and on the upper part of the body in children.[1,6–8] About half of adult patients report symptoms at the site of EM, usually mild itching, burning, or pain. In Europe, the proportion of patients with systemic symptoms (20–51%) such as fatigue, headache, myalgia, and arthralgia, which usually vary in intensity, is smaller than in the USA (up to 80%).[1,15,16] Multiple EM, defined as the presence of two or more skin lesions, is more frequent in children than in adults, and apparently a more common finding in the USA (up to 50%[1,7,8,15]) than in Europe (3–8% of adult patients with EM).[1,6,8,15–17] It represents clinical evidence of dissemination.

Borrelial lymphocytoma is a solitary bluish-red swelling with a diameter of up to a few centimeters, often arising in the vicinity of a recent or concurrent EM. Histologically, it consists of a dense polyclonal lymphocytic infiltration of cutis and subcutis. This rare manifestation is most frequently located on the earlobe in children, and on breast in adults, and is of longer duration than EM, but also resolves spontaneously.[6,17,18]

Acrodermatitis chronica atrophicans is a chronic skin manifestation of LB seen almost exclusively in Europe. ACA does not resolve spontaneously. Women are affected more often than men; both are usually older than 40 years. Previous other signs of LB are rarely reported. The onset is hardly noticeable with slightly bluish-red discoloration and edema, usually on the dorsal part of a hand or foot. The lesion is initially unilateral; later on it may become more or less symmetrical. It enlarges very slowly over months to

TABLE 24.1 The Course and Clinical Case Definitions for the Main Manifestations of Lyme Borreliosis

Clinical Sign	Onset	Duration[a]	Clinical Case Definition	Laboratory Evidence: Essential	Laboratory/Clinical Evidence: Supporting
Erythema migrans	Days to weeks[b]	Several weeks	Expanding red or bluish-red patch (≥5 cm in diameter)[c], with or without central clearing. Advancing edge typically distinct, often intensely colored, not markedly elevated.	None	Detection of *Borrelia burgdorferi* s.l. by culture and/or PCR from skin biopsy.
Borrelial lymphocytoma (rare)	Days to weeks[b]	Several weeks to months	Painless bluish-red nodule or plaque, usually on ear lobe, ear helix, nipple, or scrotum; more frequent in children (especially on ear) than in adults.	Seroconversion[d] or positive serology. Histology in unclear cases	Histology. Detection of *B. burgdorferi* s.l. by culture and/or PCR from skin biopsy. Recent or concomitant erythema migrans.
Acrodermatitis chronica atrophicans	Several months to years[b]	Unlimited (permanent)	Long-standing red or bluish-red lesions, usually on the extensor surfaces of extremities. Initial doughy swelling. Lesions eventually become atrophic. Possible nodules over bony prominences.	High level of specific serum IgG antibodies	Histology. Detection of *B. burgdorferi* s.l. by culture and/or PCR from skin biopsy.
Lyme neuroborreliosis	Weeks to months[b]	Several weeks to months (unlimited*)	In adults mainly meningo-radiculitis, meningitis, peripheral facial palsy; rarely encephalitis, myelitis; very rarely cerebral vasculitis. In children mainly meningitis and facial palsy.	Pleocytosis and demonstration synthesis of antibodies to Lyme borreliae[e]	Detection of *B. burgdorferi* s.l. by culture and/or PCR from CSF. Intrathecal synthesis of total IgM, and/or IgG, and/or IgA. Specific serum antibodies. Recent or concomitant erythema migrans.

(*Continued*)

TABLE 24.1 (Continued)

Clinical Sign	Onset	Duration[a]	Clinical Case Definition	Laboratory Evidence: Essential	Laboratory/Clinical Evidence: Supporting
Lyme arthritis	Several weeks to months[b]	Several months (unlimited*)	Recurrent attacks or persisting objective joint swelling in one or a few large joints. Alternative explanations must be excluded.	Specific serum IgG antibodies, usually in high concentrations	Synovial fluid analysis. Detection of *B. burgdorferi* s.l. by PCR and/or culture from synovial fluid and/or tissue.
Lyme carditis (rare)	Weeks to months[b]	Several weeks	Acute onset of atrio-ventricular (I–III) conduction disturbances, rhythm disturbances, sometimes myocarditis or pancarditis. Alternative explanations must be excluded.	Specific serum antibodies	Detection of *B. burgdorferi* s.l. by culture and/or PCR from endomyocardial biopsy. Recent or concomitant erythema migrans and/or neurologic disorders.
Ocular manifestations (rare)	Days to months[b]	Weeks to several months	Conjunctivitis, uveitis, papillitis, episcleritis, keratitis.	Specific serum antibodies	Recent or concomitant Lyme borreliosis manifestations. Detection of *B. burgdorferi* s.l. by culture and/or PCR from ocular fluid.

[a] *Without adequate antibiotic treatment.*
[b] *After infection.*
[c] *If <5 cm in diameter a history of tick bite, a delay in appearance (after the tick bite) of at least 2 days and an expanding rash at the site of the tick bite are required.*
[d] *As a rule, initial and follow-up samples have to be tested in parallel in order to avoid changes by inter-assay variation.*
[e] *In early cases, intrathecally produced specific antibodies may still be absent.*
* *Valid for cases of late Lyme neuroborreliosis and Lyme arthritis (rare manifestations).*
(Adapted from Stanek et al.[14])

years, the edema slowly resides, and atrophy gradually predominates; the skin becomes thin, wrinkled, and violet, with prominent underlying veins. Deformations of joints are not unusual and sensation can be impaired due to peripheral nerve involvement in the affected skin.[1,6,17]

6.2 Nervous System

Early Lyme neuroborreliosis typically presents with aseptic meningitis, and involvement of cranial and peripheral nerves.[1,6–8,17,19,20] In Europe, the leading clinical symptom is pain due to radiculoneuritis, usually in the thoracic or abdominal region, which is often belt-like and most pronounced during night, lasting for many weeks if untreated. Patients with borrelial meningitis usually have mild or intermittent headaches and only mildly expressed or absent meningeal signs; nausea and vomiting are rare. In adult European patients, there is no fever.[1,6,19] CSF shows lymphocytic pleocytosis up to several hundred cells $\times$ 10^6/L, normal or slightly raised protein concentration, and normal or mildly depleted glucose concentration. Overall, the course of borrelial meningitis resembles mild but unusually protracted viral meningitis with intermittent improvements and deterioration. Any cranial nerve can be affected; most frequent manifestation is peripheral facial palsy.[1,6–8,17,19,20] Up to 10% of European patients with untreated early Lyme neuroborreliosis develop features of disseminated encephalomyelitis resembling multiple sclerosis.[1,6] Peripheral neuritis, usually accompanying ACA, can develop.[1,6] Subtle encephalopathy has been reported predominantly in US patients.[7,8]

6.3 Heart

Lyme carditis is characterized by changing atrioventricular blocks as a result of conduction disturbances. Heart-related symptoms and electrocardiogram abnormalities usually disappear within 3–6 weeks; however, in some patients degree I atrioventricular block persists. Complete heart block would be the only life-threatening manifestation of LB necessitating insertion of a temporary heart pacemaker.[1,6–8,17,21]

6.4 Joints

Lyme arthritis consists of intermittent attacks of inflammation of one or more, predominantly large joints, most often knee, followed by elbow, ankle, and sometimes shoulder and hip. It is the most frequent clinical sign of disseminated LB in the USA[7,8] emerging in 70% of patients with untreated EM;[22] in Europe, Lyme arthritis seems to be less common. The affected joint is usually painful, swollen, and warm, with normally colored skin.[1,6–8,17,22,23] Joint inflammation typically starts acutely and lasts a few days to weeks, sometimes several months; it is usually recurrent. About 10%

of patients have longstanding arthritis with a duration of ≥1 year.[7,23] Approximately half of patients with Lyme arthritis have moderately raised erythrocyte sedimentation rate and mild leucocytosis. Leucocyte counts in synovial fluid range from 0.5 to 110×10^9/L with a predominance of polymorphonuclears.[1,6–8,23]

6.5 Eyes

Eye manifestations in LB seem to be rare and are usually associated with other signs of the disease. Eyes can be affected primarily as a result of the inflammation of eye tissue such as conjunctivitis, keratitis, iridocyclitis, retinal vasculitis, chorioiditis, and optic neuropathy (extremely rarely episcleritis, panuveitis, panophthalmitis); or secondarily due to extraocular manifestations of LB such as pareses of cranial nerves and orbital myositis.[1,6–8,17]

6.6 Other Rare Manifestations

Case reports of patients with myositis, osteomyelitis, diffuse fasciitis, eosinophilic fasciitis, and panniculitis have been interpreted as manifestations of LB.[1,6] Borrelia isolation has been reported from lesional tissue of individual patients with granuloma annulare, circumscribed scleroderma, and lichen sclerosus et atrophicus, and borrelial infection of liver, lymphatic system, respiratory tract, urinary tract, and genital organs has been reported, although the associations have not yet been well established.[1,6]

6.7 Non-Specific Symptoms

Some patients with LB report several non-specific complaints such as fatigue, nervousness, concentration difficulties, headache, myalgia, and arthralgia. It is highly speculative to interpret such symptoms as due to infection with Lyme borreliae in the absence of objective manifestations of LB.[1,6–8,16] Lyme borreliae could trigger musculoskeletal, neurocognitive, or fatigue symptoms, but these symptoms do not differ from those triggered by some other infections or by stressful physical or emotional events.[1,7,8]

7. HOW DO YOU DIAGNOSE?

Diagnosis of LB should be based on the presence or a reliable history of objective manifestations of the disease. Typical EM is the only manifestation that enables a reliable clinical diagnosis. For all other manifestations, demonstration of borrelial infection is also needed. Clinical case definitions that enable diagnosis of LB, including essential and supporting laboratory evidence, are shown in Table 24.1.[4]

TABLE 24.2 Differential Diagnoses of the Main Manifestations of Lyme Borreliosis

Manifestation of Lyme Borreliosis	Differential Diagnosis
Erythema migrans	Fungal infection (especially when lesions in inguinal or axillary regions) Erysipelas (when erythema migrans do not show central clearing) Toxic or allergic reaction (when a skin lesion develops immediately or during the first 24 h after a tick bite, it is usually not the result of borrelial infection) Bacterial cellulitis, erythema multiforme Southern tick-associated rash illness (STARI) Nummular eczema, granuloma annulare, contact dermatitis Fixed drug eruption, pityriasis rosea Parvovirus B19 infection (in children)
Borrelial lymphocytoma	Breast cancer (when lymphocytoma is located on the breast) B-cell lymphoma, pseudolymphoma
Acrodermatitis chronica atrophicans	Old skin, chilblains Chronic venous insufficiency, superficial thrombophlebitis, hypostatic eczema Arterial obliterative disease, acrocyanosis, livedo reticularis Erythromelalgia, lymphoedema, scleroderma lesions
Fibrous nodules in patients with acrodermatitis	Rheumatoid nodules, gout (tophi), erythema nodosum
Early Lyme neuroborreliosis	
Peripheral facial palsy	Other causes of facial palsy
Meningitis	Viral meningitis
Radiculitis	Mechanical radiculopathy
Late Lyme neuroborreliosis	First episode of relapsing-remitting multiple sclerosis Primary progressive multiple sclerosis
Lyme carditis	Other infectious and non-infectious causes of conduction disturbances Myopericarditis
Lyme arthritis	HLA B27-positive juvenile oligoarthritis, reactive arthritis in adults Septic arthritis, viral arthritis, psoriatic arthritis Sarcoid arthritis, early rheumatoid arthritis Seronegative spondyloarthropathies Gout, pseudo-gout

(Modified According to Stanek et al.[1] and Strle and Stanek[6])

8. HOW DO YOU DIFFERENTIATE THIS DISEASE FROM SIMILAR ENTITIES?

Principal differential diagnoses of the main clinical manifestations of LB are depicted in Table 24.2.[1,6]

9. WHAT IS THE THERAPEUTIC APPROACH?

Lyme borreliae are susceptible *in vitro* to tetracyclines, penicillins, second-generation and third-generation cephalosporins, and macrolides but resistant to fluoroquinolones, rifampicin, and first-generation cephalosporins.[1,8,24]

Although EM will eventually resolve without treatment, 10 to 21 days of oral antibiotics is recommended to prevent dissemination and development of later sequelae. Doxycycline, amoxicillin, phenoxymethylpenicillin, and cefuroxime axetil are highly effective and are the preferred agents for this manifestation, while macrolides are used as second-line antibiotics.[1,8] Doxycycline is the sole drug for which clinical trials have shown that only 10 days of treatment is effective.[25,26] Doxycycline, however, can cause photosensitivity and is contraindicated in children <8 years and in pregnant or breastfeeding women.[8] Fever, if present, should resolve within 48 h of initiation of therapy and the skin lesion usually resolves within 7–14 days. Other symptoms, such as fatigue or arthralgia, tend to improve but do not always resolve within this timeframe, lasting for more than 3 months in one-quarter of patients from the USA and in about 10% in Europe.[8,16] Extension of antibiotic treatment does not hasten relief of symptoms.[1,8,25]

Oral antibiotics are also used as first-line treatment for the other cutaneous manifestations of LB, and as initial treatment for patients with Lyme arthritis.[1,8] The preferred parenteral drug for LB is ceftriaxone because it is highly active against Lyme borreliae *in vitro*, crosses the blood–brain barrier well, and has a long serum half-life, which enables once daily dosing. Alternative choices are cefotaxime and intravenous penicillin. Parenteral antibiotic treatment is recommended for late Lyme neuroborreliosis and as an initial treatment of patients with serious heart conduction disturbances who need hospitalization. In several European countries and the USA, parenteral treatment has been the preferred management strategy also for early Lyme neuroborreliosis.[1,7,8] European studies, however, have provided convincing evidence that oral doxycycline is equally effective.[27] Similarly to other manifestations, Lyme arthritis typically also responds to antibiotics. Patients whose arthritis is improved but not resolved after an initial course of 4-week oral treatment can be retreated with a second course of oral antibiotics, reserving parenteral antibiotics for those without any substantial clinical response.[7,8] About 10% of patients in the USA, however, do not respond to antibiotic treatment. Patients with "antibiotic-refractory" Lyme arthritis (defined as persistent synovitis for at least 2 months after completion of a course of intravenous ceftriaxone or after completion of two 4-week courses of an oral antibiotic, in conjunction with negative PCR testing on synovial tissue/fluid) are not believed to be actively infected. They are usually treated with non-steroidal anti-inflammatory agents, intra-articular injections of corticosteroids, or disease-modifying antirheumatic drugs, and sometimes with arthroscopic synovectomy.[7,8]

According to guidelines, treatment for pregnant women with LB is much the same as treatment for women who are not pregnant, except that doxycycline should be avoided.[1,8] However, direct information on the efficacy of such an approach is limited. No data reliably lend support to congenital LB syndrome.[8,28]

10. WHAT ARE THE PREVENTIVE AND INFECTION CONTROL MEASURES?

LB can be prevented by avoiding tick-infested environments, by covering bare skin, and by use of tick repellents. The density of tick populations around residences can be reduced by removal of leaf litter, placing of wood chips where lawns are adjacent to forests, application of acaricides, and construction of fences to keep out deer. Daily inspection of skin to remove attached ticks is recommended. Removal is done by grasping the tick as close to the mouthparts as possible with forceps (or tweezers) and then gently pulling it out.[1,7,8] More than 96% of patients who find and remove an attached *I. scapularis* tick will not contract LB, without any other intervention, even in highly endemic geographical regions. If the tick is not found or removed, the probability of infection approaches the infection rate in the regional tick population (approximately 25% of nymphal stage *Ixodes* ticks are infected in highly endemic areas in the USA, and 10% in Europe).[29,30] In a study from the USA, one 200 mg dose of doxycycline was 87% effective in the prevention of EM at the tick-bite site.[31] Such prophylaxis should be considered for individuals in highly endemic areas of the USA who had been bitten by an *I. scapularis* tick that was attached for ≥ 36 hours.[32] In Europe, observation is recommended for *I. ricinus* tick bites, because the infection rate of ticks is lower than in the USA, and studies on the efficacy of antibiotic prophylaxis have not been done.[1] No vaccine is available to prevent LB in man.[1,8]

REFERENCES

1. Stanek G, Wormser GP, Gray J, Strle F. Lyme borreliosis. *Lancet* 2012;**379**:461–73.
2. CDC, Division of Vector-borne Infectious Diseases. Lyme Disease. <http://www.cdc.gov/ncidod/dvbid/lyme/index.htm> [accessed 28.11.13].
3. Randolph SE. To what extent has climate change contributed to the recent epidemiology of tick-borne diseases?. *Vet Parasitol* 2010;**167**:92–4.

4. Margos G, Vollmer SA, Ogden NH, Fish D. Population genetics, taxonomy, phylogeny and evolution of Borrelia burgdorferi sensu lato. *Infect Genet Evol* 2011;**11**:1545–63.
5. Anonymous. Institute of Public Health of the Republic of Slovenia. Registered communicable diseases in Slovenia in 2009. Ljubljana 2010.
6. Strle F, Stanek G. Clinical manifestations and diagnosis of Lyme borreliosis. *Curr Prob Dermatol* 2009;**37**:51–110.
7. Steere AC. Lyme disease. *N Engl J Med* 2001;**345**:115–25.
8. Wormser GP, Dattwyler RJ, Shapiro ED, et al. The clinical assessment, treatment, and prevention of Lyme disease, human granulocytic anaplasmosis, and babesiosis: clinical practice guidelines by the Infectious Diseases Society of America. *Clin Infect Dis* 2006;**43**:1089–134.
9. Gray JS. The ecology of ticks transmitting Lyme borreliosis. *Exp Appl Acarol* 1998;**22**:249–58.
10. Stanek G, Kahl O. Chemoprophylaxis for Lyme borreliosis? *Zentralbl Bakteriol* 1999;**289**:655–65.
11. Wormser GP, Brisson D, Liveris D, et al. *Borrelia burgdorferi* genotype predicts the capacity for hematogenous dissemination during early Lyme disease. *J Infect Dis* 2008;**198**:1358–64.
12. Strle K, Drouin EE, Shen S, et al. *Borrelia burgdorferi* stimulates macrophages to secrete higher levels of cytokines and chemokines than *Borrelia afzelii* or *Borrelia garinii*. *J Infect Dis* 2009;**200**:1936–43.
13. Baranton G, De Martino SJ. *Borrelia burgdorferi* sensu lato diversity and its influence on pathogenicity in humans. *Curr Prob Dermatol* 2009;**37**:1–17.
14. Stanek G, Fingerle V, Hunfeld KP, et al. Lyme borreliosis: clinical case definitions for diagnosis and management in Europe. *Clin Microbiol Infect* 2011;**17**:69–79.
15. Strle F, Nadelman RB, Cimperman J, et al. Comparison of culture-confirmed erythema migrans caused by Borrelia burgdorferi sensu stricto in New York State and Borrelia afzelii in Slovenia. *Ann Intern Med* 1999;**130**:32–6.
16. Cerar D, Cerar T, Ruzić-Sabljić E, Wormser GP, Strle F. Subjective symptoms after treatment of early Lyme disease. *Am J Med* 2010;**123**:79–86.
17. Stanek G, Strle F. Lyme borreliosis. *Lancet* 2003;**362**:1639–47.
18. Maraspin V, Cimperman J, Lotric-Furlan S, et al. Solitary borrelial lymphocytoma in adult patients. *Wien Klin Wochenschr* 2002;**114**:515–23.
19. Kristoferitsch W. Neurological manifestations of Lyme borreliosis: clinical definition and differential diagnosis. *Scand J Infect Dis* 1991;**77**:64–73.
20. Mygland A, Ljostad U, Fingerle V, Rupprecht T, Schmutzhard E, Steiner I. EFNS guidelines on the diagnosis and management of European Lyme neuroborreliosis. *Eur J Neurol* 2010;**17**:8–16.
21. Steere AC, Batsford WP, Weinberg M, et al. Lyme carditis: cardiac abnormalities of Lyme disease. *Ann Intern Med* 1980;**93**:8–16.
22. Steere AC, Schoen RT, Taylor E. The clinical evolution of Lyme arthritis. *Ann Intern Med* 1987;**107**:725–31.
23. Steere AC, Glickstein L. Elucidation of Lyme arthritis. *Nat Rev Immunol* 2004;**4**:143–52.
24. Hunfeld KP, Ruzic-Sabljic E, Norris DE, Kraiczy P, Strle F. In vitro susceptibility testing of *Borrelia burgdorferi* sensu lato isolates cultured from patients with erythema migrans before and after antimicrobial chemotherapy. *Antimicrob Agents Chemother* 2005;**49**:1294–301.

25. Wormser GP, Ramanathan R, Nowakowski J, et al. Duration of antibiotic therapy for early Lyme disease. A randomized, double-blind, placebo-controlled trial. *Ann Intern Med* 2003;**138**:697–704.
26. Stupica D, Lusa L, Ruzic-Sabljic E, Cerar T, Strle F. Treatment of erythema migrans with doxycycline for 10 days versus 15 days. *Clin Infect Dis* 2012;**55**:343–50.
27. Ljostad U, Skogvoll E, Eikeland R, et al. Oral doxycycline versus intravenous ceftriaxone for European Lyme neuroborreliosis: a multicentre, non-inferiority, double-blind, randomised trial. *Lancet Neurol* 2008;**7**:690–5.
28. Maraspin V, Cimperman J, Lotric-Furlan S, Pleterski-Rigler D, Strle F. Treatment of erythema migrans in pregnancy. *Clin Infect Dis* 1996;**22**:788–93.
29. Piesman J. Lyme borreliosis in North America. In: Gray JS, Kahl O, Lane RS, Stanek G, editors. *Lyme borreliosis: biology, epidemiology and control*. New York: CABI Publishing; 2002. p. 223–49.
30. Gern L, Humair PF. Lyme borreliosis in Europe. In: Gray JS, Kahl O, Lane RS, Stanek G, editors. *Lyme borreliosis: biology, epidemiology and control*. New York: CABI Publishing; 2002. p. 149–74.
31. Nadelman RB, Nowakowski J, Fish D, et al. Prophylaxis with single dose doxycycline for the prevention of Lyme disease after an *Ixodes scapularis* tick bite. *N Engl J Med* 2001;**345**:79–84.
32. Warshafsky S, Lee DH, Francois LK, Nowakowski J, Nadelman RB, Wormser GP. Efficacy of antibiotic prophylaxis for the prevention of Lyme disease: an updated systematic review and meta-analysis. *J Antimicrob Chemother* 2010;**65**:1137–44.

Chapter 25

Plasmodium knowlesi

Jana Preis[1] and Larry Lutwick[2]

[1]*Division of Infectious Diseases, New York Harbor Health Care System, Brooklyn Campus, and State University of New York, Downstate Medical School, Brooklyn, New York, NY, USA,*
[2]*Division of Infectious Diseases, Departments of Medicine and Biomedical Sciences, Western Michigan University Homer Stryker MD School of Medicine, Kalamazoo, MI, USA*

CASE PRESENTATION[1]

A 35-year-old Swedish man was seen in a hospital in Stockholm with 2 days of fever as high as 40°C, sweats, headache, and fatigue. He was found to be moderately leukopenic (2200/μL) and thrombocytopenic (58,000/μL) but not anemic (15.4 g/dL) or azotemic. By history, he had just returned from a 2-week holiday in Sarawak, Malaysian Borneo. He had trekked through the jungles of the Bario Highlands during the last week of the trip and did not take any antimalarial chemoprophylaxis. The symptom complex began 11 days after leaving the Highlands.

He was started on oral mefloquine therapy because of a strong suspicion of malaria. Although a rapid test for malaria was negative, malaria parasites were detected by examination of his blood smears, both thin and thick. Estimated parasitemia was 0.1%. The *Plasmodium* species could not be conclusively identified but *P. malariae* was suspected because late trophozoites were observed and the parasitized red blood cells were not enlarged. The patient improved rapidly after starting mefloquine therapy and was discharged afebrile after 2 days. Blood cultures were negative. Nadirs of hemoglobin and platelet count were 9.5 and 34,000, respectively.

PCR assays for *Plasmodium* species *falciparum*, *vivax*, *ovale*, and *malariae* were negative and because of history of travel and parasite morphology similar to *P. malariae*, a blood sample was sent for testing for *P. knowlesi* PCR and was positive.

Emerging Infectious Diseases. DOI: http://dx.doi.org/10.1016/B978-0-12-416975-3.00025-X

1. WHY THIS CASE WAS SIGNIFICANTLY IMPORTANT AS AN EMERGING INFECTION

Zoonotic malaria was considered to be extremely rare until a large focus of *P. knowlesi* infections in the Kapit Division of Sarawak, Malaysian Borneo, was described in 2004:[2] 112 were single *P. knowlesi* infections, eight were *P. knowlesi* coinfections with other *Plasmodium* species, and none were *P. malariae*.

Since then, human cases have been described in virtually all South-East Asian countries, and *P. knowlesi* is now considered the fifth species of *Plasmodium* causing malaria in humans.

P. knowlesi was first isolated and studied in detail at the Kolkata School of Tropical Medicine in India in the early 1930s, after it was noticed in a blood film from a long-tailed macaque that had been imported from Singapore. Researchers observed that all three human recipients of *P. knowlesi*-infected blood developed malaria and commented that although the fever pattern was daily, further work was required to confirm the periodicity of the fever. They also observed that the morphology of the parasites resembled that of *P. malariae*. The authors studied the morphology of the parasite in detail in non-human primate hosts and named this new malaria parasite in honor of Robert Knowles. The infection was initially thought to be non-life threatening and, in fact, replaced *P. vivax* as the fever-producing agent for the treatment of neurosyphilis in the 1930s[3] but was discontinued when deaths occurred.

2. WHAT IS THE CAUSATIVE AGENT?

Malaria is caused by protozoan parasites belonging to the genus *Plasmodium*. Over 150 species have been described to date, infecting mammals, birds, and reptiles. Despite having such a large number of hosts, in general, malaria parasites tend to be host specific. For example, humans are the natural hosts for four species, *P. falciparum*, *P. vivax*, *P. malariae*, and *P. ovale*, while long-tailed macaques (*Macaca fascicularis*) are hosts for five, *P. knowlesi*, *P. fieldi*, *P. coatneyi*, *P. cynomolgi*, and *P. inui*. Human cases have been described in virtually all South-East Asian countries, and *P. knowlesi* is now considered the fifth species of malaria in humans.

Figure 25.1 shows the morphology of *Plasmodium knowlesi* in a Giemsa-stained thin blood smear.

3. WHAT IS THE FREQUENCY OF THE DISEASE/ EPIDEMIOLOGY?

Following the description of the large focus of human *knowlesi* malaria cases in the Kapit Division of Malaysian Borneo in 2004, there have been reports

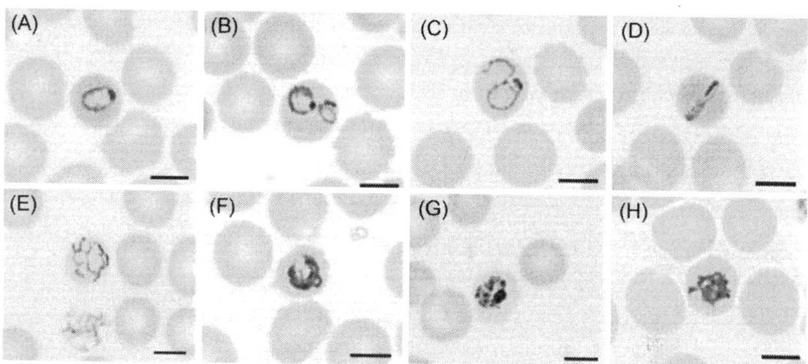

FIGURE 25.1 Morphology of *Plasmodium knowlesi* in a Giemsa-stained thin blood smear. Infected erythrocytes were not enlarged, lacked Schuffner stippling, and contained much pigment. Shown are examples of trophozoites (A–F), a schizont (G), and a gametocyte (H). Scale bars = 5 μm. *From: van Hellemond et al. Human* Plasmodium knowlesi *infection detected by rapid diagnostic tests for malaria. Emerg Infect Dis 2009;15:1478–80.*

of infections acquired in Kapit and other locations in Malaysia. Human cases are not restricted to Malaysia with reports of infections acquired in Thailand, the Philippines, Myanmar, Singapore, Vietnam, Indonesia, Brunei, and Cambodia. Transmission of *knowlesi* malaria to humans has therefore been reported in all the countries in South-East Asia except Laos. It is the most common form of human malaria in some of these areas. Visitors to South-East Asia have not been spared, with reports of adult travelers from countries including Sweden, Finland, France, Spain, the Netherlands, Taiwan, the United States, Germany, and the United Kingdom returning home with *P. knowlesi* infections following visits to Sarawak, the Philippines, Peninsular Malaysia, Brunei, and Thailand.[1]

4. HOW IS THE PARASITE TRANSMITTED?

The primary natural hosts of *P. knowlesi* that were initially identified were long-tailed (*M. fascicularis*) and pig-tailed (*Macaca nemestrina*) macaques.[4] These two macaque species are distributed throughout South-East Asia and are the most common non-human primates in this region. The prevalence of infection in some areas such as the Kapit Division of Malaysian Borneo can be very high in these macaques indicating that malaria transmission is intense among the wild macaque population.

The vectors of *knowlesi* malaria are forest-dwelling mosquitoes that belong to the *Anopheles leucosphyrus* group, which consists of 20 species, and their distribution in South-East Asia largely overlaps that of long-tailed and pig-tailed macaques. This group of mosquitoes is characterized by outdoor biting in a jungle forested habitat. *A. latens* appears to be the main vector in these forested areas, such as Kapit, but in deforested areas, other species such as *A. dirus* and *A. balabacensis* are likely vectors.[5] As humans encroach on the forested macaque habitat, Yacob et al. speculate that more anthrophilic members of the leucosphyrus group may be selected for and produce a host switch to humans for *P. knowlesi*.[6]

5. WHAT FACTORS ARE INVOLVED IN DISEASE PATHOGENESIS? WHAT ARE THE PATHOGENIC MECHANISMS?

There are no clinical signs and symptoms when the malaria parasites are developing in the preclinical phase in the liver. There is no persistent hypnozoite hepatic stage so relapses do not occur. *P. knowlesi* invasion of human red blood cells (RBCs), like *P. vivax*, is dependent on the erythrocyte chemokine receptor, known as the Duffy blood group antigen.[7]

Symptoms are associated with the cycle of the parasites in the erythrocytes. The duration of the erythrocytic cycle depends on the species of *Plasmodium*: *P. knowlesi* has the shortest cycle, approximately 24 h, while for *P. falciparum*, *P. vivax*, and *P. ovale*, it is approximately 48 h, and for *P. malariae*, it is 72 h. Therefore, if untreated, the parasite counts or parasitemia will continue to increase approximately every 24, 48, or 72 h, depending on the species. That the erythrocytic cycle of *P. knowlesi* is only 24 h underscores the need for prompt consideration of this species to prevent potentially life-threatening complications. The pathogenesis of severe *knowlesi* disease is not fully understood. A study of pretreatment cytokine concentrations at admission showed that *knowlesi* malaria patients with complicated disease had higher levels of cytokines including tumor necrosis factor alpha (TNF-α), IL-6, IL-8, IL-1ra, and IL-10 than patients with uncomplicated disease.[8] The anti-inflammatory cytokines IL-1ra and IL-10 were associated with parasitemia in *knowlesi* malaria. A small study of *ex vivo* cytoadherence[9] demonstrated that late-trophozoite- and schizont-infected erythrocytes from patients with *knowlesi* malaria have the capacity to bind to the human endothelial cell receptors' intracellular adhesion molecule 1 (ICAM-1) and vascular cell adhesion molecule (VCAM) but not to CD36.

6. WHAT ARE THE CLINICAL MANIFESTATIONS?

The symptoms of acute *knowlesi* infection are of a non-specific infectious illness similar to those seen in *falciparum* and *vivax* malaria. Fevers, chills, and rigors are the most dominant features reported, while headaches, myalgia/arthralgia, malaise, and poor appetite are also commonly present. Cough (48 and 56%), abdominal pain (31 and 52%), and diarrhea (18 and 29%) were additional symptoms noted in prospective studies of 107[10] and 130 patients,[11] respectively, presenting with acute *knowlesi* malaria. The median duration of illness prior to presentation to a healthcare facility for *knowlesi* malaria has been reported to be between 4 and 5 days. The most common examination findings reported for 107 prospectively studied *knowlesi* malaria patients were tachypnea, fever, and tachycardia; palpable liver and spleen were reported in 24 to 40%. Clinical signs of severe disease including low oxygen saturations, tachypnea, chest crackles, hypotension, and jaundice have been documented. Laboratory findings are diverse including complete blood count abnormalities, metabolic changes, and abnormal liver function tests. Hematological and biochemical parameters respond rapidly following treatment, with the exception of hemoglobin levels, serum albumin concentrations, and liver enzyme levels, which typically return to normal limits by day 28.

Severe disease is also reported in the literature: patients presented with a 3- to 7-day history of a fever associated with non-specific features that included shortness of breath, abdominal pain, and vomiting; cases tend to have high parasitemia, severe thrombocytopenia, renal failure, hypotension, jaundice, and deranged liver enzymes.

Features of severe malaria defined by laboratory findings reflect extensive studies of *P. falciparum* infections:[12] such prognostic markers include a white cell peripheral leucocytosis count of >12,000 cells/μL, a serum creatinine concentration of >265 μmol/L (≥3.0 mg/dL), a hemoglobin concentration of <7.1 g/dL, and a blood glucose level of <2.2 mmol/L (<40 mg/dL). A three-fold increase in aminotransferase enzyme levels and increased serum lactate and low bicarbonate concentrations are also associated with a poor outcome in *falciparum* malaria. In comparing severe cases of *P. knowlesi* disease and *P. falciparum* disease,[11] after adjusting for age and parasite count, higher percentages of trophozoites and schizonts were seen in severe *falciparum* disease.

Parasitemia appears to be a strong predictor of complications in *knowlesi* malaria infection. In the Barber study,[11] more than 80% of those with *P. knowlesi* parasitemia of >100,000 parasites/μL also met other criteria for severity. However, severe infection could occur with lower degrees of parasitemia—9.4 percent of *P. knowlesi* infections with parasitemia of less than 20,000 parasites/μL still met criteria for severe disease. This report found that the >100,000/μL parasite count was highly specific (100%) for severe disease but only 30% sensitive. Indeed, another study found that patients with a parasite density of >35,000/μL (or >1%) or a platelet count of <45,000/μL were at risk for severe disease.[13]

Clinically, older individuals with *P. knowlesi* infections are at greater risk for severe disease, which correlated with higher parasitemia as it does with older individuals with *falciparum* malaria. Importantly, in reviewing a total of 84 cases of severe *knowlesi* malaria, no cases of *knowlesi*-associated coma were found[11] as compared to as high as 50% of Asian adults with *falciparum* malaria. Since the pathophysiology of cerebral *falciparum* malaria seems to relate to microvascular sequestration of parasitized RBCs, cytoadherence to ICAM-1 as well as reduced RBC deformity, it is possible that the mechanisms of vascular injury in *P. knowlesi* infection may differ from *P. falciparum*.[11]

7. HOW DO YOU DIAGNOSE?

The most widely used method for detection of malaria in rural settings is microscopy, since it is a relatively cheap, rapid, quantitative, and sensitive technique. From the onset, it has been found that it is not possible to accurately identify *P. knowlesi* by microscopy, using either thin or thick smears. In the Singh report,[2] as many of a fifth of total malaria cases in the Kapit division of Malaysian Borneo had been diagnosed as *P. malariae* but the morphology was thought to be atypical. The morphological features of the early ring forms of *P. knowlesi* are identical to those of *P. falciparum*, with the possibility of double-chromatin dots, multiple infections per erythrocyte, and no enlargement of infected erythrocytes. Additionally, the later trophozoites were found to occupy less than two-thirds of the RBC cytoplasm and the trophozoite cytoplasm was compact and not amoeboid. Accurate identification of *P. knowlesi* by microscopy is a diagnostic challenge, particularly when parasitemia is low. A photomicrograph of a thin smear from a patient with *P. knowlesi* is shown in Figure 25.1.

Indeed, a more recent report[14] also found that microscopy could not reliably distinguish between *P. knowlesi*, *P. falciparum*, and *P. vivax* in a region where all three of these species coexist. Misdiagnosis could result in administration of inappropriate therapy and increase the risk of relapses or severe complications.

More specific testing is needed, especially when the affected individual is from or has visited the areas known to be endemic for *P. knowlesi* and the morphology of the parasite is atypical.

8. HOW DO YOU DIFFERENTIATE THIS DISEASE FROM SIMILAR ENTITIES?

Any febrile illnesses from endemic areas of malaria require consideration of the diagnosis of malaria. At the same time, blood and other cultures, chest X-ray, and other diagnostic modalities need to be undertaken in the appropriate circumstances.

Molecular detection methods have been developed for the accurate identification of malaria parasites, and these methods have consistently proven to be more sensitive and specific than microscopy. Singh's[2] initial PCR assay developed for the detection of *P. knowlesi* was a nested PCR assay with primers Pmk8 and Pmkr9, based on the small-subunit (SSU) rRNA genes. Indeed, using this assay, it was found that 35 of 36 samples of DNA extracted from thick smears collected in 1996 from patients in differing divisions of Sarawak were positive for *P. knowlesi*, underscoring the evidence that *P. knowlesi* has been widely distributed for some time and is not newly emergent in these areas.[15]

Immunochromatographic rapid diagnostic tests (RDTs) have been developed for detection of malaria and are particularly useful for investigations of outbreaks in rural settings where electricity is unavailable and in laboratories in developed countries where laboratory technologists are unfamiliar with detecting malaria by microscopy. The RDTs contain antibodies that are specific for histidine-rich protein 2 (HRP2) of *P. falciparum* or are specific for lactate dehydrogenase (LDH) of either *P. falciparum* or *P. vivax*. An aldolase antigen assay also exists. Barber et al.[14] reports that, based on limited data, combining the antigen testing may allow the differentiation of mono-infections of these species. A large number of these malaria RDTs are available with variable sensitivity especially at low parasitemias. A 2013 report[16] found that a pLDH-based LDH RDT performed better than an aldolase-based RDT in the diagnosis of *knowlesi* malaria of any severity at a parasite count above 1000/μL. As expected, the sensitivity was best (95%) in high parasite counts but was lower at lower degrees of parasitemia with an overall sensitivity of 74%.

9. WHAT IS THE THERAPEUTIC APPROACH?

Although patients with *knowlesi* malaria have been successfully treated with a wide range of antimalarial drugs, the optimal treatment for either uncomplicated or complicated disease is unknown. One may assume that there is little antimalarial drug resistance, as *knowlesi* malaria is primarily a zoonosis, so *P. knowlesi* parasites have not been subjected to any significant antimalarial drug pressure. Indeed, in non-human primate studies, tetracycline, clindamycin, trimethoprim, erythromycin, and artemisinins have all been shown to have an antiparasitic effect in *P. knowlesi* infections.

In mild infections in travelers, a variety of antimalarial drugs have been used with success[17] including quinine, doxycycline, mefloquine, atovaquone/proguanil, and artemether/lumefantine, and chloroquine had been used for uncomplicated *P. knowlesi* routinely.[18] In severe disease, however, a prospective study[11] found that oral artemisinin (ACTs—artemisinin-containing therapies, such as artemether-lumefantrine and artesunate-mefloquine) or intravenous artesunate improved the prognosis in *P. knowlesi* infection. It should be noted in this study that *knowlesi* was associated with a three-fold higher risk of severity than falciparum. In Sabah, Malaysia, deaths due to *P. knowlesi* were associated with initial diagnosis as *P. malariae*, which delayed appropriate therapy.[19] All cases of *P. malariae*-diagnosed infections with significant parasitemias and a link to *knowlesi*-endemic areas should be treated as severe *P. knowlesi* infections.

10. WHAT ARE THE PREVENTIVE AND INFECTION CONTROL MEASURES?

The usual modalities for the prevention of malaria are avoidance of *Anopheles* mosquitoes and administration of chemoprophylaxis. Avoidance consists of the use of mosquito repellants and minimizing outdoor activities during the *Anopheles* feeding hours (from dusk to dawn). If air conditioning is not available, mosquito netting should be used, preferably netting that is insecticide treated, and spraying on indoor walls. Chemoprophylaxis for *P. knowlesi* needs to be aimed at species with multidrug resistance such as what might occur with *P. falciparum*.

As with the other species of malaria, investigations have researched the possibility of a vaccine for *P. knowlesi*. The area of the *P. knowlesi* merozoite Duffy blood group antigen-binding site that binds to the human erythrocyte is a cysteine-rich domain known as region II of the alpha protein. Blocking the binding by glycosylating the Duffy antigen blocked binding of *P. vivax* merozoites (and presumably *P. knowlesi* as well) to rhesus erythrocytes.[7] In 2002, antibodies raised against the receptor-binding domain of *P. knowlesi* (as a recombinant protein) was reported to inhibit erythrocyte invasion.[20] Subsequently, an apical membrane antigen (AMA1) of *P. knowlesi* produced in recombinant yeast showed promise when given to rhesus macaques in an adjuvanted form[21] in the control of parasitemia. Studies in Malian children with a similar *P. falciparum* vaccine, however, only produced protection when the AMA antigen in the infecting *P. falciparum* strain was homologous to the vaccine protein.[22] The lack of protection may be overcome by using a number of AMA1 alleles in a single immunization.[23]

REFERENCES

1. Bronner U, Divis PCS, Färnert A, Singh B. Swedish traveller with *Plasmodium knowlesi* malaria after visiting Malaysian Borneo. *Malar J* 2009;**8**:15.
2. Singh B, Kim Sung L, Matusop A, et al. A large focus of naturally acquired *Plasmodium knowlesi* infections in human beings. *Lancet* 2004;**363**:1017–24.
3. Coatney G, Collins W, Warren M, Contacos P. *CD-ROM: the primate malarias [original book published 1971]. Division of Parasitic Diseases, ed.* Atlanta, GA: CDC; 2003.
4. Lee KS, Divis PC, Zakaria SK, et al. *Plasmodium knowlesi*: reservoir hosts and tracking the emergence in humans and macaques. *PLoS Pathog* 2011;**7**:e1002015.
5. Cox-Singh J. Zoonotic malaria: *Plasmodium knowlesi*, an emerging pathogen. *Curr Opin Infect Dis* 2012;**25**:530–6.
6. Yakob L, Bonsall MB, Yan G. Modelling knowlesi malaria transmission in humans: vector preverence and host competence. *Malar J* 2010;**9**:329.
7. Chitnis CE, Chaudhuri A, Horuk R, Pogo AO, Miller LH. The domain on the Duffy blood group antigen for binding *Plasmodium vivax* and *P. knowlesi* malarial parasites to erythrocytes. *J Exp Med* 1996;**184**:1531–6.
8. Cox-Singh J, Singh B, Daneshvar C. Anti-inflammatory cytokines predominate in acute human *Plasmodium knowlesi* infection. *PLoS One* 2011;**6**:e20541.

9. Fatih FA, Angela Siner A, Atique A, et al. Cytoadherence and virulence—the case of *Plasmodium knowlesi* malaria. *Malar J* 2012;**11**:33.
10. Daneshvar C, Davis TME, Cox-Singh J, et al. Clinical and laboratory features of human knowlesi infection. *Clin Infect Dis* 2009;**49**:852–60.
11. Barber BE, William T, Grigg MJ, et al. A prospective comparative study of knowlesi, falciparum, and vivax malaria in Savah, Malaysia: high proportion with severe disease from *Plasmodium knowlesi* and *Plasmodium falciparum* but no mortality with early referral and artesunate therapy. *Clin Infect Dis* 2013;**56**:383–97.
12. World Health Organisation. Management of severe falciparum malaria: a practical handbook. Available at: <http://www.who.int/malaria/docs/hbsm.pdf>.
13. Willman M, Ahmed A, Siner A, et al. Laboratory markers of disease severity in *Plasmodium knowlesi* infection: a case control study. *Malar J* 2012;**11**:363.
14. Barber BE, William T, Grigg MJU, Yeo TW, Anstey NM. Limitations of microscopy to differeniate *Plasmodium* species in a region co-endemic for *Plasmodium falciparum, Plasmodium vivax* and *Plasmodium knowlesi*. *Malar J* 2013;**12**:8.
15. Lee K-S, Cox-Singh J, Brooke G, et al. *Plasmodium knowlesi* from archival blood films: further evidence that human infections are widely distributed and not newly emergent in Malaysian Borneo. *Int J Parasitol* 2009;**39**:1125–8.
16. Barber BE, William T, Grigg MJ, et al. Evaqluation of the senstivity of a pLDH-based and an aldolase-based rapid diagnostic test for diagnosis of uncomplicated and severe malaria caused by PCR-confirmed *Plasmodium knowlesi, Plasmodium falciparum*, and *Plasmodium vivax*. *J Clin Microbiol* 2013;**51**:1118–23.
17. Tanizaki R, Ujiie M, Kato M, et al. First case of *Plasmodium knowlesi* infection in a Japanese traveller returning from Malaysia. *Malar J* 2013;**12**:128.
18. Daneshvar C, Davis TM, Cox-Singh J, et al. Clinical and parasitological response to oral chloroquine in uncomplicated human *Plasmodium knowlesi* infections. *Malar J* 2010;**9**:238.
19. Rajahram GS, Barber BE, William T, et al. Deaths due to *Plasmodium knowlesi* malaria in Sabah, Malaysia; association with reporting as *Plasmodium malariae* and delayed parenteral artesunate. *Malar J* 2012;**11**:284.
20. Singh AP, Puri SK, Chitnis CE. Antibodies raised against receptor-binding domain of *Plasmodium knowlesi* Duffy binding protein inhibit erythrocyte invasion. *Mol Biochem Parasitol* 2002;**121**:21–31.
21. Hamid MMA, Remarque EJ, van der Duivenvoorde LM, et al. Vaccination with *Plasmodium knowlesi* AMA1 formulated in the novel adjuvant Co-vaccine HT protects against blood-stage challenge in rhesus macaques. *PLoS One* 2011;**6**:e20547.
22. Thera MA, Doumbo OK, Coulibaly D, et al. A field trial to assess a blood-stage malaria vaccine. *N Engl J Med* 2011;**365**:1004–13.
23. Miura K, Herrera R, Diouf A, et al. Overcoming allelic specificity by immunization with five allelic forms of *Plasmodium falciparum* apical membrane antigen 1. *Infect Immun* 2013;**81**:1491–501.

Chapter 26

Measles

Alpay Azap[1] and Filiz Pehlivanoglu[2]
[1]*Ankara University, Infectious Diseases and Clinical Microbiology Department, Ankara, Turkey,* [2]*Haseki Training and Research Hospital, Infectious Diseases and Clinical Microbiology Clinic, Istanbul, Turkey*

CASE PRESENTATION

A 38-year-old female patient was admitted to the Haseki Training and Research Hospital's emergency department with complaints of respiratory distress. The patient had had fever, influenza-like symptoms, and malaise for 3 days. On her admission, she had a body temperature of 39°C, and maculopapular rash on the face, neck, and upper trunk. The WBC count was 9.500/mm^3, ALT was 51 U/L, AST was 48 U/L, GGT was 70 U/L, and CRP was high, 18 times higher than the upper limit. The patient had no history of contact with persons with fever and rash. There was no underlying comorbid condition. She had not received MMR vaccine. Measles IgM antibodies were positive with ELISA. She developed acute respiratory distress syndrome (ARDS) on the second day of her admission. She was intubated and transferred to the intensive care unit. She died 2 days later.

Besides a case presentation, we analyzed 82 adult measles cases applied to Haseki Training and Research Hospital in 2012 and 2013. The diagnoses were established based on measles IgM positivity by ELISA. Of these cases, 68% were hospitalized. The mean age of the cases was 27, between 18 and 41 years, and 66% were female. Seven pregnant cases were aged between 18 and 30 years. Only 23% of the patients reported history of contact with a patient with typical eruptions. Seven percent of the patients had no fever at admission to hospital. The most common symptoms were conjunctivitis (38%), LAP (29%), Koplik's spots (18%), viral pneumonia (20%), and diarrhea (24%), respectively. All patients had maculopapular eruptions. Hospitalization ranged from 2 to 9 days, average 5 days. Leukocyte count was between 2400 and 13.500/mm^3, average 5.228. Alanin transferase values ranged between 8 and 581 U/L, and 42% of the patients had ALT values

Emerging Infectious Diseases. DOI: http://dx.doi.org/10.1016/B978-0-12-416975-3.00026-1

above normal limit. Forty percent of the patients were given symptomatic treatment, 28% received antibiotics, 5% received only vitamin A, and 27% received both antibiotics and vitamin A.

1. BRIEF JUSTIFICATION ON WHY THIS CASE WAS IDENTIFIED AS EMERGENT

Measles or rubeola is a highly infectious disease that caused epidemics throughout history, probably for more than 5000 years.[1] It was endemically seen in crowded populations as a childhood disease. Although the clinical course of measles is often not fatal, it has high fatality in populations not previously exposed to the virus and in undernourished children living in developing or poor countries. Prior to the license of Enders' effective vaccine in 1963, many deadly measles epidemics have been seen all over the world. One of these epidemics struck the islands of Fiji in 1875 and killed as many as one-quarter of the Fijian population in 3 months.[1] Over the past few decades, the World Health Organization (WHO) has launched a number of programs in order to control and eliminate measles. Despite great efforts, measles elimination could not be achieved even in Europe and the east Mediterranean region. Since 2007, many cases of measles have been seen in France, Germany, England, Netherland, Norway, Italy, Bulgaria, Russia, and Turkey. During 2011, measles outbreaks were reported in 36 of 53 member states in Europe, with a total of more than 26,000 measles cases.[2] Thus, the WHO Regional Committee for Europe recommitted to the goal of "eliminating measles in Europe by 2010" and revised the target date for elimination from 2010 to 2015.[3]

2. WHAT IS THE CAUSATIVE AGENT? (TAXONOMY AND DESCRIPTION OF THE AGENT)

Measles virus (MV) is a member of the genus *Morbillivirus* of the family Paramyxoviridea.[4] It is an enveloped, non-segmented, single-stranded, negative-sense RNA virus. Measles virions are seen as pleomorphic spheres with a diameter of 100–250 nm on electron microscopy. Measles virus encodes at least eight structural proteins, which have letter names: F, C, H, L, M, N, P, and V. Of these proteins, H (hemagglutinin) has a role in the attachment of the virus to host cells, and F (fusion) is involved in the spread of the virus from one cell to another.[4]

Measles virus is closely related to the viruses causing diseases in animals such as canine and phocine distemper and rinderpest viruses. It is stated that MV adapted to humans when humans first began to domesticate animals in Mesopotamia around 3000 BC.[1] Today, wild MV is pathogenic only for primates.

3. WHAT IS THE FREQUENCY OF THE DISEASE? (PREVALENCE, INCIDENCE, BURDEN, AND IMPACT OF THE DISEASE)

Measles is one of the most contagious diseases, with a secondary attack rate of >90% in susceptible household contacts.[5] Prior to the availability of measles vaccine, measles infected over 90% of children before they reached 15 years of age. These infections were estimated to cause more than 2 million deaths and between 15,000 and 60,000 cases of blindness annually worldwide.[6] The case fatality rate for measles is highest in infants aged under 12 months. In developed countries, the case fatality rate for measles is 0.05–0.1 per 1000 cases, whereas in some developing countries it is still high among children, reaching 5–6%.[7] WHO estimated that in the year 2000, 535,000 children died because of measles. This number accounts for 5% of all deaths under five mortality. With the help of vaccination efforts, estimated global measles mortality decreased 74% from 535,000 deaths in 2000 to 139,300 in 2010.[7]

Measles has important complications affecting the respiratory tract and central nervous system (CNS). Approximately 30% of cases have one or more complications.[7] The most common respiratory complications are otitis media and pneumonia. Pneumonia may be caused by the virus itself or by bacterial superinfection and accounts for 60% of measles deaths in infants.[4] Acute encephalitis is as rare as one per 1000–2000 cases, but is a serious complication of measles ranging from mild to severe, and has a high proportion of neurologic sequelae. A less common, one per 100,000 cases, but very serious complication is subacute sclerosing panencephalitis (SSPE), a mortal degenerative CNS disease that occurs secondary to persistent infection with a defective measles virus.[5] Other complications of measles are blindness, stomatitis, diarrhea, subclinical hepatitis, appendicitis, ileocolitis, pericarditis, myocarditis, hypocalcemia, and thrombocytopenia.[4,5]

4. WHAT ARE THE TRANSMISSION ROUTES?

Humans are the only natural host for sustaining measles virus transmission. Infected people are usually contagious from 4 days before to 4 days after rash onset (period of infectivity). Transmission occurs primarily person to person via large respiratory droplets. The virus remains infective in droplet nuclei form in air in closed areas for up to 2 hours after a person with measles has occupied the area.[8] This fact may account for the increased incidence of measles in the winter. Airborne spread of measles in physicians' offices has been observed.[4]

5. WHICH FACTORS ARE INVOLVED IN DISEASE PATHOGENESIS? WHAT IS THE PATHOGENIC MECHANISM?

Measles virus enters the body via respiratory epithelium at any point from the nose to the lower parts of the respiratory tract. The H (hemagglutinin) and F (fusion) proteins that present on the virion surface initiate infection of susceptible cells. The H is responsible for interaction of the virus with specific MV receptors on susceptible cells and is an important determinant of cell tropism. The F protein is responsible for fusion of the viral envelope and cellular plasma membrane to initiate infection. Antibodies to these proteins can neutralize virus infectivity.[4,9] After replication in the respiratory epithelium, MV reaches the local lymph nodes from which it will spread to blood, spleen, lymphatic tissue, lung, thymus, liver, skin, conjunctiva, intestine, and bladder. Infection of the entire respiratory mucosa accounts for the classic signs of measles: cough and coryza. A few days after generalized involvement of the respiratory tract has occurred, Koplik's spots appear on buccal mucosa and are followed by the development of skin rash. The appearance of the measles rash coincides temporally with the appearance of serum antibody, suggesting that the skin and mucous membrane manifestations of disease are due to hypersensitivity of the host to the virus. The main component of the immune system that is responsible for the fight against MV is cell-mediated immunity. Patients with deficiencies in cell-mediated immunity may develop measles giant cell (Hecht's) pneumonia without a rash after an exposure to MV or after vaccination.[4,9]

On the other hand, measles itself is associated with immune suppression lasting for several weeks. Loss of delayed-type hypersensitivity responses and increased susceptibility to secondary infections seen in measles cases are the results of MV-induced immunosuppression. Various mechanisms for MV-induced immunosuppression have been suggested, including lymphopenia, type 2 skewing of cytokine responses, and suppression of lymphocyte proliferation.[9]

6. WHAT ARE THE CLINICAL MANIFESTATIONS?

Typical measles infection can be subdivided into the following clinical stages: incubation, prodromal phase, exanthem, and recovery.[10] The incubation period for measles is 10–14 days but can range from as few as 7 days to as many as 21 days. It is slightly longer in adults than in children. There are no signs or symptoms of measles during this time. Malaise, fever, anorexia, conjunctivitis, and respiratory symptoms such as cough and coryza (three "c" signs) are seen in the prodromal phase, which usually lasts for 2 or 3 days, maximum 8 days.[4] Koplik's spots, bluish-gray particles on an erythematous base, occur on buccal mucosa approximately 48 hours prior to the characteristic skin rash. This enanthema is pathognomonic for measles but

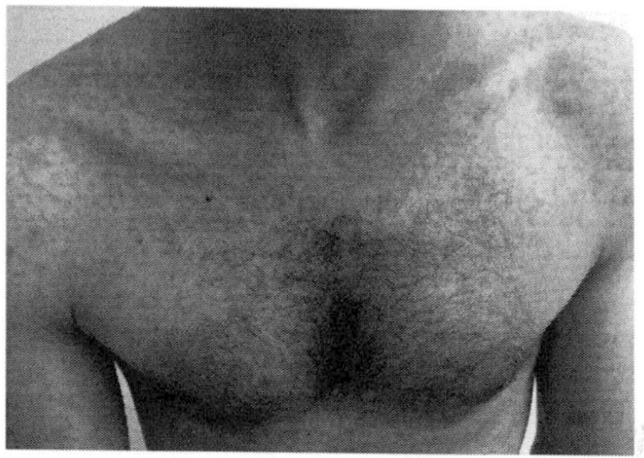

FIGURE 26.1 Maculopapular rash on a patient with measles.

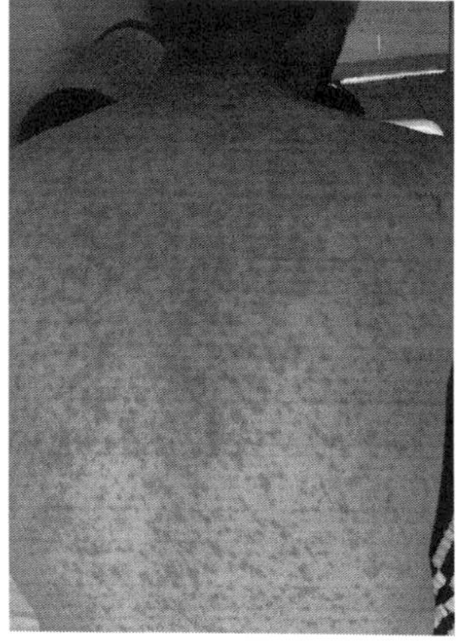

FIGURE 26.2 Maculopapular rash on the back.

does not appear in all patients. Koplik's spots usually disappear in 12 to 72 hours.[10] The exanthema of measles is a maculopapular rash beginning on the face and spreading cephalocaudally and centrifugally to involve the neck, upper trunk, lower trunk, and extremities (Figures 26.1 and 26.2). The rash becomes confluent, especially on the face and the neck. After 3 to 4 days, the rash begins to fade and turn into brownish macular lesions with fine desquamation (Figure 26.3). The rash usually lasts 5 to 7 days. Several days after the rash occurs fever resolves and the patient feels better. Cough may be the last symptom to disappear and may persist for 1 to 2 weeks after measles infection.

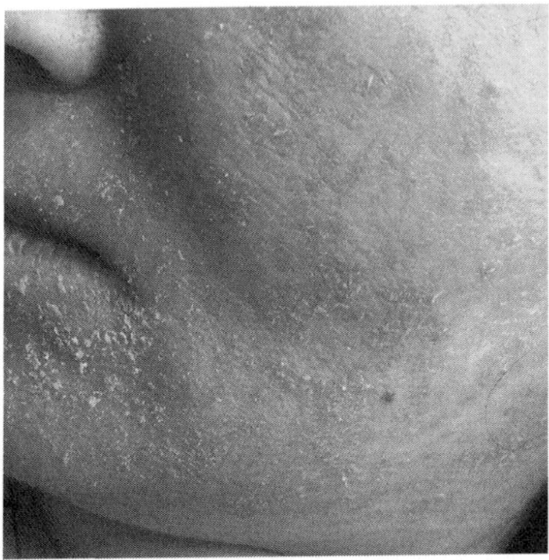

FIGURE 26.3 Desquamation on the face.

6.1 Modified Measles

In patients who have some degree of immunity to virus, a modified and mild form of measles can occur.[4,5] These patients are people who recently received immunoglobulin products or young infants who have maternal antibodies against measles. The incubation period usually lasts longer than classical measles (17–21 days). The clinical course may vary and classical prodromal signs such as Koplik's spots may not appear and the rash may be different in this form of measles.

6.2 Atypical Measles

This is a rare form of measles that occurs when a person previously vaccinated with killed-virus measles vaccine is infected with wild-type measles. This type of vaccine has not been used since 1967. Atypical measles has a shorter prodrome (1 to 2 days) followed by a rash that begins on distal parts of extremities and spreads centripetally, sparing the head, neck, and face. The rash may be petechial, urticarial, vesicular, or maculopapular. The clinical course is more severe than classical measles with a higher rate of lung and CNS involvement. Patients with atypical measles do not transmit the disease to other people.[4,5]

7. HOW DO YOU DIAGNOSE?

Laboratory confirmation of measles requires either measles virus detection by polymerase chain reaction (PCR) or detection of measles-specific

IgM antibodies by serological testing. Generally, a previously susceptible person exposed to either vaccine or wild-type measles virus first mounts an IgM response and then an IgG response. Approximately 70% of measles cases are IgM positive at 0–2 days after the rash onset, and 90% are positive 3–5 days after rash onset. IgM antibody levels peak after 7–10 days and then decline, being rarely detectible after 6–8 weeks. IgG antibody levels peak within 3 weeks and persist long after the infection.[7] False-positive IgM test results may sometimes occur due to cross-reacting IgM antibodies to viruses (e.g., Epstein–Barr virus [EBV], human parvovirus B19), rheumatoid factor or other autoantibodies, and polyclonal stimulation of IgM response by EBV.[7]

PCR is the most useful test in the first few days of illness when serology was negative. The test is also useful as a test for measles infection in severely immunocompromised patients in whom antibody response is unreliable. Measles virus can be isolated from urine, nasopharyngeal aspirates, heparinized blood, or throat swabs. Clinical specimens should be obtained within 7 days, and not more than 10 days, after onset of rash.[7,8] WHO recommends IgM antibody detection by ELISA as the standard test for routine measles surveillance.[7]

8. HOW DO YOU DIFFERENTIATE THIS DISEASE FROM SIMILAR ENTITIES?

Accurate diagnosis is important because of the potential effect of measles on patients as well as on their contacts. Patients with measles frequently present with fever and rash. Measles should be considered in the differential diagnosis of fever and generalized maculopapular rash. The differential diagnosis of measles is very broad and includes rubella, parvovirus B19 infection, rickettsial, enteroviral, and adenoviral infections, human herpes virus 6 infection, human herpes virus 7 infection, infectious mononucleosis, Kawasaki's disease, meningococcemia, toxic shock syndrome, dengue, HIV, secondary syphilis, drug reactions, cutaneous lupus erythematosus, and inflammatory bowel disease.[11,12] Medical history, physical examination, and laboratory methods are helpful to differentiate measles from other diseases.

A history of contact to a person with suspected or proven measles is an important clue for diagnosis and is usually the case during an ongoing epidemic. Detection of prodromal signs of fever, conjunctivitis, cough, and coryza in a patient who has not been vaccinated before suggests measles. In such a patient, buccal mucosa should be examined for Koplik's spots. Koplik's spots are pathognomonic for measles but they do not exist in all patients with measles. The localization and appearance of rash are typical but not pathognomonic. Measles may be confirmed in the laboratory by molecular or serologic tests.

9. WHAT IS THE THERAPEUTIC APPROACH?

Treatment of measles mainly involves supportive therapy such as antipyretics and fluids as indicated.[4,5] Prompt treatment of bacterial superinfections with appropriate antimicrobials is important, but prophylactic antibiotics to prevent superinfections are not recommended. Although there are limited data indicating the beneficial effect of intravenous ribavirin in the treatment of measles pneumonia, the efficacy of ribavirin (intravenously or by aerosol) for treatment of severe measles is unproven.[13]

Vitamin A decreases the severity of measles especially in malnourished children suffering vitamin A deficiency. Vitamin A should be administered orally for 1 day at a dose of 100,000 IU for children 6 months to 1 year old and 200,000 IU for children older than 1 year. Administration of vitamin A has been reported to reduce serologic conversion in vaccine recipients and should therefore be avoided at or after immunization. Side effects of vitamin A are transient vomiting and headache.[4]

10. WHAT ARE THE PREVENTIVE AND INFECTION CONTROL MEASURES?

10.1 Prevention

A very safe, cheap, and effective live attenuated vaccine against measles has been available since the 1960s. Two doses of vaccine (first very early in the second year of life and second in early childhood) may provide long-lasting (life-long) immunity in 95% of vaccinated individuals.[4] Vaccination is not recommended for infants younger than 12 months because of the risk of suppression of immunity induction by residual transplacentally acquired antibodies. However, in situations or regions in which the incidence of natural measles before the age of 1 year is high, live measles vaccine may be given at 6 to 9 months of age but should be routinely followed by an additional two doses given after 12 months of age.[4]

Live measles vaccine is contraindicated in pregnant women and in persons who have cell-mediated immunodeficiency. Women who have received measles vaccine should avoid becoming pregnant for 28 days after vaccination. Since serious hypersensitivity reactions to measles vaccine have been observed in persons allergic to egg protein, people with a history of anaphylactic reactions after egg ingestion should be vaccinated only with extreme caution.[4]

Passive immunization against measles is also possible by immunoglobulin administration. Immunoglobulin is recommended for the following risk groups[5,8]:

- People with malignant disease, particularly if they are receiving chemotherapy, radiotherapy, or both.

- People with cell-mediated immunodeficiency, including patients with AIDS.
- Infants younger than 1 year if they are in household contact with a patient, including newborns whose mothers have measles.
- Pregnant women.

Immunoglobulin must be administered within 6 days after an exposure at a dosage of 0.25 mL/kg body weight, with a maximum of 15 mL intramuscularly. The recommended dose of IG for immunocompromised people is 0.5 mL/kg of body weight (maximum 15 mL) intramuscularly.[8] Immunoglobulin should not be used to control measles outbreaks.[5]

Susceptible people who are exposed to measles, with the exception of young infants, pregnant women, and immunocompromised persons, may be protected by administration of live vaccine in the first 72 hours of exposure as an alternative to immunoglobulin.[4]

10.2 Infection Control Measures

Healthcare facilities are important sites for acquisition of measles, especially in developed countries where measles is re-emerging. Waiting areas in outpatient clinics and emergency departments serve as sites of amplification in many measles outbreaks.[14] Young physicians are usually not familiar with measles, especially in countries with very low incidence, thus delayed diagnosis of a case is common in many instances. Delay in diagnosis causes implementation of inadequate infection control measures and spread of the infection not only in the community but among susceptible healthcare workers and patients as well.[14] It is important to increase measles awareness among healthcare workers by training them to recognize patients who may have measles. Applying a clear case definition is essential for early recognition and timely implementation of appropriate isolation precautions.[15]

10.3 Managing Patients with Measles[15,16]

Patients with laboratory confirmed measles and patients meeting the clinical case definition for measles should be managed with airborne precautions during the "period of infectivity." If the patient is immunocompromised (including pregnant women), airborne precautions should continue for the entire duration of illness. Airborne precautions are as follows:

- Airborne infection isolation room (room with negative pressure) should be available.
- Fitted N-95 masks should be worn by all non-immune healthcare workers and visitors who enter the room.
- Single use gown and gloves should be worn by staff if there is a possibility of contact with body fluids and secretions.

- Until placed in a room with negative pressure, the patient should wear a surgical mask.
- Healthcare workers and visitors without evidence of immunity should not be permitted to enter the rooms of patients with measles.
- If a negative pressure room is temporarily unavailable, the patient should be managed in a single room with the door closed.
  - The patient should wear a surgical mask.
  - The room should not be used for at least 2 hours after being vacated by the infectious patient and should be cleaned prior to being reoccupied.

10.4 Managing the Infection Control among Healthcare Workers[14–17]

The risk of acquisition of measles is higher for healthcare workers than for the general population.[15] Morbidity of measles in healthcare workers is also high since the complications of disease occur more frequently in adults.

- The immune status of all staff (clinical and non-clinical) in every healthcare facility should be determined prior to exposure. Nursing and medical students should also be screened and vaccinated if needed before they start their probation.
- All staff born before 1957 are considered immune.
- All staff born after 1957 are considered immune if they have documentation of having received two doses of measles-mumps-rubella (MMR) vaccine, and have had laboratory confirmed measles or documented laboratory evidence of immunity.
- All non-immune staff should be vaccinated with MMR unless contraindicated (e.g., pregnancy).
- If a patient with measles is admitted to a unit then all non-immune staff should be redeployed to other work areas until they have documented evidence of having received at least one dose of MMR vaccine.
- Exposed staff should be offered post-exposure prophylaxis as appropriate (see above).
- Exposed non-immune staff should be relieved of direct patient contact from day 5 to day 21 post-exposure regardless of whether they have received vaccination or immunoglobulin.

10.5 Managing Non-Immune Exposed Patients

- Non-immune exposed patients should be managed with airborne precautions until 21 days post-exposure or 4 days after development of rash regardless of whether they have received vaccination or immunoglobulin.

- Immunocompromised patients who have been exposed should be managed with airborne precautions until 21 days post-exposure or for the duration of illness regardless of whether they have received immunoglobulin.

REFERENCES

1. Dobson M. *Disease: the extraordinary stories behind history's deadliest killers*. London: Quercus History; 2007. p. 140–5.
2. World Health Organization. *Wkly Epidemiol Rec* 2011;**86**:557–64.
3. Regional Committee for Europe resolution EUR/RC60/R12 on renewed commitment to elimination of measles and rubella and prevention of congenital rubella syndrome by 2015 and sustained support for polio-free status in the WHO European Region. Copenhagen, WHO Regional Office for Europe. <http://www.euro.who.int/__data/assets/pdf_file/0016/122236/RC60_eRes12.pdf>; 2010.
4. Gershon AA. Measles virus (rubeola). In: Mandell GL, Bennett JE, Dolin R, editors. *Principles and practice of infectious diseases*. 7th ed. New York: Churchill Livingstone; 2010. p. 2229–36.
5. Sabella C. Measles: not just a childhood rash. *Cleve Clin J Med* 2010;**77**:207–13.
6. World Health Organization Global Measles and Rubella Strategic plan: 2012–20, Geneva. <http://www.who.int/immunization/newsroom/Measles_Rubella_StrategicPlan_2012_2020.pdf>; 2012.
7. World Health Organization, Surveillance Guidelines for Measles, Rubella and Congenital Rubella Syndrome in the WHO European Region. Copenhagen, December 2012.
8. Measles. Epidemiology and prevention of vaccine-preventable diseases. The pink book: course textbook—12th ed., Second Printing, May 2012. p. 173–92. <http://www.cdc.gov/vaccines/pubs/pinkbook/meas.html>.
9. Griffin DE. Measles virus-induced suppression of immune responses. *Immunol Rev* 2010;**236**:176–89.
10. Perry RT, Halsey NA. The clinical significance of measles: a review. *J Infect Dis* 2004;**189**(Suppl. 1):S4.
11. Black JB, Durigon E, Kite-Powell K, de Souza L, Curli SP, Afonso AMS, et al. Seroconversion to human herpesvirus 6 and human herpesvirus 7 among Brazilian children with clinical diagnoses of measles or rubella. *Clin Infect Dis* 1996;**23**:1156–8.
12. Asaria P, MacMahon E. Measles in the United Kingdom: can we eradicate it by 2010? *BMJ* 2006;**333**:890–5.
13. Forni AL, Schluger NW, Roberts RB. Severe measles pneumonitis in adults: evaluation of clinical characteristics and therapy with intravenous ribavirin. *Clin Infect Dis* 1994;**19**:454–62.
14. Maltezou HC, Wicker S. Measles in health-care settings. *Am J Infect Dis* 2013;**41**:661–3.
15. Botelho-Nevers E, Gautret P, Biellik R, Brouqui P. Nosocomial transmission of measles: an updated review. *Vaccine* 2012;**30**:3996–4001.
16. Best E, Voss L, Roberts S, Freeman J. Measles—Infection Control Definitions & Guidelines. <http://www.adhb.govt.nz/starshipclinicalguidelines/Measles%20Infection%20Control%20and%20Guideline.htm>; 2011.
17. Kutty P, Rota J, Bellini W, Redd SB, Barskey A, Wallace G. Chapter 7: Measles. In: CDC, VPD surveillance manual, 6th ed.; 2013.

Chapter 27

Pertussis

Larry Lutwick[1] **and Jana Preis**[2]
[1]*Division of Infectious Diseases, Departments of Medicine and Biomedical Sciences, Western Michigan University Homer Stryker MD School of Medicine, Kalamazoo, MI, USA,*
[2]*Division of Infectious Diseases, New York Harbor Health Care System, Brooklyn Campus, and State University of New York, Downstate Medical School, Brooklyn, New York, NY, USA*

CASE PRESENTATION[1]

A 19-day-old female infant was admitted to a neonatal ICU in Ankara, Turkey. A product of an unremarkable gestation and vaginal delivery at 39 weeks, she began to have symptoms of an upper respiratory infection at the age of 15 days. The illness abruptly worsened 3 days later with cough, cyanosis, periods of apnea, and progressive respiratory failure precipitating her admission. The mother, who had received DPT (diphtheria, pertussis, and tetanus) vaccination as a child, reported a mild cough beginning in the 2 weeks prior to delivery. Upon arrival, the infant was having apneic episodes with cyanosis and her O_2 saturation dropped to 36% during a coughing spasm. An initial white blood cell count was 27,200/μL with 60% lymphocytes and a chest X-ray did not reveal any infiltrates.

Despite ampicillin and gentamicin, she deteriorated further and endotracheal intubation was necessary but was discontinued on day 2 of her admission. Nasopharyngeal washings were tested by PCR and were negative for respiratory syncytial virus, influenza A and B, adenovirus, parainfluenza virus, and coronavirus. Based on the infant's symptom complex and lymphocytosis, she had erythromycin therapy added on day 4. Nasopharyngeal aspirates obtained before erythromycin treatment was begun were positive for pertussis by PCR and by culture on Bordet–Gengou agar. The infant's serum antibody against pertussis toxin was undetectable. The infant slowly improved with fewer bouts of paroxysmal coughing and she was discharged on hospital day 10. The mother was the only contact who reported cough and was the only contact that was seropositive for pertussis although her culture was negative.

Emerging Infectious Diseases. DOI: http://dx.doi.org/10.1016/B978-0-12-416975-3.00027-3

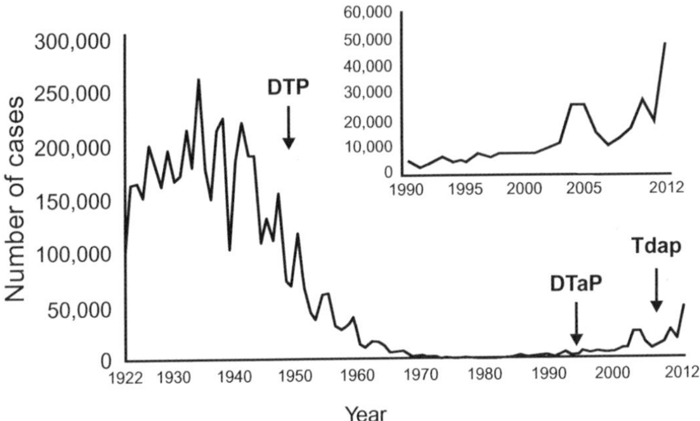

FIGURE 27.1 The USA reported pertussis cases: 1922–2012. *From: CDC.*

1. BRIEF JUSTIFICATION ON WHY THIS CASE WAS IDENTIFIED AS EMERGENT

As shown in Figure 27.1 from the Centers for Disease Control and Prevention (CDC)[2] and further discussed in the section on frequency and epidemiology, the numbers of reported cases of pertussis in the USA and other parts of the developed world began to increase from 10,000 to 25,000 in the first decade of the 21st century. In 2012, almost 50,000 cases were reported, the highest number of cases since 1955 with many infected infants requiring hospitalization, including at least 18 deaths. Outbreaks were also occurring in Europe, Australia, and Japan. It is this dramatic increase, which appears to be tied to the introduction of the safer vaccine (the pertussis paradox[3]), that has made this infection an emerging (actually reemerging) disease, which has its highest risk of significant morbidity and mortality in infants.

2. WHAT IS THE CAUSATIVE AGENT?

The primary cause of pertussis, *Bordetella pertussis*, is one of eight species of the *Bordetella* genus. Among the group, *B. pertussis*, *B. parapertussis*, and *B. bronchiseptica* are the most studied. *B. bronchiseptica* causes kennel cough in dogs and asymptomatic carriage can occur in many animals, but is not a usual human pathogen. The organism appears to be the evolutionary ancestor of both *B. pertussis* and *B. parapertussis*. The organisms are highly trophic for the cilia of the mucosa of the respiratory system and from this location a significant number of toxins are produced, which participate in the disease pathogenesis.

B. pertussis is a small, strictly aerobic, Gram-negative coccobacillus that is quite fastidious, requiring specific media for its isolation. The morphology of the organism is not distinguishable from other *Bordetellae* or from *Haemophilus* species. *B. pertussis* is non-motile, oxidase positive, catalase positive, and is relatively inert biochemically. The organism is only isolated from humans.

3. WHAT IS THE FREQUENCY OF THE DISEASE?

Pertussis (whooping cough) remains the most commonly reported vaccine-preventable disease in the United States in children younger than 5 years. Pertussis immunization as part of the DPT vaccine was introduced in the United States in the late 1940s and, as can be seen in Figure 27.1, the number of reported cases dropped slowly over the decades of the 1950s and 1960s (with periodic bumps characteristic of the infection) to a nadir of 1010 cases in 1976. That number is >99% lower than the number of cases in 1947. The numbers of reported cases remained low but in the 2000–4500 range through the mid-1990s. Because of issues of reactogenicity with DPT related to the whole cell pertussis component, which became a political issue as the vaccine was inaccurately blamed for permanent brain damage among other things, the pertussis pediatric immunization schedule was gradually changed to an acellular one. This vaccine utilizes acellular pertussis toxoids, which were much better tolerated and seemed to produce a similar immune response as compared to DPT. Unlike DPT, the vaccine with a smaller amount of pertussis toxoids (Tdap) could be tolerated in adults.

In 2010, according to CDC, the US pertussis rate reached 27,550 cases (the highest number since 1959), with 27 related deaths.

In 2011, according to statistics from CDC, adolescents (aged 11–19 years) and adults together accounted for 47% of pertussis cases, while children aged 7–10 years accounted for 18% of cases.

During 2012, 48,277 cases of pertussis were reported to CDC. The incidence rate of pertussis among infants exceeds that of all other age groups. The second highest rates of disease are observed among children 7–10 years old. Rates also increased in adolescents 13 and 14 years of age. Eighteen pertussis-related deaths during 2012 have been reported to CDC. Almost all of the deaths occurred among infants younger than 3 months of age. During 2012, increased pertussis cases or outbreaks were reported in a majority of states. Forty-nine states and Washington, DC reported increases in disease in 2012 compared with 2011.

CDC has estimated that 5–10% of all cases of pertussis are recognized and reported. In studies, 12–32% of adults with prolonged (1–4-week) cough have been found to have pertussis. It is estimated worldwide to be 48.5 million cases, with a mortality rate of nearly 295,000 deaths per year. The case-fatality rate among infants in low-income countries may be as high as 4%.

4. HOW ARE THE BACTERIA TRANSMITTED?

Pertussis is a highly contagious respiratory infection, particularly in the household setting. Indeed, direct inoculation with as few as 140 organisms can cause disease in susceptible children.[4] In a review, attack rates for unvaccinated children in the household setting ranged from 64 to 86% (average 76%) but was much lower (range 0–36% in classroom contract studies).[5] Clearly, transmission required repeated or prolonged exposure and/or close contact. Airborne transmission via respiratory droplets had been postulated but not clearly proven until 2012 when Merkel's laboratory[5] demonstrated airborne transmission between infected and naïve baboons.

4.1 Household

Transmission of the infection to high risk infants tends to occur from inside the family unit. An Australian review[6] reported that 39% of the time the mother was the source, 16% fathers, 5% grandparents, and both siblings and non-family sources had very heterogeneous rates. In as many as 52% of circumstances, no source was identified. In a Dutch review,[7] the estimated relative infections of mothers were 3.9 and fathers 0.44. A report from Korea came to similar conclusions.[8] Clearly, the mother appears the major player in neonatal transmission and should be at the head of the line if selective, rather than universal, family "cocooning" of the infant is used. Even an entire family vaccination may miss potential exposures as 37% of contacts of English infants aged < 10 weeks were non-household individuals lasting > 15 minutes.[9]

4.2 Nosocomial

Perhaps the worst scenario for nosocomial transmission of *B. pertussis* is a neonatal nursery. In 2003, such an outbreak[10] (prior to the acellular vaccine use) was introduced by a symptomatic nurse who was not diagnosed with the infection despite a characteristic illness. The nurse had made multiple health-care visits but the disease was detected only after a 2-month-old premature infant developed pertussis and four other nurses were subsequently diagnosed with pertussis, probably transmitted by the nurse prior to diagnosis. Azithromycin prophylaxis was recommended to all infants in the unit during the time of the illness of the index nurse. Seventy-two infants received post-exposure prophylaxis (PEP) as well as 72 healthcare workers (HCW). No other infant cases occurred but a resident physician who cared for the ill infant and declined prophylaxis did develop pertussis. A more recent outbreak among oncology nurses also showed transmission among HCW and none from the HCW to patients.[11] With the availability of acellular pertussis vaccine boosters, HCW, especially those with exposure to infants, should be vaccinated.

5. WHICH FACTORS ARE INVOLVED IN DISEASE PATHOGENESIS? WHAT ARE THE PATHOGENIC MECHANISMS?

As reviewed by Preston,[12] de Gouw et al.,[13] and Hewlett,[14] *B. pertussis* produces a cadre of factors involved in pathogenesis. The major adhesion factors enabling close contact of the organism and the respiratory epithelium are filamentous hemagglutinin (FHA), fimbriae, and pertactin (prn). Both FHA and prn are included in the antigen profile of the acellular vaccine. The toxins produced by the organism include:

- pertussis toxin (Ptx), a complex hexameric protein that is an ADP-ribosyltransferase;
- adenylate cyclase toxin (ACT), which is post-translationally modified to be able to facilitate apoptosis and cytotoxicity;
- a type 3 secretion system effector protein such as BteA that induces rapid, non-apoptotic cell death; and
- tracheal cytotoxic (TCT), a disaccharide tetrapeptide derived from the organism's cell wall, which causes ciliostasis and damages respiratory epithelial cells.

Ptx is a major part of the forms of the acellular pertussis vaccine and is felt to play a major role in the disease but it cannot be a *sine qua non* for the disease as *B. parapertussis* can produce a very similar disease despite not producing Ptx. *B. parapertussis* does contain the gene for Ptx but does not express it because of mutations in the promoter gene.

B. pertussis contains several regulatory systems controlling the expression of the virulence genes in response to environmental signals; the most notable of these is bvgASR. The organism also contains several iron acquisition systems.

In addition to these varied adherence and toxic factors, the organism can successfully persist in the human host by its ability to "interfere with almost every aspect of the immune system, from the inhibition of complement and phagocyte-mediated killing to the suppression of T- and B-cell responses".[13] Further understanding of the immune system modifications facilitated by the organism may assist in developing more effective vaccines.

6. WHAT ARE THE CLINICAL MANIFESTATIONS?

Pertussis generally has an incubation period of 7–10 days but it is important to realize that more than a fifth of secondary cases in a German household milieu presented more than 4 weeks following the primary case onset,[15] which may reflect a longer incubation period or exposure later in the primary case's illness. As reviewed by Cherry and Heininger[16] and Mattoo and Cherry,[17] there are many factors that impact on the pertussis illness including age and previous immunization status of the individual, size of inoculum, antimicrobial therapy, presence of passively acquired specific antibody, and genetic factors of both the host and the organism.

Pertussis is an acute infection of the respiratory tract and the classical illness most frequently occurs as a primary infection in young, unimmunized children. The clinical course of the 6–12-week process is divided into three stages:

1. The catarrhal phase is characterized by insidious onset of mild upper respiratory symptoms similar to rhinovirus infections including low-grade fever, coryza, sneezing, and a mild, occasional cough. During the 1–2 weeks of this stage, the cough gradually becomes more severe.
2. The paroxysmal phase manifests as spasmodic coughing episodes, or paroxysms of as many as 10 or more coughs without an inspiration. These spasms sometimes are followed by a long inspiratory whooping sound and/or by posttussive vomiting. Paroxysmal attacks occur more frequently at night and may be precipitated by eating. Cyanosis can occur during paroxysms and the spasmodic cough may have many complications including cerebral hypoxia, subcutaneous emphysema, subconjunctival hemorrhage, umbilical or inguinal hernia, rib fracture, severe alkalosis, and seizures. Young children and infants may appear quite distressed and exhausted following an episode. Remarkably, the child can appear well between attacks. This stage usually lasts 2–6 weeks, but may persist for up to 10 weeks. Neither fever nor pharyngitis is common in pertussis unless secondary bacterial superinfections intervene.
3. In the third or convalescent phase, recovery is gradual with paroxysms subsiding initially in frequency and then in severity and the cough may disappear in 2–3 weeks. Paroxysmal episodes may return with other respiratory infections.[17]

It is important to be aware that mild or even asymptomatic infection can occur especially in previously immunized children, adolescents, and adults, but even in unimmunized children, as many of 5% of apparently healthy infants had polymerase chain reaction (PCR) evidence of pertussis.[18] The milder illness in older children, adolescents, and adults is manifest by chronic cough lasting 3–4 weeks or longer. In a report of a group of college students who were not clinically diagnosed with pertussis but found to have laboratory evidence of the illness,[19] the mean length of cough was 3 weeks. The only two features differentiating pertussis from non-pertussis in the study were non-productive cough and less likely previous use of antimicrobials. Classical pertussis can occur in adults and it has been observed that adults immunologically primed from previous infection were more likely to have typical pertussis than those primed by immunization.[17]

Clearly, much of the severe morbidity and almost all of the mortality from *B. pertussis* infection occur in those younger than 6 months. As many as 63% of infants younger than 6 months of age with pertussis require hospitalization.[20] In this CDC report from infants with pertussis in 1997–2000, there were 11.8% with pneumonia, 1.4% seizures, 0.2% encephalopathy, and 0.8% died. This compares with parallel numbers of 28, 8.6, 0.7, 0.1, and <0.1% in infants 6–11 months of age.

Neonatal pertussis is observed to be especially severe with as much as a 3% risk of death.[17] Symptoms can be substantially different with periods of apnea and sometimes hypoxia-induced seizures usually the most common manifestation of infection. The cough is present, but so weak that it may be unrecognized. In these children with so-called malignant pertussis, leukocytosis, particularly with white blood cell (WBC) counts of 30,000 to 100,000, and severe pulmonary hypertension are ominous signs for mortality.[21] In a study comparing neonatal pertussis to other neonatal respiratory infections,[22] pertussis-positive neonates had longer hospital stays, less fever, more apnea and cyanosis spells, required more days of supplemental O_2 in the hospital, and represented all the infants discharged on respiratory supportive care.

7. HOW DO YOU DIAGNOSE?

Figure 27.2 shows the timing of pertussis diagnostic studies.

Symptom complexes such as a spasmodic cough without fever or chronic cough for more than 3 weeks should suggest the diagnosis of *B. pertussis* infection and appropriate tests should be performed. A video demonstrating

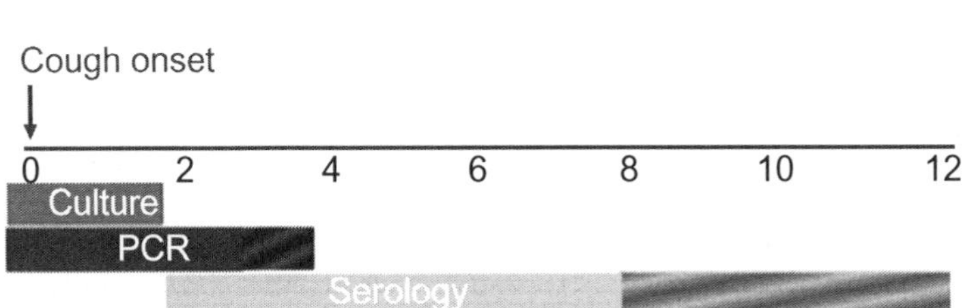

FIGURE 27.2 The timing of pertussis diagnostic studies. *From: CDC.*

the proper technique of obtaining specimens for culture and PCR can be found at: https://www.youtube.com/watch?v=DVJNWefmHjE&feature=related.

7.1 Culture

Isolation of *B. pertussis* from nasopharyngeal specimens (not pharyngeal) remains the gold standard for diagnosis. Culture is 100% specific. The degree of sensitivity (12–60%) is modified by length of illness, vaccination status, and patient age,[23] and is highest during the initial catarrhal phase. Isolation requires special transport and culture material such as Bordet–Gengou agar. Positivity can require 3–7 days of incubation.

7.2 PCR

PCR for *B. pertussis* offers a higher sensitivity than most techniques but there is no standardized testing and may be positive with other *Bordetella* species. Results are obtained within 1 day. The assays are much more sensitive late in disease and can be still positive if only non-viable organisms are present. PCR is a supplement to, not a replacement for, culture.

A direct fluorescent antigen stain is available but is not generally recommended for diagnostic use and serologies are also available. Antibodies against Ptx and filamentous hemagglutinin are also available. Anti-pertussis toxin (anti-Ptx) may take 3 weeks to become detectable and anti-FHA can help distinguish different *Bordetella* species, but neither test is US FDA approved.[23]

8. HOW DO YOU DIFFERENTIATE THIS DISEASE FROM SIMILAR ENTITIES?

Especially early in pertussis illness, prior to or early in the spasmodic cough, a variety of respiratory viruses such as the rhinovirus, respiratory syncytial virus, parainfluenza virus, adenovirus, and influenza may be confused with pertussis. Specific tests for these agents along with pertussis can help but it needs to be remembered that pertussis can co-infect with a respiratory virus and diagnosing a respiratory virus does not exclude the possibility of pertussis.[24] Especially in populations currently immunized with the acellular vaccine, mild pertussis can be difficult to differentiate from *B. parapertussis* and other viral or bacterial respiratory infections.[25]

Infection with *B. parapertussis* can cause a disease similar to pertussis but parapertussis tends to be milder and of shorter duration than pertussis. It appears that in some areas such as Scandinavia, the two infections can be of equal frequency but generally diagnosed less often and the ratio of subclinical to clinical illness is much higher in parapertussis.[26] In culture, *B. parapertussis* is less fastidious, is oxidase negative, urea positive, and produces a brown pigment on heart infusion agar.[23] Using different primers for PCR, pertussis (with IS481) and parapertussis (IS1001) can be differentiated, although IS1001 is present in *B. bronchiseptica*.[23]

Serologically, parapertussis does not express Ptx so that Ptx antibodies can distinguish the illnesses of pertussis and parapertussis but the antibody responses to FHA and prn are of similar magnitude.[27]

B. holmesii, a more recently described organism, had been initially described as a cause of bacteremia but can cause a symptom complex similar to *B. pertussis* and, like *B. parapertussis*, can co-circulate in a population. *B. holmesii* contains IS481 as *B. pertussis* but does not produce an anti-PT response.[28] It appears that current vaccination of adolescents and adults does not affect the incidence of *B. parapertussis* or *B. holmesii* as much as *B. pertussis*.[25,28]

9. WHAT IS THE THERAPEUTIC APPROACH?

With increasing incidence and widespread community transmission of pertussis, extensive contact tracing and broad-scale use of PEP among contacts may not be an effective use of limited public health resources. While antimicrobial agents may prevent pertussis if given prior to symptom onset, there are no data to indicate that widespread use of PEP among contacts effectively controls or limits the scope of pertussis outbreaks. If used, PEP use should be targeted to persons at high risk of developing severe pertussis and to persons who will have close contact with those at high risk of developing severe pertussis.

A 2007 Cochrane Review[29] looked at 13 trials with 2197 participants: 11 trials investigated treatment and two trials investigated prophylaxis. Short-term macrolide antimicrobial use (azithromycin for 3 to 5 days, or clarithromycin or erythromycin for 7 days) were as effective as long term (erythromycin for 10 to 14 days) in eradicating *B. pertussis* from the nasopharynx and had fewer side effects. Trimethoprim/sulfamethoxazole for 7 days was also effective. There were no differences in clinical outcomes or microbiological relapse between short- and long-term antimicrobial use. Indeed, while therapy is effective in eliminating *B. pertussis* from patients with the disease, rendering them non-infectious, treatment does not alter the subsequent clinical course of the illness. The review also found that contact prophylaxis of contacts older than 6 months of age with antimicrobials did not significantly improve clinical symptoms or the number of cases developing culture-positive *B. pertussis*.

Rarely, *B. pertussis* has been found to be resistant to macrolides. Most of the resistant isolates are reported from the USA[30] but a 2012 report from France also documented resistance.[31]

10. WHAT ARE THE PREVENTIVE AND INFECTION CONTROL MEASURES?

Vaccination has been the primary way of prevention of pertussis. As previously noted, the number of cases of pertussis in the industrialized world has increased in the wake of introduction of the acellular pertussis vaccine (DTaP in primary immunization of children and Tdap in adults). The small letters d and p in the adult vaccine represents a lower amount of immunogen given.

As reviewed by Cherry,[32,33] the pertussis reemergence may well be related to increased awareness of the disease, changes in case definition, increased incidence of non-pertussis *Bordetellae* and the greater availability of PCR as a diagnostic test. However, several other issues exist regarding this rise in the incidence of pertussis, including lower potency of DTaP contributing to a more rapidly waning immunity and potential genetic drift of strains of *B. pertussis*.

Clearly, the current formulation of the acellular pertussis vaccine is less potent than its whole cell ancestor. Whether measuring specific antibodies to proteins in the vaccine during the primary sequence[34] or assessing pertussis incidence after the sequence ends at age 5,[35] immunity to pertussis promptly waned. Indeed, pertussis incidence and risk ratios rose in the three study states[35] to as high as 8.8 times and 3.9 times in Minnesota between year 1 and year 6 after completion of the sequence. Currently, three-antigen and five-antigen component acellular pertussis vaccines are available in the USA. The five component vaccine had greater efficacy in a head-to-head comparison.[36] Interestingly, the relative risk of pertussis was 8.57 times higher in those with a five-dose aP schedule as compared to those who had received at least one dose of whole cell vaccine,[37] while those with six doses of aP had a relative risk of 3.55 as compared to the one or more whole cell group. Whether changing the balance of antigens in the vaccine or adding different antigens is unclear. In a response to the waning immunity observed following the primary immunization sequence of the acellular pertussis vaccine, further immunizations were recommended first for adolescents, then for adults 19–64 years old, and then for adults 65 years and older. Finally, in order to best "cocoon" the as yet unborn baby, recommendations now exist for aP vaccination of the pregnant woman in the 2nd or 3rd trimester.[38]

Evolutional changes have been observed in *B. pertussis* that may have had a role in the organism's ability to persist in a highly immunized population. Despite the genetically monomorphic nature of the pathogen without horizontal acquisition of new genes, Mooi and colleagues have observed strains carrying a mutation in the gene for the Ptx gene promoter, which mediates increased production that was associated with pertussis resurgence[39] and other genetic differences between surface proteins including Ptx, prn, and fimbriae.[40] Additionally, Lan's Australian group[41,42] have divided

B. pertussis isolates into six clusters based on single nucleotide polymorphisms (SNPs). The reemergence of pertussis seemed to coincide with the emergence of SNP cluster 1 of strains carrying prn and Ptx genes (prn2 and ptxP3) able to evade acellular pertussis vaccine-induced selective pressure.

It is likely that waning immunity related to the acellular pertussis vaccine and selective pressure of this vaccine have worked in tandem with better recognition of the illness to cause this "pertussis paradox". It may well be the case that a better vaccine may be needed to control this resurgence and better protect the susceptible infant.

REFERENCES

1. Armangil D, Tekinalp G, Yurdakõk M, Yalçin E. Maternal pertussis is hazardous for a newborn:a case report. *Turk J Pediat* 2010;**52**:206–10.
2. <http://www.cdc.gov/pertussis/surv-reported.html>; [accessed 1.10.13].
3. Allen A. The pertussis paradox. *Science* 2013;**341**:454–5.
4. MacDonald H, MacDonald EJ. Experimental pertussis. *J Infect Dis* 1933;**53**:328–30.
5. Warfel JM, Beren J, Merkel TJ. Airborne transmission of Bordetella pertussis. *J Infect Dis* 2012;**206**:902–6.
6. Wiley KE, Zuo Y, Macartney KK, McIntyre PB. Sources of pertussis infection in young infants: a review of key evidence informing targeting of the cocoon strategy. *Vaccine* 2013;**31**:618–25.
7. de Greeff SC, de Melker HE, Westerhof A, et al. Estimation of household transmission rates of pertussis and the effect of cocooning vaccination strategies on infant pertussis. *Epidemiology* 2012;**23**:852–60.
8. Kwon HJ, Yum SK, Choi UY, et al. Infant pertussis and household transmission in Korea. *J Korean Med Sci* 2012;**27**:1547–51.
9. van Hoek AJ, Andrews N, Campbell H, et al. The social life of infants in the context of infectious diseases transmission; social contacts and mixing patterns of the young. *PLoS One* 2013;**8**:e76180.
10. Bryant KA, Humbaugh K, Brothers K, et al. Measures to control of pertussis in a neonatal intermediate care nursery after exposure to a healthcare worker. *Infect Control Hosp Epidemiol* 2006;**27**:541–5.
11. Baugh V, McCarthy N. Outbreak of *Bordetella pertussis* among oncology nurse specialists. *Occupat Med* 2010;**60**:401–5.
12. Preston A. *Bordetella pertussis*: the intersection of genomics and pathobiology. *Can Med Assn J* 2005;**173**:55–62.
13. de Gouw D, Diavatopoulos DA, Bootsma HJ, Hermans PWN, Mooi FR. Pertussis: a matter of immune modulation. *FEMS Microbiol Rev* 2011;**35**:441–74.
14. Hewlett EL. A commentary on the pathogenesis of pertussis. *Clin Infect Dis* 1999;**28**: s94–8.
15. Heininger U, Cherry JD, Stehr K, et al. Comparative study of Lederle Takeda acellular pertussis component DTP (DTaP) vaccine and Lederle whole-cell component DTP vaccine in German children after household exposure. *Pediatrics* 1998;**102**:546–53.
16. Cherry JD, Heininger U. Pertussis and other *Bordetella* infections. In: Feigin RD, Cherry JD, Demmler GJ, Kaplan S, editors. *Textbook of pediatric infectious diseases*. 5th ed. Philadelphia, PA: WB Saunders; 2004. p. 1588–608.

17. Mattoo S, Cherry JD. Molecular pathogenesis, epidemiology, and clinical manifestations of respiratory infections due to *Bordetella pertussis* and other *Bordetella* subspecies. *Clin Microbiol Rev* 2005;**18**:326–82.
18. Heininger U, Kleeman WJ, Cherry JD, Group SS. A controlled study of the relationship between *Bordetella pertussis* infection and sudden deaths in German infants. *Pediatrics* 2004;**114**:e9–15.
19. Mink CM, Cherry JD, Christenson P, et al. A search for *Bordetella pertussis* infection in university students. *Clin Infect Dis* 1992;**14**:464–71.
20. CDC. Pertussis—United States, 1997–2000. *MMWR* 2002;**51**:73–6. Age distribution and incidence of reported cases. <http://www.cdc/gov/nip/ed/slides/pertussis8p.ppt>.
21. Berger JT, Carcillo JA, Shanley TP, et al. Critical pertussis illness in children: a multicenter prospective cohort study. *Pediatr Crit Care Med* 2013;**14**:356–65.
22. Castagnini LA, Munoz FM. Clinical characteristics and outcomes of neonatal pertussis: a comparative study. *J Pediatr* 2010;**156**:498–500.
23. Leber AL, Salamon DP, Prince HE. Pertussis diagnosis in the 21th century: progress and pitfalls, part I. *Clin Microbiol Newslett* 2011;**33**:111–5.
24. Ferronato AE, Gilio AE, Viera SE. Respiratory viral infections in infants with clinically suspected pertussis. *J Pediatr (Rio J)* 2013;**89**:549–53.
25. Liese JG, Renner C, Stojanov S, et al. Clinical and epidemiological picture of *B. pertussis* and *B. parapertussis* infections after introduction of acellular pertussis vaccines. *Arch Dis Child* 2003;**88**:684–7.
26. Lautrop H. Epidemics of parapertussis. 20 years' observation in Denmark. *Lancet* 1971; **i**:1195–8.
27. Bergfors E, Trollfors B, Taranger J, et al. Parapertussis: differences and similarities in incidence, clinical course, and antibody responses. *Int J Infect Dis* 1999;**3**:140–6.
28. Rodgers L, Martin SW, Cohn A, et al. Epidemiologic and laboratory features of a large outbreak of pertussis-like illnesses associated with cocirculating *Bordetella holmesii* and *Bordetella pertussis*—Ohio, 2010–2011. *Clin Infect Dis* 2013;**56**:322–31.
29. Altunaji S, Kukuruzovic R, Curtis N, Massie J. Antibiotics for whooping cough (pertussis). *Cochrane Database Syst Rev* 2007;**18**: CD004404.
30. Bartkus JM, Juni BA, Ehresmann K, et al. Identification of a mutation associated with erythromycin resistance in *Bordetella pertussis*: implications for surveillance of antimicrobial resistance. *J Clin Microbiol* 2003;**41**:1167–72.
31. Guillot S, Descours G, Billet Y, et al. Macrolide-resistant *Bordetella pertussis* infection in newborn girl, France. *Emerg Infect Dis* 2012;**18**:966–8.
32. Cherry JD. Epidemic pertussis in 2012—the resurgence of a vaccine-preventable disease. *N Engl J Med* 2012;**367**:785–7.
33. Cherry JD. Who do pertussis vaccines fail? *Pediatrics* 2012;**129**:968–70.
34. Guerra FA, Blatter MM, Greenberg DR, et al. Safety and immunogenicity of a pentavalent pertussis vaccine compared with licensed equivalent vaccines in US infants and toddlers and persistence of antibodies before a preschool booster dose: a randomized, clinical trial. *Pediatrics* 2009;**123**:301–12.
35. Tartof SY, Lewis M, Kenyon C, et al. Waning immunity to pertussis following 5 doses of DTaP. *Pediatrics* 2013;**131**:e1047–52.
36. Olin P, Rasmussen F, Gustafsson L, et al. Randomised controlled trial of two-component, three-component, and five-component acellular pertussis vaccines compared with whole-cell pertussis vaccine. *Lancet* 1997;**350**:1569–77.

37. Witt MA, Arias L, Katz PH, Truong ET, Witt DJ. Reduced risk of pertussis among persons ever vaccinated with whole cell pertussis vaccine compared to recipients of acellular pertussis vaccines in a large US cohort. *Clin Infect Dis* 2013;**56**:1248–54.
38. CDC. Updated recommendations for use of tetanus toxoid, reduced diphtheria toxoid, and acellular pertussis vaccine (Tdap) in pregnant women—Advisory Committee on Immunization Practices (ACIP). *MMWR* 2013;**62**:131–5.
39. Mooi FR, van Loo IHM, van Gent M, et al. *Bordetella pertussis* strains with increased toxin production associated with pertussis resurgence. *Emerg Infect Dis* 2009;**15**:1206–13.
40. Mooi FR. *Bordetella pertussis* and vaccination: the persistence of a genetically monomorphic pathogen. *Infect Genet Evol* 2010;**10**:36–49.
41. Lam C, Octavia S, Bahrame Z, et al. Selection and emergence of pertussis toxin promoter ptxP3 allele in the evolution of *Bordetella pertussis*. *Infect Genet Evol* 2012;**12**:492–5.
42. Octavia S, Sintchenko V, Gilbert G, et al. Newly emerging clones of *Bordetella pertussis* carrying prn2 and ptxP3 alleles implicated in Australia pertussis epidemic in 2008–2010. *J Infect Dis* 2012;**205**:1220–4.

Chapter 28

Buruli Ulcer (Atypical Mycobacteria)

Dr. Gene Khai Lin Huang[1] and Prof. Dr. Paul Johnson[2]

[1]*Infectious Diseases Registrar, Infectious Diseases Department, Austin Health, Melbourne, Australia,* [2]*Deputy Director, Infectious Diseases Department, Austin Health, Melbourne, Australia; Professor, University of Melbourne, Department of Microbiology and Immunology, Department of Medicine, Austin & Northern Clinical School, Melbourne, Australia; Director, World Health Organization Collaborating Centre for Mycobacterium ulcerans, Melbourne, Australia*

CASE PRESENTATION

A 19-year-old, previously healthy man living in regional Victoria was referred to the Austin Hospital in Melbourne with a non-healing ulcer of his right knee. The patient had been managed at a regional health service for the preceding 3 months receiving multiple courses of oral and intravenous antibiotics without effect. He remained systemically well with no fever or constitutional symptoms.

Examination of his right leg revealed extensive woody induration and erythema extending from knee to ankle with a 2 centimeter undermined ulcer just above his right knee. His hemoglobin was 12.8 g/L (12.0–18.0 g/L), with an elevated white cell count of 13.1×10^9/L ($4.0–11.0 \times 10^9$/L), neutrophilia of 9.4×10^9/L ($2.0–7.5 \times 10^9$/L), and his C-reactive protein was 17.1 mg/L (<3 mg/L). Imaging studies including MRI and ultrasound identified soft tissue edema without evidence of collection or underlying osteomyelitis.

Swabs of the ulcer had cultured *S. aureus* and *E. faecalis*. A subsequent biopsy identified dermal necrosis and subcutaneous inflammation, with abundant extracellular acid fast bacilli (AFB) on auramine-rhodamine staining. Polymerase chain reaction (PCR) testing for *Mycobacterium ulcerans* was positive on two occasions and *M. ulcerans* was isolated from broth culture and solid media at 6 weeks' incubation. The patient had previously traveled to the Bellarine Peninsula, a known endemic area for *M. ulcerans*, in the 6 months prior to the development of the ulcer.

Emerging Infectious Diseases. DOI: http://dx.doi.org/10.1016/B978-0-12-416975-3.00028-5

Serological tests for human immunodeficiency virus, hepatitis B and C were negative. There was no evidence of vasculitis, erythema nodosum, or pyoderma gangrenosum on histology of his biopsy.

Immediate surgery was considered but the extent of involvement was too large and it was not clear where the limits of the infection were anatomically. Oral rifampicin (600 mg) and moxifloxacin (400 mg) were commenced and there was initial improvement in the induration and swelling of his leg.

Four weeks following commencement of his antibiotics, the patient was readmitted to hospital with increasing pain, fevers to 38.7°C, disruption of the skin integrity of his right thigh and lower leg with increased edema, and profuse serous exudate. His white cell count was 14.9×10^9/L ($4.0-11.0 \times 10^9$/L), with an increased C-reactive protein of 69.8 mg/L (<3 mg/L). He was commenced on intravenous (IV) piperacillin-tazobactam 4.5 g three times daily to cover for superimposed infection and also oral prednisolone 50 mg daily for a presumed paradoxical inflammatory reaction to residual mycobacteria. Ultrasound identified liquefaction of subcutaneous fat and wound cultures grew scanty mixed enteric flora. A repeat PCR for *M. ulcerans* was positive, and AFB were again seen on auramine-rhodamine staining on a wound swab. Despite this, mycobacterial cultures remained negative after prolonged 12 weeks' incubation, in keeping with a diagnosis of paradoxical inflammation.

The introduction of steroid therapy and IV antibiotics was associated with rapid resolution of fevers, inflammatory markers, and swelling; however, the patient did go on to require a limited surgical debridement of his lateral knee, followed by negative-pressure wound therapy and subsequent split skin grafting for delayed wound healing. Medical therapy with oral rifampicin and moxifloxacin was continued for a total duration of 3 months with eventual microbiological cure by culture, although swabs of small areas of spontaneous discharge of pus did remain PCR positive for at least 3 months following completion of antibiotic therapy.[1]

This case illustrates three important clinical features of Buruli ulcer: first, although the classic presentation is a slowly progressive undermined ulcer that will not heal, it may also present as an atypical cellulitis and the true diagnosis is only suspected when treatment fails and part of the involved area breaks down and ulcerates. Second, diagnosis is very difficult when patients present outside an endemic area. Third, once considered, PCR rapidly confirms the diagnosis and the correct treatment can begin.

1. WHY THIS CASE WAS SIGNIFICANTLY IMPORTANT AS AN EMERGING INFECTION

Buruli ulcer (*M. ulcerans* infection) is the third most common mycobacterial pathogen affecting humans.[2] Although it is not a fatal infection, it can result in significant morbidity and mortality. In West Africa, where the disease is endemic, more than 25% of those affected are left with permanent disability.[3] It has been classified by the World Health Organization (WHO) as one of a group of important neglected tropical diseases.[4]

2. WHAT IS THE CAUSATIVE AGENT?

M. ulcerans is a slow growing environmental mycobacterium closely related to *Mycobacterium marinum*, sharing >97% of its nucleotide identity.[5] Unlike *M. marinum*, which causes slowly progressive granulomatous lesions with intracellular bacteria seen on histology, *M. ulcerans* causes a progressive necrotizing ulcerative disease of skin, soft tissue, and sometimes bone.

M. ulcerans grows preferentially at 30–32°C on Löwenstein–Jensen media with no growth and variable thermo-tolerance at temperatures of 37°C.[6] This is in contrast to *M. tuberculosis*, which grows at higher temperatures. Histopathology is characteristic with large numbers of extracellular AFB and dermal necrosis during the acute phase, with a relative lack of inflammatory cells. Over time, a degree of host response develops and persistent infection results in granulomatous reaction with relatively few organisms.[7]

Buruli ulcer is a geographically restricted infection. The environmental reservoir of *M. ulcerans* has yet to be convincingly established. Cases often cluster around bodies of water and have a focal distribution with a short distance between endemic and non-endemic areas.[8] *M. ulcerans* has been isolated from wildlife, domestic mammal species,[9] and aquatic insects.[10] DNA has been identified by PCR from various environmental specimens[11] including mosquitoes.[12]

3. WHAT IS THE FREQUENCY OF THE DISEASE?

M. ulcerans was first identified in Australia in 1948 when a new mycobacterial species was isolated from six patients presenting with unusual skin lesions.[13] The name Buruli ulcer is derived from the Buruli region near the Nile River in Uganda, where in 1961 a large number of cases were reported.[14] Buruli ulcer has since been described in more than 30 countries worldwide with the greatest frequency in West and sub-Saharan Africa (Figure 28.1) where the incidence can be as high as 150.8 per 100,000 in the most endemic districts.[15] In some areas this exceeds that of leprosy or tuberculosis.[16] Rates of disease may be underestimated in parts of West Africa where local cultural beliefs and the resultant stigma may lead to a reluctance to seek medical attention.[17,18]

Cases have occurred in other areas including the Americas, Asia, the Western Pacific, and particular regions of Australia.[2,19–22] Buruli ulcer has a number of names based on the geographical location where cases have been

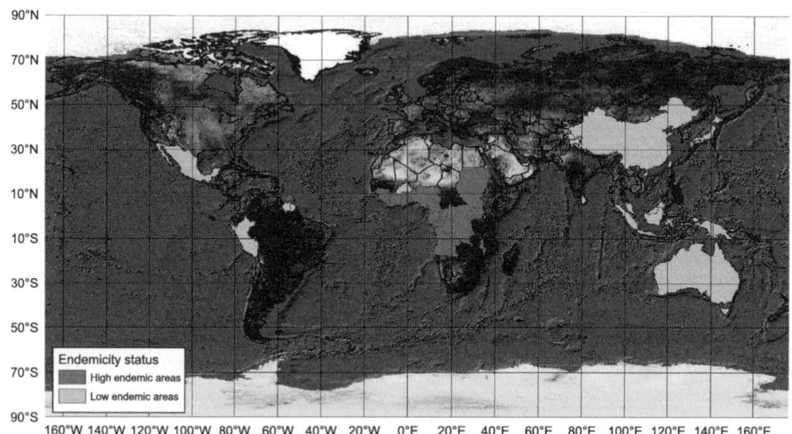

FIGURE 28.1 Distribution of *M. ulcerans* worldwide. The boundaries and names shown and the designations used on this map do not imply the expression of any opinion whatsoever on the part of the World Health Organization concerning the legal status of any country, territory, city, or area or of its authorities, or concerning the delimitation of its frontiers or boundaries. Dotted lines on maps represent approximate border lines for which there may not yet be full agreement. *Data source: WHO/Global Buruli Ulcer Initiative Map Production: Public Health Mapping & GIS Communicable Diseases (CDS) World Health Organization.*

identified. These include the Bairnsdale ulcer, the Daintree ulcer, and the Kumasi ulcer.[23] Buruli ulcer is the standard internationally accepted name preferred by the WHO.

4. HOW ARE THE BACTERIA TRANSMITTED?

The exact route of transmission of *M. ulcerans* is uncertain, although several hypotheses exist. *M. ulcerans* infection is thought to be acquired through environmental contact resulting in inoculation of skin with the organism[23] with only one reported case of transmission following a human bite.[24] Routes of transmission may vary between different geographic regions.

Proximity to water sources has long been identified as a risk factor for infection, particularly in Africa (Figure 28.2). Additional risk factors include wading or swimming, poor wound care, and having uncovered skin.[23,25,26] Aquatic insects have been hypothesized to be potential vectors in transmission of disease to humans in some areas.[10,23,27] In support of this theory, *M. ulcerans* has been isolated from aquatic insects in West Africa and has also been shown to be transmissible to mice via biting aquatic insects of the family Naucoridae.[10]

In Australia, where endemic areas are relatively well defined, being bitten by mosquitoes has been shown to be independently associated with risk of

FIGURE 28.2 Locals having a bath in one of the local waterways in Cameroon. Living in proximity to and in contact with water is a known risk factor for Buruli ulcer although the exact mechanism of transmission is not yet known. *From: Pouillot R, Matias G, Wondje CM, et al. Risk factors for Buruli ulcer: a case control study in Cameroon. PLoS Negl Trop Dis 2007;1:e101.*

Buruli ulcer.[25] In these regions, mosquitoes themselves,[12,28] as well as possums (small tree dwelling marsupials) and their excreta,[9] have been shown to be PCR positive. The exact link and sequence of transmission between humans, mosquitoes, and possums has not yet been elucidated.

5. WHAT FACTORS ARE INVOLVED IN DISEASE PATHOGENESIS? WHAT ARE THE PATHOGENIC MECHANISMS?

The clinical and histopathological presentation is mediated by a polyketide cytotoxin produced by *M. ulcerans* called mycolactone[29] (Figure 28.3). Mycolactone diffuses into surrounding soft tissues inducing necrosis, cellular apoptosis, and inhibition of the local immune response. This results in the characteristic tissue destruction with a lack of inflammation.[30] The genes for the production of mycolactone are encoded by a virulence plasmid pMUM, unique to the species.[31] Isolates from different geographic regions produce distinct patterns of mycolactone variants.[32]

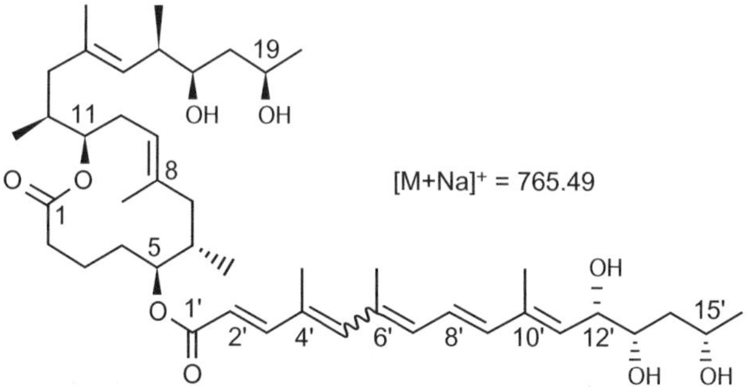

FIGURE 28.3 Chemical structure of mycolactone A/B showing the lactone core ring and polyketide side chains. *From: Sarfo FS, Phillips RO, Rangers, B, et al. Detection of mycolactone A/B in Mycobacterium ulcerans-infected human tissue. PLoS Negl Trop Dis 2010;4:e577.*

6. WHAT ARE THE CLINICAL MANIFESTATIONS?

Given the subacute nature of the disease, there is often a long delay from first symptoms to presentation. In a recent publication from Australia, where endemic areas have been carefully identified, authors were able to identify patients with Buruli ulcer who had a single visit exposure to an endemic area.[33] They calculated the mean incubation period as 4.5 months.

Most patients initially notice a papular lesion, nodule or plaque, which subsequently ulcerates.[34] The resulting ulcer is typically painless and necrotic with undermined edges and gradually extends into the subcutaneous fat layer (Figure 28.4). There is a notable lack of systemic symptoms. Lesions are most common on limbs and exposed areas, although they may occur anywhere. A subgroup of patients may present with acute onset non-pitting edema of a body region such as a limb, the face, or the abdominal wall with or without a focal ulcer.[3,35]

If left untreated, Buruli ulcer causes slowly progressive tissue destruction. Although healing may occur with eventual development of host immunity, there is often significant scarring and resultant disability.

7. HOW DO YOU DIAGNOSE?

Laboratory tools that assist in establishing the diagnosis of Buruli ulcer include staining for AFB, histopathology, culture, and PCR. In settings with limited access to laboratory services, diagnosis is often made on clinical grounds.

AFB may be seen with Ziehl–Neelsen stain of a swab or tissue sample; however, this is non-specific. Culture of the organism is slow and insensitive, and if relied upon, can lead to a significant delay in diagnosis.

In 1997, an Australian group developed an *M. ulcerans*-specific PCR that identifies insertion sequence IS2404, an 1109 base pairs sequence that is

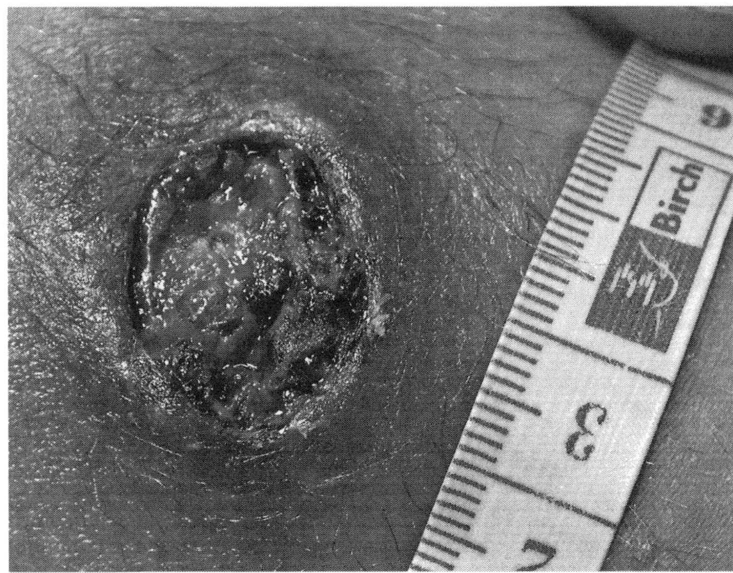

FIGURE 28.4 Buruli ulcer on the lower limb of a 10-year-old girl in Victoria, Australia. Note the tissue necrosis with undermined edges. *Photo: Dr. Gene Khai Lin Huang with written permission obtained.*

repeated at high copy numbers throughout the *M. ulcerans* genome.[36] This PCR can be performed directly on an ulcer swab or fresh tissue from a biopsy and has very high sensitivity and specificity. A negative PCR in the context of a positive smear for AFB suggests an alternate diagnosis.[37]

8. HOW DO YOU DIFFERENTIATE THIS DISEASE FROM SIMILAR ENTITIES?

Differential diagnoses to consider include necrotizing fasciitis, venous or arterial ulcer, diabetic ulcer, leprosy, yaws, cutaneous leishmaniasis, tropical phagedenic ulcer, vasculitic ulcer, erythema nodosum, and pyoderma gangrenosum.

9. WHAT IS THE THERAPEUTIC APPROACH?

The goal of therapy is two-fold, to treat the infection and also to repair and minimize tissue damage and disability.

The traditional approach to Buruli ulcer had been centered on surgical resection with wide margins, which is satisfactory for small lesions.[38] It was thought that antibiotics were ineffective and two early randomized studies performed in Africa showed no benefit.[39,40] Larger lesions, or infections that involve critical structures, have always presented a major challenge.

In 2005, a pilot study established that the combination of oral rifampicin and intramuscular streptomycin was able to sterilize early lesions in humans.[41] Following this study, antibiotic therapy has increasingly become the cornerstone of treatment, with less recourse to surgery.[4,37]

The approach to therapy depends on the presentation as well as the size and stage of the lesion, as outlined in the WHO classification categories:

- Category I: single lesion <5 cm in diameter. Most are curable with antibiotic therapy alone (or excision with wide margins).
- Category II: single lesion 5–15 cm in diameter. Some curable with antibiotic therapy alone.
- Category III: single lesion >15 cm in diameter, multiple lesions, lesions at a critical site (eye, breast, genitalia), and osteomyelitis. Usually require surgery in addition to antibiotics but with a conservative and delayed surgical approach.

With regards to choice of antibiotics, the WHO recommends the combination of rifampicin and streptomycin or rifampicin plus another oral agent (clarithromycin or moxifloxacin) for a minimum of 8 weeks' duration.[4] The Australian approach has been in keeping with these recommendations but with a preference for an oral regimen (i.e., rifampicin plus a fluoroquinolone or clarithromycin) for 2 months' duration.[37,42–44] It has been observed that significant healing continues to occur after completion of antibiotic therapy.[42,45,46]

The temperature-dependent growth characteristics of *M. ulcerans*[6] have been utilized for therapeutic purposes with several small case series demonstrating that continuous heat applied to the skin at temperatures of 38–39°C can be used to treat Buruli ulcer.[47,48]

Paradoxical reactions in the form of worsening or progressive ulceration during antibiotic therapy are not uncommon and typically occur several weeks after the commencement of antibiotics.[46,49] It is important not to misinterpret this as treatment failure. Steroid therapy (0.5–1 mg/kg prednisolone daily) in this setting may be of benefit.[1,50]

10. WHAT ARE THE PREVENTIVE AND INFECTION CONTROL MEASURES?

Given the uncertainty surrounding the transmission and environmental reservoirs of *M. ulcerans* and lack of human to human cases, the main focus of public health efforts has been on early diagnosis and treatment. It is reasonable to recommend the use of protective clothing while working outside in endemic areas, minimising mosquito exposure and attention to wound care.

Probably the best public health message is to be aware that the Buruli ulcer may present several months after exposure, that the diagnosis is difficult if not considered by doctors unfamiliar with the condition, and that morbidity is minimized by early diagnosis.

REFERENCES

1. Trevillyan JM, Johnson PD. Steroids control paradoxical worsening of Mycobacterium ulcerans infection following initiation of antibiotic therapy. *Med J Aust* 2013;**198**:443–4.
2. Asiedu K, Scherpbier R, Raviglione M. *Buruli ulcer: Mycobacterium ulcerans infection.* Geneva: World Health Organization; 2000.
3. Johnson PD, Stinear TP, Pamela LC, et al. Buruli ulcer (*M. ulcerans* infection): new insights, new hope for disease control. *PLoS Med* 2005;**2**:e108.
4. WHO. Treatment of *Mycobacterium ulcerans* disease (Buruli ulcer) [homepage on the internet]. c2012 [cited; 2013 Oct 20]. Available from: <http://apps.who.int/iris/bitstream/10665/77771/1/9789241503402_eng.pdf>.
5. Stinear TP, Seemann T, Pidot S, et al. Reductive evolution and niche adaptation inferred from the genome of *Mycobacterium ulcerans*, the causative agent of Buruli ulcer. *Genome Res* 2007;**17**:192–200.
6. Eddyani M, Portaels F. Survival of Mycobacterium ulcerans at 37° C. *Clin Microbiol Infect* 2007;**13**:1033–5.
7. Hayman J, McQueen A. The pathology of Mycobacterium ulcerans infection. *Pathology* 1985;**17**:594–600.
8. World Health Organization. Buruli ulcer: progress report, 2004–2008. *Wkly Epidemiol Rec* 2008;**83**:145–56.
9. Fyfe JA, Lavender CJ, Handasyde KA, et al. A major role for mammals in the ecology of Mycobacterium ulcerans. *PLoS Negl Trop Dis* 2010;**4**:e791.
10. Marsollier L, Robert R, Aubry J, et al. Aquatic insects as a vector for Mycobacterium ulcerans. *Appl Environ Microbiol* 2002;**68**:4623–8.
11. Vandelannoote K, Durnez L, Amissah D, et al. Application of real-time PCR in Ghana, a Buruli ulcer-endemic country, confirms the presence of *Mycobacterium ulcerans* in the environment. *FEMS Microbiol Lett* 2010;**304**:191–4.
12. Lavender CF, Fyfe JA, Azuolas J, et al. Risk of Buruli ulcer and detection of Mycobacterium ulcerans in mosquitoes in Southeastern Australia. *PLoS Negl Trop Dis* 2011;**5**:e1305.
13. MacCallum P, Tolhurst JC, Buckle G, Sissons H. A new mycobacterial infection in man. *J Pathol Bacteriol* 1948;**60**:93–122.
14. Clancey J, Dodge R, Lunn HF. Study of a mycobacterium causing skin ulceration in Uganda. *Ann Soc Belg Med Trop* 1962;**4**:585–90.
15. Amofah G, Bonsu F, Tetteh C, et al. Buruli ulcer in Ghana: results of a national case search. *Emerg Infect Dis* 2002;**8**:167.
16. Debacker M, Aguiar J, Steunou C, et al. Mycobacterium ulcerans disease (Buruli ulcer) in rural hospital, Southern Benin, 1997–2001. *Emerg Infect Dis* 2004;**10**:1391.
17. Stienstra Y, van der Graaf WT, Asamoa K, van der Werf TS. Beliefs and attitudes toward Buruli ulcer in Ghana. *Am J Trop Med Hyg* 2002;**67**:207–13.
18. Aujoulat I, Johnson C, Zinsou C, Guédénon A, Portaels F. Psychological aspects of health seeking behaviours of patients with Buruli ulcer in Southern Benin. *Trop Med Int Health* 2003;**8**:750–9.

19. Steffen CM, Smith M, McBride WJ. Mycobacterium ulcerans infection in North Queensland: the "Daintree ulcer." *ANZ J Surg* 2010;**80**:732–6.
20. Francis G, Whitby M, Woods M. Mycobacterium ulcerans infection: a rediscovered focus in the Capricorn Coast region of central Queensland. *Med J Aust* 2006;**185**:179–80.
21. Radford AL. Mycobacterium ulcerans in Australia. *Aust N Z J Med* 1975;**5**:162–9.
22. Johnson PD, Veitch MG, Leslie DE, Flood PE, Hayman JA. The emergence of Mycobacterium ulcerans infection near Melbourne. *Med J Aust* 1996;**164**:76.
23. Merritt RW, Walker ED, Small PL, et al. Ecology and transmission of Buruli ulcer disease: a systematic review. *PLoS Negl Trop Dis* 2010;**4**:e911.
24. Debacker M, Zinsou C, Aguiar J, Meyers WM, Portaels F. First case of Mycobacterium ulcerans disease (Buruli ulcer) following a human bite. *Clin Infect Dis* 2003;**36**:e67–8.
25. Quek TY, Athan E, Henry MJ, et al. Risk factors for Mycobacterium ulcerans infection, Southeastern Australia. *Emerg Infect Dis* 2007;**13**:1661–6.
26. Debacker M, Portaels F, Aguiar J, et al. Risk factors for Buruli ulcer, Benin. *Emerg Infect Dis* 2006;**12**:1325.
27. Portaels F, Elsen P, Guimaraes-Peres A, Fonteyne PA, Meyers WM. Insects in the transmission of Mycobacterium ulcerans infection. *Lancet* 1999;**353**:986.
28. Johnson PD, Azuolas J, Lavender CJ, et al. Mycobacterium ulcerans in mosquitoes captured during outbreak of Buruli ulcer, southeastern Australia. *Emerg Infect Dis* 2007;**13**:1653.
29. George KM, Chatterjee D, Gunawardana G, et al. Mycolactone: a polyketide toxin from Mycobacterium ulcerans required for virulence. *Science* 1999;**283**:854–7.
30. George KM, Pascopella L, Welty DM, Small PLA. Mycobacterium ulcerans toxin, mycolactone, causes apoptosis in guinea pig ulcers and tissue culture cells. *Infect Immun* 2000;**68**:877–83.
31. Stinear TS, Mve-Obiang A, Small PL, et al. Giant plasmid-encoded polyketide synthases produce the macrolide toxin of *Mycobacterium ulcerans*. *Proc Natl Acad Sci USA* 2004;**101**:1345–9.
32. van der Werf TS, Stinear T, Stienstra Y, van der Graaf W, Small P. Mycolactones and Mycobacterium ulcerans disease. *Lancet* 2003;**362**:1062.
33. Trubiano JA, Lavender CJ, Fyfe JA, Bittmann S, Johnson PD. The incubation period of Buruli ulcer (Mycobacterium ulcerans infection). *PLoS Negl Trop Dis* 2013;**7**:e2463.
34. Hayman J. Clinical features of Mycobacterium ulcerans infection. *Australas J Dermatol* 1985;**26**:67–73.
35. Jenkins GA, Smith M, Fairley M, Johnson PD. Acute oedematous Mycobacterium ulcerans infection in a farmer from far north Queensland. *Med J Aust* 2002;**176**:180–1.
36. Ross BC, Marino L, Oppedisano F, Edwards R, Robins-Browne RM, Johnson PD. Development of a PCR assay for rapid diagnosis of Mycobacterium ulcerans infection. *J Clin Microbiol* 1997;**35**:1696–700.
37. Johnson PD, Hayman JA, Quek TY, et al. Consensus recommendations for the diagnosis, treatment and control of Mycobacterium ulcerans infection (Bairnsdale or Buruli ulcer) in Victoria, Australia. *Med J Aust* 2007;**186**:64.
38. Debacker M, Aguiar J, Steunou C, Zinsou C, Meyers WM, Portaels F. Buruli ulcer recurrence, Benin. *Emerg Infect Dis* 2005;**11**:584.
39. Revill WD, Morrow RH, Pike MC, et al. A controlled trial of the treatment of Mycobacterium ulcerans infection with clofazimine. *Lancet* 1973;**2**:873–7.
40. Fehr H, Egger M, Senn I. Cotrimoxazol in the treatment of Mycobacterium ulcerans infection (Buruli ulcer) in West Africa. *Trop Doct* 1994;**24**:61–3.

41. Etuaful S, Carbonnelle B, Grosset J, et al. Efficacy of the combination rifampin-streptomycin in preventing growth of Mycobacterium ulcerans in early lesions of Buruli ulcer in humans. *Antimicrob Agents Chemother* 2005;**49**:3182.
42. Gordon CL, Buntine JA, Hayman JA, et al. All-oral antibiotic treatment for buruli ulcer: a report of four patients. *PLoS Negl Trop Dis* 2010;**4**:e770.
43. O'Brien DP, McDonald A, Callan P, et al. Successful outcomes with oral fluoroquinolones combined with rifampicin in the treatment of Mycobacterium ulcerans: an observational cohort study. *PLoS Negl Trop Dis* 2012;**6**(1):e1473.
44. O'Brien DP, McDonald A, Callan P, et al. Successful outcomes with oral fluoroquinolones combined with rifampicin in the treatment of Mycobacterium ulcerans: an observational cohort study. *PLoS Negl Trop Dis* 2007;**6**:e1473.
45. Chauty A, Ardant MF, Adeye A, et al. Promising clinical efficacy of streptomycin-rifampin combination for treatment of buruli ulcer (Mycobacterium ulcerans disease). *Antimicrob Agents Chemother* 2007;**51**:4029–35.
46. Nienhuis WA, Stienstra Y, Abass KM, et al. Paradoxical responses after start of antimicrobial treatment in Mycobacterium ulcerans infection. *Clin Infect Dis* 2012;**54**:519–26.
47. Meyers WM, Shelly WM, Connor DH. Heat treatment of Mycobacterium ulcerans infections without surgical excision. *Am J Trop Med Hyg* 1974;**23**:924–9.
48. Junghanss T, Um Boock A, Vogel M, Schuette D, Weinlaeder H, Pluschke G. Phase change material for thermotherapy of Buruli ulcer: a prospective observational single centre proof-of-principle trial. *PLoS Negl Trop Dis* 2009;**3**:e380.
49. O'Brien DP, Robson ME, Callan PP, McDonald AH. Paradoxical immune-mediated reactions to Mycobacterium ulcerans during antibiotic treatment: a result of treatment success, not failure. *Med J Aust* 2009;**191**:564–6.
50. Friedman ND, McDonald AH, Robson MW, O'Brien DP. Corticosteroid use for paradoxical reactions during antibiotic treatment for Mycobacterium ulcerans. *PLoS Negl Trop Dis* 2012;**6**:e1767.

Index

Note: Page numbers followed by "*f*", "*t*" and "*b*" refers to figures, tables and boxes respectively.

N

O

P

Q

R

Edwards Brothers Malloy
Ann Arbor MI. USA
October 31, 2014